P9-DTY-535

Progress in SEPARATION AND PURIFICATION

Volume 3

A Wiley-Interscience Series

Progress in SEPARATION AND PURIFICATION

VOLUME 3

Edited by

EDMOND S. PERRY

Research Laboratories
Eastman Kodak Company
Rochester, New York

and

CAREL J. VAN OSS

Department of Microbiology
Immunochemistry Laboratory
State University of New York at Buffalo
Buffalo, New York

Wiley–Interscience, A Division of John Wiley & Sons, Inc.
New York • London • Sydney • Toronto

Library of Congress Catalogue Card Number: 67-29539

ISBN—0-471-68100-8

Printed in the United States of America

10 9 8 7 6 5 4 3 2 1

Preface to Volume 3

Progress in Separation and Purification was launched in January 1968 and Volume 2 appeared in March 1969. The present Volume 3 continues to carry forth our original objectives: to have a forum for special papers to provide the expert as well as the newcomer to the field with a current awareness of the progress being made in this expanding segment of science.

The original intention for the Series was to compile volumes from submitted papers irrespective of their subject matter. Thus Volume 1 contains articles describing a wide range of methods and techniques. Volume 2, however, is a collection of papers allied to a common interest and originally presented together as a symposium. In this instance, the symposium chairman assumed the editorship of the volume while we assisted in getting it into print. Such a procedure can at times bring new thoughts and ideas to the Series through the editor-pro-tem. We think that this procedural variant has been a valuable adjunct to the Series and intend to continue the practice occasionally in future volumes.

The present volume, like Volume 1, contains a group of articles on a variety of subjects within the field of separation and purification. Again, the authors have presented their particular expertise in a manner to benefit the novice and expert alike.

As always, we shall be happy to receive suggestions and recommendations for improving this Series, and inquiries about publication in the Series are always welcomed. Finally, to the contributors who have made Volume 3 possible, we extend our sincere gratitude.

Edmond S. Perry
Carel J. van Oss

June, 1970

Preface to Volume 1

Through the ages man has been aware of the importance of separating the valuable from the less valuable in nature's mixtures. This trait has persisted; it is difficult to single out an area of science of a science-based industry where separations and purifications do not play an important role. Yet, in spite of this heritage, only within recent times has the science of separations been recognized in its individuality and has finally been accorded independent identity. The beginning of this era is difficult to pinpoint, but it is approximately the time of the second world war. Certainly, the science and technology generated in the struggle to separate the fissionable isotopes, to isolate and purify marketable quantities of antibiotics, the discovery of gas-phase chromatography, and the industries based on solid-state phenomena were important factors in leading the way for the science of separations and purifications as we know it today.

In the intervening twenty-five to thirty years, there has occurred an extensive development in this field. The literature has grown immensely; it is still expanding at a rapid rate for both the science and the technology of the subject, and a forum is needed for specialists as well as those new to the field or those working in related areas to keep up with the development. *Progress in Separation and Purification* is devoted to this purpose. Its broadcast objective is to provide its readers with a high degree of current awareness on the progress being made in the large and complex field. We hope that the Series will help the practitioner to keep abreast of the ever growing literature by providing him with authoritative summaries on significant new developments and critical evaluations of new methods, apparatus, and techniques. The organization and condensation of the literature which is now dispersed throughout the chemical, biological, and nuclear sciences will also help to bring to the science and technology of separations its rightful status.

Fast and expeditious reporting are particular objectives of this Series. We plan to publish volumes at intervals commensurate with the procurement of articles. Manuscripts will be processed as received, and a volume will be issued when sufficient material has been assembled. The choice of subject and the manner in which ideas and opinions are expressed are essentially left to the discretion of the authors. Our only request of them is to render a service which will be of value to the reader and to the science and technology of separation and purification.

The nine articles in this first volume fall within the spirit of this liberal policy. Each author is an acknowledged leader in the field and has written on a subject with which he has had intimate experience. The random order of appearance of the articles in the volume has helped to expedite publication. We shall be happy to receive suggestions and recommendations for improving the service this Series purports to provide, and we invite inquiries for publication in the Series from authors.

My sincere gratitude goes to the authors of this first volume who were willing to embark with me on this new venture. Special thanks are due Dr. Arnold Weissberger, who suggested this undertaking and has provided advice and counsel in getting it under way.

Edmond S. Perry

May, 1968

Contributors to Volume 3

J. CALVIN GIDDINGS, *Department of Chemistry, University of Utah, Salt Lake City, Utah*

MARJORIE M. GOODMAN, *Research and Development Division, Smith Kline and French Laboratories, Philadelphia, Pennsylvania*

NORMAN N. LI, *Corporate Research Laboratories, Esso Research and Engineering Company, Linden, New Jersey*

BERNARD LOEV, *Research and Development Division, Smith Kline and French Laboratories, Philadelphia, Pennsylvania*

ROBERT B. LONG, *Corporate Research Laboratories, Esso Research and Engineering Company, Linden, New Jersey*

H. K. LONSDALE, *Chemistry Department, Gulf General Atomic Incorporated, San Diego, California*

F. W. H. M. MERKUS, *Head, Pharmaceutical Department, Roman Catholic Hospital, Sittard, The Netherlands*

MARCUS N. MYERS, *Department of Chemistry, University of Utah, Salt Lake City, Utah*

CAREL J. VAN OSS, *Department of Microbiology, Immunochemistry Laboratory, State University of New York at Buffalo, Buffalo, New York*

T. A. SULLIVAN, *U. S. Department of the Interior, Bureau of Mines, Boulder City Metallurgy Research Laboratory, Boulder City, Nevada*

ROKUS A. DE ZEEUW, *Laboratory for Pharmaceutical and Analytical Chemistry, State University, Groningen, The Netherlands*

Contents

Vapor-Programmed Thin-Layer Chromatography, Development and Applications

Rokus A. de Zeeuw

Laboratory for Pharmaceutical and Analytical Chemistry, State University, Groningen, The Netherlands

I. INTRODUCTION

During the past ten years thin-layer chromatography (TLC) has become a very useful and versatile tool in analytical chemistry. Despite some confusion at the beginning, the performance of the technique has developed into a uniform procedure that is discussed in the many books on the subject. It is remarkable, however, that the theoretical aspects of the TLC processes are still not fully understood.

We have been using the general procedures of TLC for several years in pharmaceutical science for the analyses of drugs with reasonable success. However, when investigating the TLC behavior of hypnotics and sedatives we observed two striking phenomena: a more efficient separation with the aid of the so-called unsaturated chamber and, furthermore, complete disappearance of a given separation in the sandwich chamber; the normal chamber provided complete resolution.

These two phenomena, for which the existing theories could not provide a suitable explanation, led to the investigations described in this chapter. After examining the influence of solvent vapor in TLC, a much better insight could be gained into the different processes affecting the migration of the spots, thus allowing a simple explanation of the two phenomena mentioned. Moreover, the better understanding of the TLC processes resulted in the development of a chamber type by which a new thin-layer chromatographic technique became possible, this technique we call vapor-programmed TLC. The development of the new apparatus is described in detail and examples of its capabilities are presented here. The results indicate that vapor-programmed TLC is a valuable contribution to the existing possibilities of TLC.

II. EXPERIMENTAL DATA

The chromatographic experiments described in this chapter were carried out according to the standard procedures and apparatus described below, unless stated otherwise in the text.

Adsorbents

(*a*) Silica gel GF 254 (E. Merck), 30 g/60 ml demineralized water are carefully mixed and stirred for 1 min in a mortar to give a homogenous suspension for the preparation of 5 plates.

(*b*) Aluminum oxide GF 254 (E. Merck), 30 g/60 ml demineralized water are carefully mixed and stirred for 10 min in a mortar to give a homogenous, slightly viscous suspension for the preparation of 5 plates.

The preparation of the suspensions can also be done with distilled water.

Spreading apparatus: Desaga.

Plate size: 20 × 20 cm.

Layer thickness: 0.25 mm when spread.

Plate drying: After preparation 15 min in air, then 30 min at 110°C in an oven with a fan. Cooling and storage is done in a desiccator over blue silica gel.

Demineralized water: Conductivity $<0.4 \times 10^{-4}\ \Omega^{-1}\ m^{-1}$.

Solvents: "Pro Analysi" grade (E. Merck). N-chambers contained 100 ml of solvent, S-chambers contained 20 ml of solvent. Solvent mixtures were prepared immediately before development and were used once. The compositions of these mixtures are given by volume.

Substances: The following 0.2% w/v solutions were used in the chromatographic experiments: hypnotics and sedatives in chloroform, color dyes in benzene, sulfonamides in acetone, and local anaesthetics in 96% ethanol (as HCl salts). The identities of the drugs were established by melting points and, if possible, by examining their identity reactions as described in the various pharmacopoeiae. The purity of the substances was examined chromatographically. All components proved to be more than 99.0% pure, except cyclobarbital which showed a slight decomposition. The azo-dyes were available as a solution in benzene (Desaga).

Sample load: 0.005 ml with the aid of 0.01 ml micropipets (Desaga), corresponding with 10 μg solute. Mixtures of substances are composed in a way that 0.005 ml contains 10 μg of each component.

Starting points: 1.5–2.5 cm from the bottom edge of the plate.

Length of run: 10 or 15 cm over the starting points.

Temperature: 21° ± 1°C.

Relative humidity: 27–51%. In this range reproducibility of the chromatograms was observed. The experiments with azo-dyes were done at a relative humidity of 26–30%.

Development chambers: Normal tank chambers (N-chambers), 21 × 21 × 9 cm (Desaga type) and sandwich chambers (S-chambers), 20 × 20 × 0.1 cm (Camag type). In the sandwich chamber a plain glass plate was used to cover the adsorbent plate and no special arrangements were made for saturation procedures.

Detection: Ultraviolet light of 254 nm by means of a Universal lamp, Sylvania Germicidal G 8 T, with UV-G5 filter (Camag).

Photography: In UV light of 254 nm with two lamps (as under Detection) on either side of the plate, exposure 15 sec, aperture 5.6, distance 70 cm. Camera: Asahi Pentax SV, Super Takumar 1:1.8/55 lens with 49 mm ghostless filter (Asahi). Film: Agfacolor CT 18 diapositive (Agfa).

Saturated chambers: N-chambers lined with filter paper, with 100ml of solvent. After a saturation time of 30–45 min the plate is introduced.

Unsaturated chambers: N-chambers without filter paper, with 100 ml of solvent. The plate is placed into the chamber immediately after introduction of the solvent.

Glass troughs: 19 × 1.5 × 1.5 cm.

Balances: Mettler K7 and Mettler BCH.

Special conditions: The spotting period, that is, the time between removal of the plate from the desiccator and the introduction into the chamber, is at least 15 min.

Plates for N-chambers are stripped 0.5 cm wide at the side edges; plates for S-chambers are stripped 0.5 cm at the side edges and the top edge to ensure suitable fitting of the spacer-frame; plates for the vapor-programming chamber are stripped 0.5 cm at the side edges and the bottom edge.

III. THE ROLE OF SOLVENT VAPOR IN TLC*

A. Introduction

In TLC, development of the plate is usually performed after saturation of the chamber with solvent vapor. This procedure is recommended in every book on TLC in order to obtain more reproducible R_f values (30) and to avoid the appearance of edge effects as described by Demole (9) and Stahl (34). Although this procedure has proved to give suitable results in many instances, some authors recommend the use of unsaturated chambers, particularly in the separation of multicomponent mixtures of closely related substances (26,1,46,29). In our experiments with hypnotics and sedatives we also obtained improved separation by using an unsaturated chamber (39). This observation was recently confirmed by Hermans and Kamp (16).

A suitable explanation of these observations was not readily available. Von Arx and Neher (1) presumed that in saturated chambers the solvent

* In this section the influence of adsorbed water vapor on the adsorbent is left out of consideration for the sake of clearness. If experiments are done under standardized relative humidity and according to our working procedure, this water vapor influence will be constant.

ascends too fast, thus preventing sufficient selectivity in the adsorption processes by the adsorbent. Zinkel and Rowe (46) ascribed the improved separations to the prolonged time of run due to solvent evaporation from the plate, in combination with the occurrence of gradient development. These presumptions were not tested experimentally and the absence of a theoretical explanation led us, therefore, to a more detailed investigation into the various TLC processes. In our opinion, the use of single-component solvents, which are generally preferred in the more theoretical studies, would be incorrect because an explanation of the phenomena herewith is not valid for multicomponent solvents. Therefore, we started to work with a binary solvent system because the processes with these systems are easily applicable to both single- and multicomponent systems. A further reason to start with a binary solvent is that in practical TLC analysis multicomponent solvents are more frequently used than single-component solvents.

B. Examination of Some TLC Processes

In TLC the solvent is ascending in a "dry" adsorbent. It is well known that during this process multicomponent solvents can "demix" on the plate (as in frontal analysis for instance). The front of the ascending liquid contains a single component A which is followed by a zone of a binary mixture $A + B$, then by a zone of a ternary mixture $A + B + C$, and so on, with the components having an increasing affinity for the adsorbent in the order $A < B < C$. The demixing causes a stepwise gradient elution which was applied by Niederwieser and Brenner (28) under the name "polyzonal TLC."

In normal TLC, using volatile multicomponent solvents, the situation is, however, much more complex because apart from the adsorption of the solvent, liquid-vapor equilibria play a role as well as the adsorption of solvent components from the gas phase. If, for example, we take a binary mixture of chloroform and ether as the solvent we may conceive the following processes:

1. By capillary action of the porous adsorbent, solvent ascends and by a process comparable to frontal analysis a certain zone of pure chloroform will be formed, followed by the binary mixture.

2. In the dry part of the plate adsorption of solvent vapor will take place and, ether being more strongly adsorbed than chloroform, the adsorbate will mainly consist of ether.

3. In the wet part of the plate, which is already covered by the ascending solvent, absorption of vapor as well as evaporation of solvent will take place.

As a consequence of the adsorption in the dry part and the absorption in the wet part of the plate, the front zone will not be pure chloroform. Furthermore, it will be obvious that the extent to which processes *2* and *3* will affect the development depends on many factors, such as vapor pressure and relative affinity of the solvent components for the adsorbent, the geometry of the chamber, the temperature and, especially, whether or not the chamber has been saturated before development.

C. Experiments with Vapors

Because our main interest lies in the differences between unsaturated chambers and saturated chambers, the experiments are aimed at elucidating the influence of these parameters. In order to obtain data on the amount of adsorbed vapor and on the velocity of this process, the increase in plate weight was measured when a dry plate was brought into a saturated or unsaturated vapor atmosphere of one of the solvent components. This was done in a plastic box in which a Mettler K7 top-balance was placed. In the case of unsaturation, the plate and 4 glass troughs filled with 20 ml of a solvent component each were brought into the box at the same instant. In the case of saturation the 4 filled troughs were placed in the box first, followed by the plate after 30 min. In both cases vapor adsorption was measured over a 30-min period.

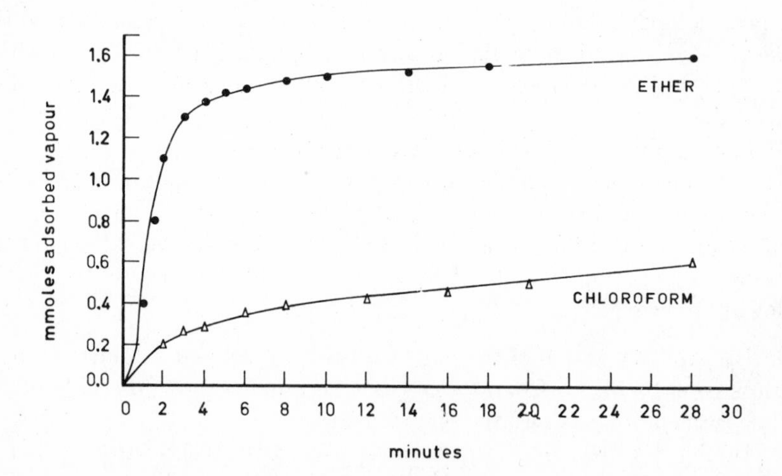

Fig. 1. Vapor adsorption of ether and chloroform by silica gel plates from an unsaturated atmosphere. Plate size 20 × 20 cm, layer thickness 0.25 mm. Weight measurement with a Mettler K 7 top balance as described in the text. Figures 1–12 are reproduced from R. A. de Zeeuw, "The role of solvent vapour in thin layer chromatography," "*J. Chromatog.*, **32,** 43 (1968), with permission of the publisher.

Evaporation of adsorbed vapor was measured by placing a dry plate into a saturated vapor atmosphere for 5 min (e.g., an N-chamber with a trough filled with a solvent component) and by observing the decrease in weight on a Mettler BCH. This was done by hanging the plate mounted in a wire frame onto the balance. The adsorbed vapor could freely evaporate in the balance cabinet. Due to the large volume of this cabinet the evaporated components produced only a small partial pressure which, in turn, had little influence on the evaporation process.

All weight measurements were repeated 5 times, each with different plates. In all cases identical adsorption and evaporation curves were obtained with the corresponding values showing less than 7% variation. Corrections for changes in upward pressure could be neglected.

The adsorption of chloroform and ether vapor from unsaturated and saturated atmospheres are given in Figures 1 and 2. The amounts of adsorbed vapor have been expressed in millimoles rather than in milligrams because of the differences in molecular weight. It can be seen, especially in the beginning, that the amount of adsorbed vapor from saturated atmospheres is higher. It should be noted that this period of about 20 min is usually an important part of the time needed for development. The evaporation curves of ether and chloroform vapor are shown in Figure 3. From these three figures it is obvious that, for this adsorbent (silica gel), the adsorption of ether is much stronger than that of chloroform. It will be clear that the shape of the curves in Figures 1, 2, and 3 are dependent on the geometry of the space in which they were measured, particularly in the

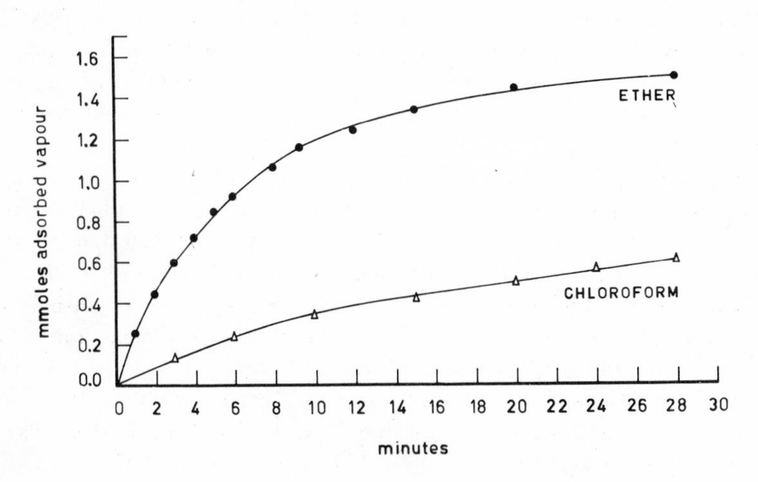

Fig. 2. Vapor adsorption of ether and chloroform by silica gel plates from a saturated atmosphere (details as in Fig. 1).

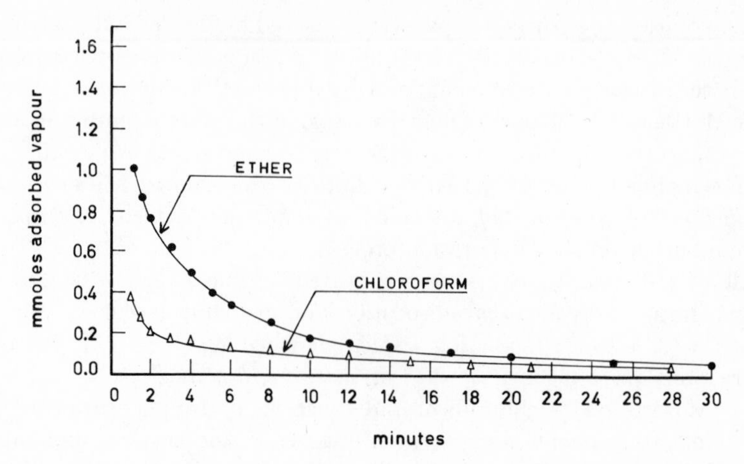

Fig. 3. Evaporation of adsorbed ether- and chloroform vapor from silica gel plates after a 5 min vapor uptake in a saturated atmosphere. Weight measurement with a Mettler BCH balance as described in the text.

beginning. Thus, the shapes in a normal chromatography tank may be different from those mentioned above, but the maximum adsorption values will remain the same.

The vapor pressures of ether and chloroform at 20°C are 435 and 154 mm, respectively (36). The amounts of ether and chloroform vapor being adsorbed by the plate at these vapor pressures and in a plastic box having a volume of about 10 l are approximately 58 and 67 cm³ respectively. It is clear, then, that only very small quantities of the available vapors are adsorbed. For 20 × 20 cm plates, for example, the volume of vapor adsorbed is less than that contained in the 1 cm layer lying just above the entire plate surface.

The next chromatographic experiments were all done with a series of hypnotics numbered as follows: 1 = heptobarbital, 2 = phenobarbital, 3 = allobarbital, 4 = hexobarbital, 5 = methylphenobarbital, 6 = bromisoval. Also, a mixture of the components 3 + 4 + 5 + 6 was used.

Figure 4 illustrates the separation in a saturated chamber with chloroform-ether (75 + 25). The separation of the mixture 3–6 is incomplete and the spreading of the spots is limited to the lower part of the plate. In an unsaturated chamber the chromatogram of Figure 5 is obtained using the same solvent composition. Comparing Figures 4 and 5 it is clear that the selectivity of the separation shown in Figure 5 is much better. The mixture 3–6 is completely separated and the spread of the spots is enlarged. Furthermore, it can be seen that the location of the spots in Figure 5 is higher, due

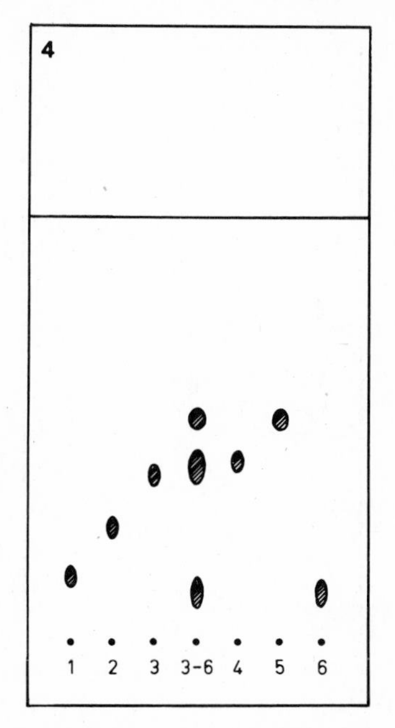

Fig. 4. Separation of hypnotics with chloroform-ether (75 + 25) in a saturated chamber. 1 = heptobarbital; 2 = phenobarbital; 3 = allobarbital; 4 = hexobarbital; 5 = methylphenobarbital; 6 = bromisoval; 3–6 = mixture of 3 + 4 + 5 + 6.

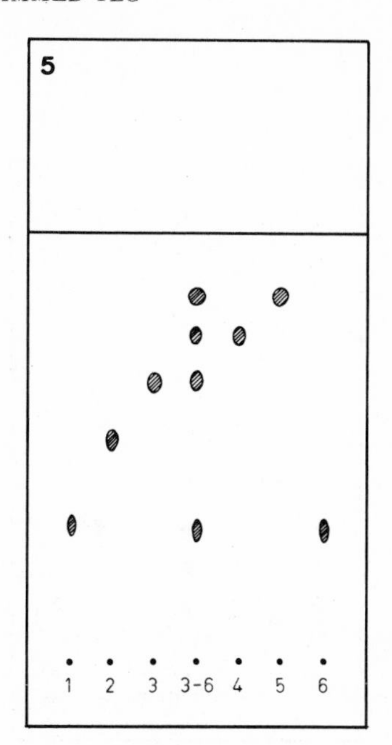

Fig. 5. Separation of hypnotics with chloroform-ether (75 + 25) in an unsaturated chamber; numbering as in Figure 4.

to solvent evaporation from the plate during the run. Hence, more solvent is needed to complete the 10 cm run, resulting in a longer duration of the development and a higher location of the spots.

Figure 6 shows the separation with chloroform and in Figure 7 ether is used as solvent, both in saturated chambers. The more polar character of ether results in higher R_f values but a closer examination of Figures 6 and 7 shows two differences in the separation sequence. In Figure 7, with ether as solvent, allobarbital (spot 3) moves faster than hexobarbital (spot 4) whereas bromisoval (spot 6) moves slower than heptobarbital, phenobarbital, and allobarbital (spots 1, 2, and 3).

With regard to Figures 4 and 5 we may conclude that in unsaturated chambers the influence of ether is smaller, since hexobarbital moves faster than allobarbital and bromisoval has the same migration rate as heptobarbital. This can be explained as follows. Ether is more polar than chloroform and the adsorbent has a greater affinity for ether vapor. However, in

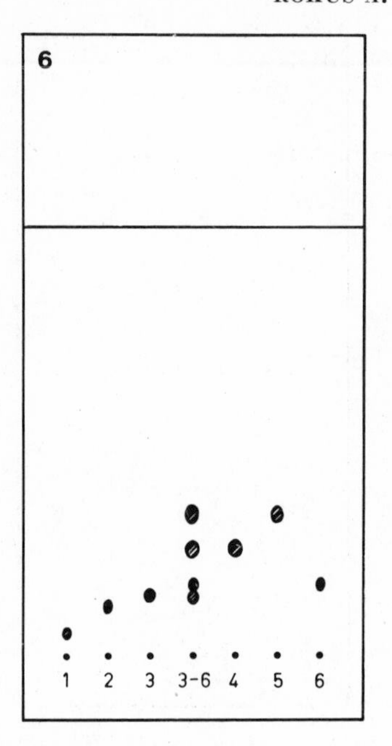

Fig. 6. Separation of hypnotics with chloroform in a saturated chamber; numbering as in Figure 4.

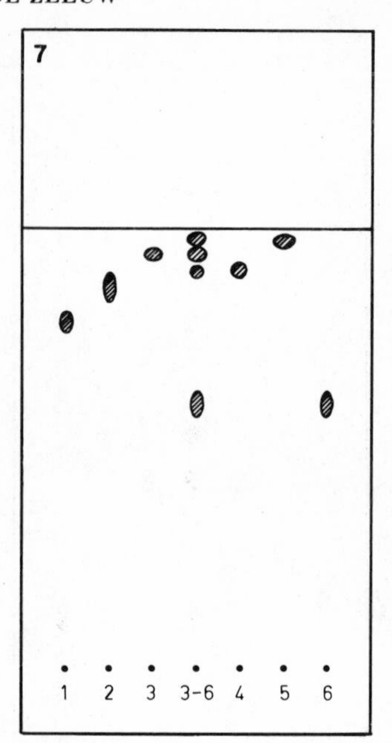

Fig. 7. Separation of hypnotics with ether in a saturated chamber; numbering as in Figure 4.

unsaturated chambers the amount of ether vapor available is less than in saturated chambers, especially in the beginning of the run and this results in a smaller ether vapor adsorption. This is in full agreement with the adsorption data of Figures 1 and 2.

From these experiments it becomes obvious that the amount of adsorbed ether vapor greatly determines the separation since the amount of ether in the solvent has not been changed. It should be noted that at the same time there is also an adsorption of the chloroform vapor but, as a consequence of the greater affinity of ether, this process will be of minor importance. Moreover, the effects of chloroform can be considered to be our base line from which the ether influence is examined. Therefore, it does not make a great difference if we replace the chloroform in the solvent by a less polar compound—benzene, for example. We should only take into account that benzene is less strongly adsorbed than chloroform. Accordingly the amount of adsorbed ether vapor will be higher. This is clearly shown in Figures 8 and 9. With benzene-ether (75 + 25) in unsaturated chambers the mixture 3–6 does not separate due to the greater influence of the ether. Further-

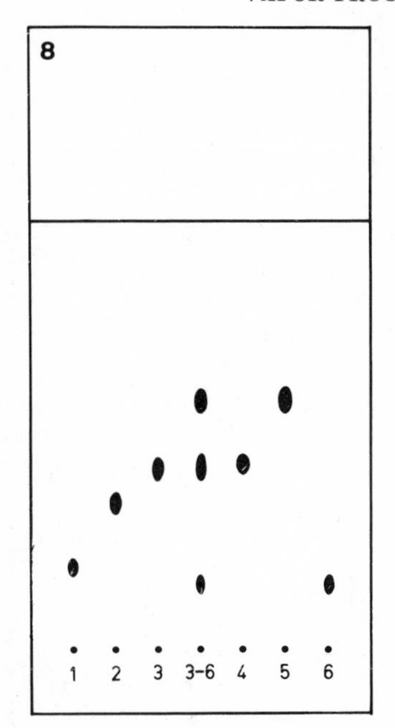

Fig. 8. Separation of hypnotics with benzene-ether (75 + 25) in an unsaturated chamber; numbering as in Figure 4.

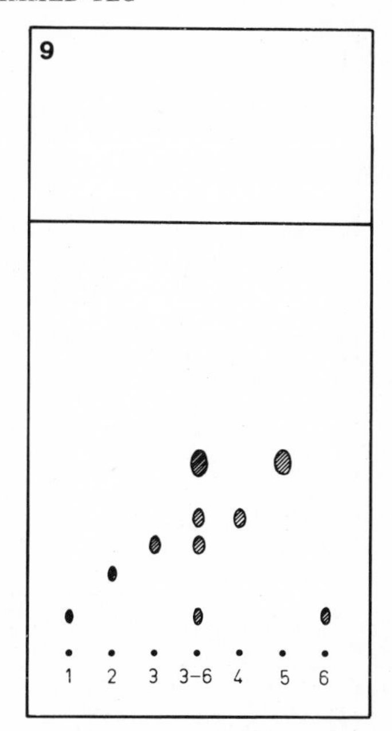

Fig. 9. Separation of hypnotics with benzene-ether (85 + 15) in an unsaturated chamber; numbering as in Figure 4.

more, bromisoval has a lower R_f value than heptobarbital. The mixture is fully separated, however, when the solvent ratio is changed to benzene-ether (85 + 15), with heptobarbital and bromisoval having the same migration rate. On the other hand, when using benzene-ether (85 + 15) in saturated chambers a chromatogram is obtained that is almost identical to the one in Figure 8 with the only difference being lower R_f values.

The role of the ether vapor can also be demonstrated using the following experiments. The separation of Figure 10 was obtained by developing the plate in an unsaturated chamber with chloroform only, but during development a trough containing 10 ml ether was present at the bottom of the chamber. The chloroform, the trough with ether, and the plate were placed in the chamber quickly one after the other. As no ether is present in the solvent, any influence of the ether can only be due to its vapor. This influence is evident if Figure 10 is compared with Figure 6, in which we also used chloroform as the solvent, but without a trough of ether. It can also be observed that the separation in Figure 10 is almost the same as in Figure 5, where we used chloroform-ether (75 + 25) as solvent.

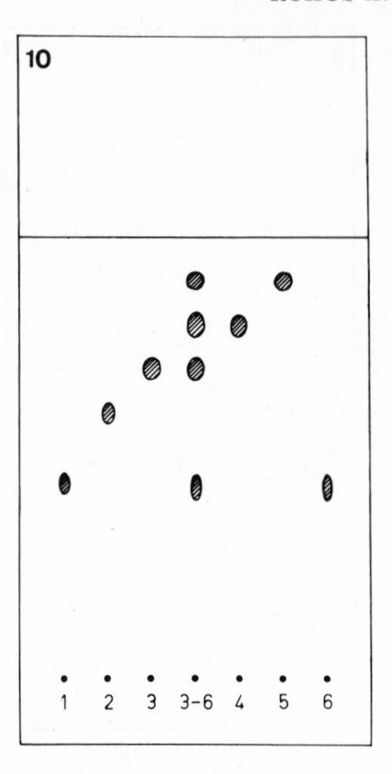

Fig. 10. Separation of hypnotics with chloroform in the presence of a trough with ether; unsaturated chamber; numbering as in Figure 4.

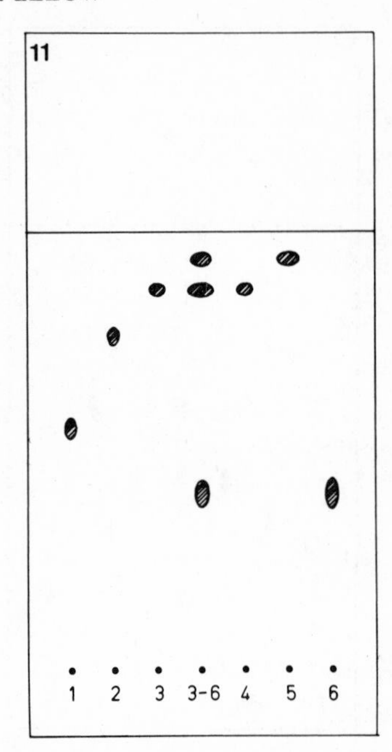

Fig. 11. Separation of hypnotics; presaturation of the plate with vapor by a trough with ether for 5 min, followed by development with chloroform; numbering as in Figure 4.

In the next experiment the amount of ether vapor available in the chamber was increased by putting the plate and the trough with ether together in the chamber for presaturation. After a 5-min presaturation time the solvent chloroform was added by means of a small tube inserted through the cover of the chamber. The separation is shown in Figure 11. The influence of ether is greater here because allobarbital and hexobarbital are not separated and bromisoval travels slower than heptobarbital. Presaturation with ether vapor for 30 min, followed by development in an S-chamber with chloroform resulted in the chromatogram of Figure 12. The effect of ether here apparently becomes so large that the separation is comparable with that in Figure 7 when we used ether only as solvent. For this experiment an S-chamber was preferred because it gives:

(*a*) decreased evaporation of adsorbed ether from the plate;
(*b*) decreased removal of adsorbed ether by chloroform vapor;
(*c*) decreased ether absorption by the solvent.

It becomes obvious from this experiment that quite small amounts of ether are responsible for great changes in the separation. After 30-min presaturation the plate has adsorbed about 1.5 mM of ether vapor, which is about 112 mg. Measurements with the solvent chloroform-ether (75 + 25) showed that 3.10 g was needed to wet 11.5 cm of the plate, in which about 415 mg of ether can be found. However, the influence of these 415 mg on the separation, which can be seen in Figure 5, is far less than the influence of the adsorbed 112 mg ether vapor which causes the separation of Figure 12. These data once again confirm our presumption that the amount of adsorbed ether vapor determines the nature of the separation to a large extent, whereas the amount of ether present in the initial solvent is of minor importance.

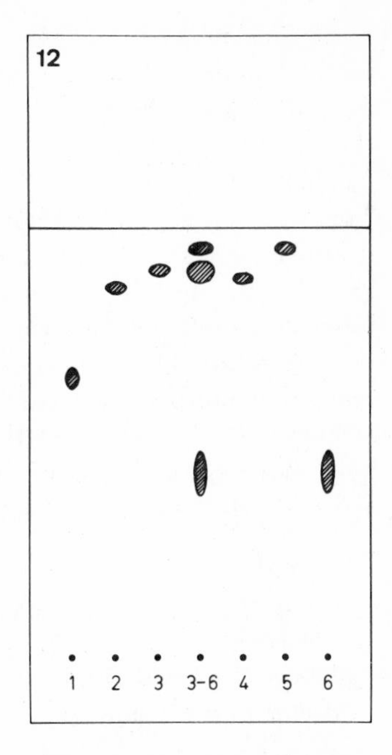

Fig. 12. Separation of hypnotics; presaturation of the plate with vapor by a trough with ether for 30 min, followed by development with chloroform in an S-chamber; numbering as in Figure 4.

It should be remembered that the influence of vapor is not restricted to vapor adsorption on the dry plate. In addition, there is also absorption of vapor by the solvent on the wet part of the plate. In our opinion however, the influence of the latter process is rather small in comparison to the influence of vapor adsorption.

D. Conclusions

The above investigations have shown that solvent vapor plays a very important role in TLC and, furthermore, it is clear that the different processes involved are generally rather complex. Therefore, it may be useful to summarize the different processes that take place during development in a scheme particularly for those processes which can be influenced by the investigator. We are aware of the shortcomings of this scheme but it might be valuable for further investigations. It should also be noted that transitions are given in the scheme and not chemical reactions.

The chromatographically important processes start at time 0 with the introduction of the solvent in the chamber (the beginning of vapor generation). The next step is the introduction of the plate which immediately begins to adsorb the generated vapor. The degree of adsorption will depend on the affinity of the vapor components for the adsorbent and on the rate of chamber saturation. The development of the plate then starts at moment I with the ascending of the solvent through the dry adsorbent. This will cause a partial adsorption of the solvent by the adsorbent (*cf.* frontal analysis). The adsorbed part of the solvent, together with the underlying adsorbent will now act as the stationary phase S. Although there are other interpretations existing for the stationary phase in TLC we believe that there are sufficient reasons to justify this term in our concept. The remaining nonadsorbed part of the solvent forms the mobile phase M, whereby Klinkenberg and Bayle (24) and Smit and Van den Hoek (32) have shown the existence of an equilibrium between both phases, so $S \leftrightarrows M$. We also make use of the following abbreviations:

Solvent = L
Vapor = V
Adsorbent = A
Vapor-impregnated adsorbent = iA
Mobile and stationary phase influenced by vapor = M^+ and S^+

The first transition at moment 0 is:

$$0^1 \qquad L \rightarrow V_0$$

When the plate is introduced adsorption will occur:

$$0^2 \qquad V_0 + A \rightarrow iA_0$$

The development of the plate then starts at moment I:

$$\mathrm{I}^1 \qquad L + iA \rightarrow S_\mathrm{I} + M_\mathrm{I}$$

At the same time vapor equilibria play a role:

$$\mathrm{I}^2 \qquad M_\mathrm{I},\ V_0, \text{ and } L \rightarrow V_\mathrm{I}$$

$$\mathrm{I}^3 \qquad V_\mathrm{I} + iA_0 \rightarrow iA_\mathrm{I}$$

$$\mathrm{I}^4 \qquad V_\mathrm{I} + M_\mathrm{I} + S_\mathrm{I} \rightarrow M_\mathrm{I}^+ + S_\mathrm{I}^+$$

The last four transitions describe the equilibria that tend to form between the solvent, the vapor phase, the adsorbent, and the mobile and stationary phase. The next transition represents the further ascending of the solvent:

$$\mathrm{I}^5 \qquad M_\mathrm{I}^+ + iA_\mathrm{I} \rightarrow S_\mathrm{II} + M_\mathrm{II}$$

At the same time, however, new supply of solvent is necessary and, moreover, new equilibria will tend to form between the various phases (II^1–II^4). Transition II^5 again represents the further ascending of the solvent, followed by transition III^1, and so on:

II^1 $\quad L + S_I \rightarrow S_I{}^1 + M_I{}^1$

II^2 $\quad M_{II}, M_I{}^1, V_I$, and $L \rightarrow V_{II}$

II^3 $\quad V_{II} + iA_I \rightarrow iA_{II}$

II^4 $\quad V_{II} + M_{II} + S_{II} \rightarrow M_{II}{}^+ + S_{II}{}^+$
$\quad V_{II} + M_I{}^1 + S_I{}^1 \rightarrow M_I{}^{1+} + S_I{}^{1+}$

II^5 $\quad M_{II} + iA_{II} \rightarrow S_{III} + M_{III}$
$\quad S_{II}{}^+ + M_I{}^{1+} \rightarrow S_{II}{}^1 + M_{II}{}^1$

III^1 $\quad L + S_I{}^{1+} \rightarrow S_I{}^2 + M_I{}^2$ . . . etc.

As it may be difficult to obtain a sufficiently clear insight in the various transitions, the illustrations of Figure 13 can be helpful. When using multicomponent solvents there will be no equilibrium during development between V, M, and S in both unsaturated and saturated chambers. When using a single-component solvent in the unsaturated chamber, there is again no equilibrium between V, M, and S in the beginning of the run, although the qualitative composition of the phases will be the same. Only the use of a single-component solvent in a fully saturated chamber will provide equilibrium during development between V, M, and S.

The transitions in the scheme and in the drawings once again underline the important role of solvent vapor. The adsorbed vapor can be considered the preliminary stationary phase because the adsorbate will mainly consist of the more polar solvent component(s). When the ascending solvent covers the vapor-impregnated areas this preliminary phase will be completed to a normal stationary phase. We further presume that solute separation will take place by interactions of the solute molecules with the stationary phase, with the mobile phase being the transport medium. So, the character of the stationary phase is very important for the separation and the experiments described above have clearly shown that this character can be highly influenced by vapor adsorption. For example, small amounts of adsorbed ether vapor on a plate which is developed with chloroform as solvent are causing such marked effects that we might think that ether was used as development solvent instead of chloroform.

In general, it can be concluded that the influence of the more polar solvent components upon the separation is mainly due to its vapor. Hence, it is possible to replace a great many multicomponent solvents by a system

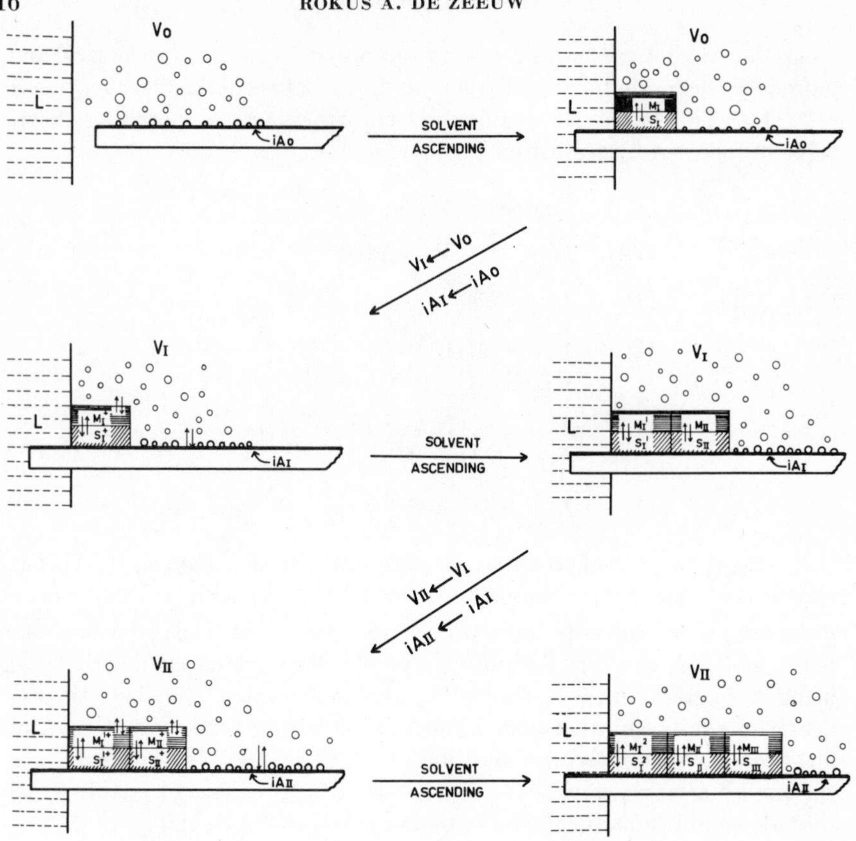

Fig. 13. Schematic illustration of the main transitions taking place during development of a plate in TLC. L = solvent; V = vapor; iA = vapor-impregnated adsorbent; S = stationary phase; M = mobile phase.

of a single-component solvent and one or more troughs with the more polar liquid components. The use of troughs also enables the investigation of the vapor effects of liquid components which cannot always be used in solvent systems. Components such as water and ammonia, which frequently cause solvent demixings, can then be investigated up to high vapor concentrations.

It was also shown that the best separations are not a priori obtained with saturated chambers, but that optimal conditions have to be established experimentally, especially with regard to the influence of solvent vapor. The use of unsaturated chambers or presaturation of the plate with one or more vapor components or the use of troughs containing liquids can be valuable.

It is well known that TLC is often compared with column chromatography (*cf.*, the name "open-column chromatography"). Moreover, many workers have tried to transfer TLC separations to columns for preparative purposes. However, the role of solvent vapor in TLC makes the assumed similarity very doubtful, since there is little or no possibility for vapor impregnation in column chromatography in the same characteristic way as is found in TLC. Nor are dry columns, such as described by Dahn and Fuchs (6) and Loev and Snader (26), comparable with TLC because the vapor impregnations take place in quite a different way. This explains why many TLC separations with multicomponent solvents cannot be obtained on columns. Recently, Van Dijk and Mijs (10) came to the same conclusions after investigating the vapor influence of benzene-methanol mixtures. Their results are in full agreement with our explanations (40,41).

Finally, there are the principles of the separation process in TLC. Until now TLC was generally considered to be based on adsorption chromatography (*cf.* 22). However, the experiments described above have shown that, particularly with the more polar solvents, a stationary phase of adsorbed solvent molecules is present. With single-component solvents this stationary phase also exists because part of the solvent is bound by the adsorbent, and this adsorbed part has different properties from the non-adsorbed part of the solvent. Consequently, the presence of both a stationary and a mobile phase allows partition processes to take place. On the other hand, these partition processes are, of course, influenced by the adsorption forces which are present within the stationary phase, so that it seems better to describe the TLC processes as a mixture of adsorption and partition chromatography. The extent to which one of these will dominate the other will be dependent on the polarity of the solvent components and on the adsorbent. With low-polar solvents, such as hexane or benzene, the stationary phase will be of little importance and adsorption chromatography will be predominant. However, with highly polar solvents, such as ethanol, methanol, or water, the stationary phase will be much stronger and then partition chromatography will be the main factor in the separation. When using solvents with medium polarity—chloroform, ether, acetone, dioxane—an intermediate form of adsorption-partition will be present.

The occurrence of partition chromatography can also be noticed in the linearity of a great many "adsorption isotherms" in TLC and by the fact that round spots are often obtained together with constancy of R_f values with increasing load. Moreover, Brenner *et al.* (5) have shown that for other more polar solvent systems the additivity rule of Martin (27) is valid. This includes a linear relationship between the R_M value $R_M = \log [(1/R_f) - 1]$ and the number of CH_2 groups in homologous series in separations based on partition chromatography.

IV. THE INFLUENCE OF WATER VAPOR

A. Introduction

After having observed the important role of solvent vapor in the separations it will also be necessary to consider the influence of water vapor. Although water is not generally used as a solvent component, the presence of water vapor in the ambient room atmosphere should be seriously taken into account. Even small amounts of water vapor can have great influence due to the great affinity of water for the various adsorbents used. In column chromatography this problem has been studied for many years, and Brockmann and Schodder (4) introduced the work with adsorbents of well-defined activity grades, based on the amounts of adsorbed water to the adsorbent. This provided more reproducible results.

Subsequently, it will be clear that in TLC, with various adsorbents being applied to the plate by means of an aqueous suspension, much attention was paid to the activation of the adsorbent layer. This has led to a great many procedures to obtain plates with a fixed activity, that is, with a reproducible amount of adsorbed water. In general, the preparation of the plates after application of the adsorbent layer includes a heating period for a certain time in an oven to remove excess water, followed by cooling and storage in a desiccator or a drying cabinet in the presence of a drying agent.

In this section we will consider the activation procedure in more detail. The experiments are done with silica gel, which is most often used as adsorbent, but the principles also apply to other adsorbents.

B. The Activation Process

The adsorption properties of silica gel for water as well as for other substances are mainly determined by the presence of surface hydroxyl groups (17). Beside these silanol groups siloxane linkages, which are far less adsorption-active, will be present at the surface. The number of available silanol groups and the other properties of the adsorbent, such as pore diameter, particle size, and surface area, are strongly determined by the manufacturing processes, thus it is obvious that one should carefully distinguish between silica gels of different manufacturers. Frequently, these products, even though they have identical names, show completely different adsorption properties.

The predominant activity of the silanol groups is responsible for the strong adsorption of water, which is bound by means of hydrogen bonding. This physisorbed water prevents the adsorption of less polar substances, thus deactivating the adsorbent. According to Snyder (33), physisorbed water can be removed by heating between 150 and 250°C, while heating over 250°C results in dehydration of the silanol groups, thus forming siloxane linkages. Rehydration of siloxane bonds does not occur at an appreciable rate in atmospheres below saturation with water vapor (23). So, activation of the TLC plates between 150 and 250°C yields the most active adsorption without decreasing the silanol groups.

An enormous variety of activation procedures have been evolved for TLC. These range from 100° to 250°C in temperature and from 20 min to several hours in duration. However, 30 min at 110°C seems to be the most accepted method. In order to maintain the high activity the plates are usually stored over a drying agent.

The next step is the application of the solutes and here, in particular, some major factors have often been overlooked or have had insufficient attention. Namely, the activated plates are removed from the dry atmosphere of the desiccator or storage cabinet and then spotted in free contact with the ambient laboratory atmosphere. During this period the activated plate will immediately adsorb water vapor from the ambient atmosphere, the quantity being dependent on the duration of the spotting period and on the relative humidity of the atmosphere. The rate of adsorption is very fast as can be seen in Figure 14, which illustrates the water vapor uptake of a 0.25 mm silica gel plate of 20 × 20 cm at a relative humidity of 50%. After about 3 min, half the initial activity has disappeared and in about 15 min the maximum water vapor uptake is ob-

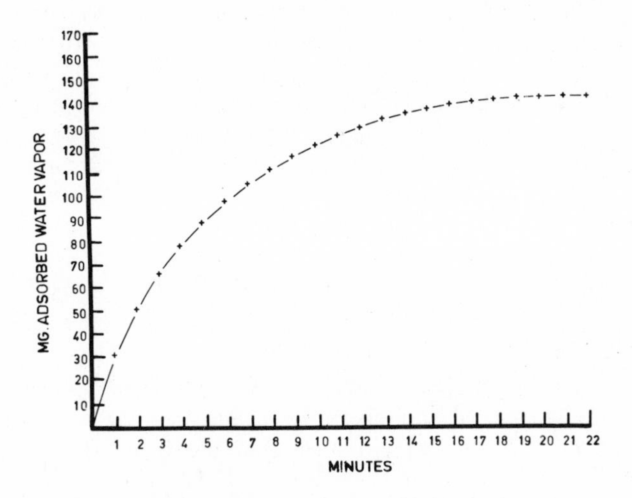

Fig. 14. Adsorption of water vapor at an activated silica gel plate in a relative humidity of 50%. The plate was activated for 30 min at 110°C then cooled and stored in a desiccator over calcium oxide. Plate size 20 × 20 cm, layer thickness 0.25 mm. Weight measurement on a Mettler BCH balance.

tained. Thus, during the spotting period a great deal of the initial activity is lost and, furthermore, it should be pointed out that if the duration of the spotting period is not standardized, TLC plates of varying activity will result. Subsequently, this will influence the separation to various extents so that decreased reproducibility may occur. There are two solutions for this problem:

(*a*) working with plates of maximum activity;
(*b*) working under standardized humidity conditions.

It will be clear that complete exclusion of water in TLC is extremely difficult. Moreover, such a technique would be very expensive and the simplicity of the normal TLC procedures would be compromised. In some instances, however, experiments have been carried out on plates with the highest possible activity. This is done by reactivation of the plates after spotting, by heating in an oven (*cf.* 38). After cooling over a drying agent development is started immediately. In this way only water vapor available in the development chamber and water vapor adsorbed during transport from the desiccator to the chamber can play a role during development.

We have investigated the reproducibility of the separations obtained with this procedure. The results indicated however, that considerable differences in R_f values could be obtained and we, therefore, have studied in more detail the effects of small humidity variations in separations of hypnotics. This was done by varying the amounts of water vapor available during the development of six different plates loaded with the following substances: (1) heptobarbital; (2) phenobarbital; (3) allobarbital; (4) hexobarbital; (5) methylphenobarbital; and (6) bromisoval and (3–6), a mixture of 3 + 4 + 5 + 6. After spotting the following steps were carried out, but under different conditions for each plate (details are given in Table I).

(*a*) Reheating of the spotted plate in an oven with a fan for 30 min at 110°C to expel adsorbed water.

(*b*) Cooling for 24 hr in a desiccator, with or without a drying agent.

(*c*) Development in unsaturated chambers at various humidities with chloroform. These humidities were established by means of troughs containing a drying agent. High humidities were obtained by using a trough with water. The troughs were placed in the chamber 24 hr before development and were allowed to remain in the chamber during development.

As a result of the differences in development procedures, the effect of increasing humidity on the separation of the hypnotics can be observed. Knowledge of the capacities of the drying agents (37) and the volumes of the desiccator (26 l) and the chamber (3.6 l) enables us to calculate the

TABLE I
Procedures of Plate Development at Various Humidities

Procedure	Reheating	Drying Agent During Cooling	Contents of Trough in Development Chamber
A	30 min at 110°C	P_2O_5	P_2O_5
B	30 min at 110°C	P_2O_5	H_2SO_4
C	30 min at 110°C	P_2O_5	CaO
D	30 min at 110°C	P_2O_5	none
E	30 min at 110°C	none	none
F	30 min at 110°C	none	H_2O

various amounts of water involved in the separations. These are listed in Table II.

The separations obtained by the different procedures are shown in Figures 15 through 20. The R_f values of the spots are listed in Table III. With increasing amounts of water a rise in R_f values can be noticed which is normal under these conditions and which is quite distinct in procedure D. The increase in R_f values can be explained by a decrease in activity. However, in procedures E and F with still increasing amounts of water there is

TABLE II
Amounts of Water Involved in the Different Separation Procedures

Procedure	A	B	C	D	E	F
Water in mg	<0.001	<0.05	<1	~48	~400	~450

TABLE III
R_f Values Obtained by the Different Separation Procedures

Substance	Procedure ($R_f \times 100$)					
	A	B	C	D	E	F
Heptobarbital	5	5	6	9	8	6
Phenobarbital	10	10	12	18	14	11
Allobarbital	12	13	15	26	17	13
Hexobarbital	18	23	29	70	27	24
Methylphenobarbital	23	31	37	84	35	31
Bromisoval	13	17	20	25	17	18

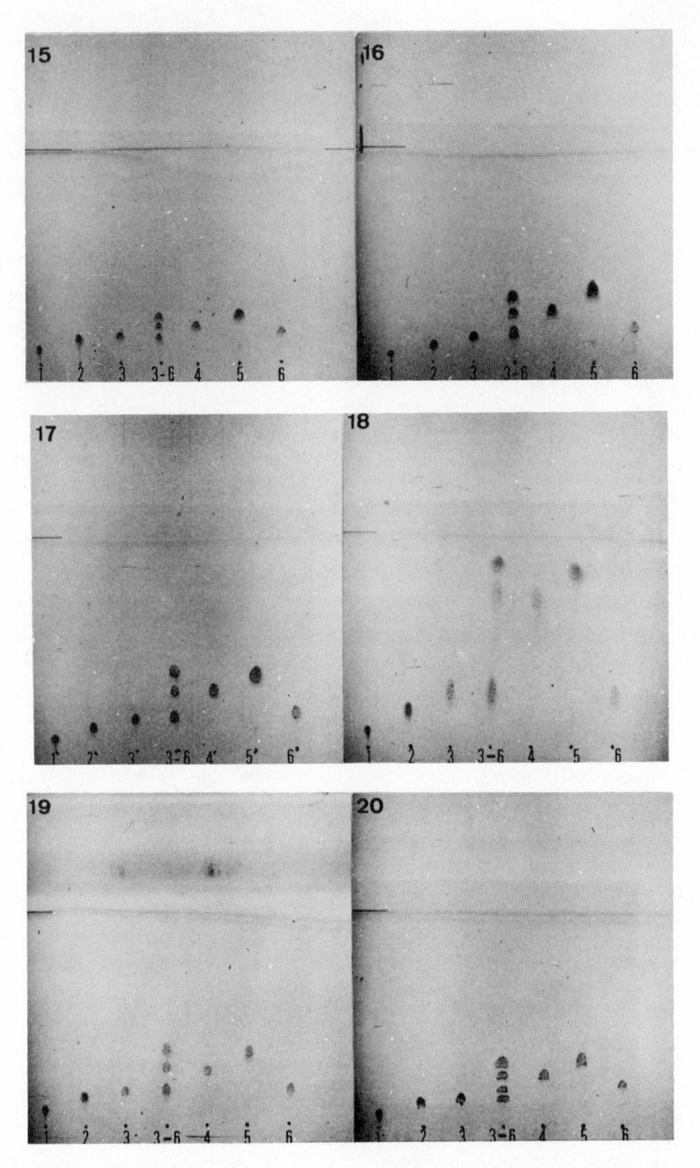

Figs. 15–20. Influence of increasing humidities on the separation of hypnotics. Development with chloroform according to procedures *A* though *F* as described in the text. Photography in UV light of 254 nm after activation with ammonia vapor. 1 = heptobarbital; 2 = phenobarbital; 3 = allobarbital; 4 = hexobarbital; 5 = methylphenobarbital; 6 = bromisoval; 3–6 = mixture of 3 + 4 + 5 + 6. Reproduced from R. A. de Zeeuw, "Influence of humidity variations in the thin layer chromatography of hypnotics," *J. Chromatog.*, **33,** 227 (1968), with permission of the publishers.

a very sharp fall in the R_f values. Hence, the R_f values in procedure F at a relative humidity of nearly 100% become so low that they are comparable to those obtained with the rather dry method B. These observations are surprising indeed because one would not expect a decrease of R_f values with increasing humidity. The effects cannot be ascribed to the presence of a small quantity of ethanol in the solvent (1% for stabilization) because the same results were obtained with chloroform free from ethanol. We have no theoretical explanation available for this phenomenon, nor do we know whether this effect can also be found with other substances and/or solvent systems. A possible explanation might be found in the assumption of a certain transition from adsorption chromatography to partition chromatography.

The results underline, however, that reactivation of the plates after spotting still remains dangerous. Because small humidity variations can cause marked changes in the R_f value, the procedure is too irreproducible for general use. Moreover, one must carefully examine every combination of solutes and solvents to ascertain if a certain reproducibility under various humidity conditions is not an accidental one. For example, in our investigation we found similar R_f values with the "dry" method B and the "wet" method F, thus indicating a very small humidity effect, but a more systematic study clearly showed us that this was not the case.

C. Working under Standardized Humidity Conditions

Working under well-defined humidity conditions has had the attention of various investigators. Badings (2) carried out separations after having equilibrated the plate in an atmosphere of nitrogen or air of constant relative humidity in the development chamber. Geiss *et al.* (13) used a special climatized development chamber with which parts of the plate or the whole plate can be climatized over different saturated salt solutions. Drost and Reith (11) cooled the plates after activation in a climatized cabinet with a constant relative humidity of 60%. Spotting is also done in this cabinet, followed by development in the usual way. The first two methods have the disadvantage of being rather complicated and time-consuming. The last method is elegant and simple but one should realize that optimal separations are not always obtained at a relative humidity of 60% and that it could be necessary to use other relative humidities. Furthermore, it should be remembered that when using a certain humidity in the cabinet, the development chamber should also be brought to the same relative humidity to avoid difficulties.

However, in our opinion the best solution for working under standardized humidity conditions is found by using a controlled humidity room. The plates are then allowed to equilibrate in the ambient room atmosphere at a given relative humidity and at a given temperature, which will result in the maximum water vapor uptake under these conditions, so that fixed amounts are obtained. Thus, no precautions are required during the spotting period, there is no difference in relative humidity inside and outside the development chamber, the plates can be stored without drying agents, and a whole range of desired humidities can easily be established by setting a single knob. Moreover, it becomes very easy to establish the most suitable relative humidity and temperature to give optimal separation. With this system we could obtain very reproducible R_f values, indeed.

For those laboratories not equipped with a controlled humidity environment, we advocate a system which we have tested and found to provide suitable results. The method requires the daily registration of the relative humidity and, by using a suitable reference mixture, one can establish whether the separations remain reproducible. This determines in which humidity range reproducibility is observed. Thus, no problems arise when the work is done within this reproducibility range but if, on the other hand, work is done outside that range one must be warned that R_f values may have changed. In those cases, however, the use of the reference mixture can be very helpful in the identification procedure. It becomes clear that when publishing R_f values, the relative humidity under which the work was done must be reported. In the past this has not been done routinely and some of the confusion arising in trying to verify reported R_f values in other laboratories may indeed be due to the lack of a knowledge of the relative humidity at which the original values were established.

This method provided fairly reproducible R_f values if the time between the removal of the plate from the desiccator and the introduction into the chamber was sufficiently long. For silica gel plates we used an acclimatization period of at least 15 min and it can be seen from Figure 14 that this allows for the maximum water vapor uptake. During this period there is ample time for spotting. The proposed method can also be used for nonactivated plates, provided that the plates are allowed to dry for a sufficiently long period after preparation to achieve equilibration with the ambient relative humidity.

When looking for humidity ranges in which a particular separation shows reproducibility, it can be noticed that these ranges increase in proportion as the polarity of the solutes and solvents increase. This means that with nonpolar substances and solvents—azo dyes developed with benzene, for example—only a small reproducibility range is available and, accordingly, marked humidity effects may occur. With medium polar systems the ranges increase, for example, when working with barbiturates and chloroform-ether solvents we found reproducible results in the range of 20–50%

relative humidity, while with the more polar system chloroform-isopropanol-25% ammonia (45 + 45 + 10) reproducibility was observed between 20 and 80% relative humidity.

These observations can be explained by the fact that solvent vapor in the development chamber tends to displace the adsorbed water vapor. The rate of displacement is dependent on the polarity and the quantity of the solvent vapor components. Nonpolar solvents, such as benzene, are less effective and, since there is a great difference in polarity between water and benzene, even small variations in the amount of adsorbed water can cause great changes in the R_f values. When working with suitable amounts of highly polar solvents, such as isopropanol and ammonia, the adsorbed water will be displaced to a great extent, the equilibrium being mainly dependent on the high concentration of solvent vapor. Differences in the amount of initially adsorbed water are thus leveled down. Moreover, the possible effect of different amounts of adsorbed water vapor is much smaller because the character of the solvent system is much more polar. Thus, reproducibility can be observed over a wide relative humidity range.

V. THE USE OF UNSATURATED CHAMBERS

A. Introduction

It was shown in Section III that separations in TLC are very dependent on the adsorption of solvent vapor. It was also observed that in separations of hypnotics with chloroform-ether (75 + 25) as solvent the use of unsaturated chambers was more effective than saturated chambers. The unsaturated chamber provided a more efficient separation in combination with improved spreading of the spots. However, if these two improvements were due only to a decreased ether vapor adsorption as compared to the saturated chamber, then it should be possible to obtain the same results by using a solvent with a decreased percentage of ether in the saturated chamber. Despite many experiments this could not be obtained and, more importantly, the spread of the spots could not be improved. So there probably was a second factor which made the results in the unsaturated chamber superior to those obtained in the saturated chamber.

B. The Vapor Gradient in Unsaturated Chambers

An example of the obvious advantages of the unsaturated chamber over the saturated one can be seen in Figure 21, in which a selection of barbiturates is developed over a distance of 17 cm. In (*a*) the separation is given with chloroform-acetone (90 + 10) in saturated chambers. The R_f values are rather low, with all spots remaining at the lower part of the plate.

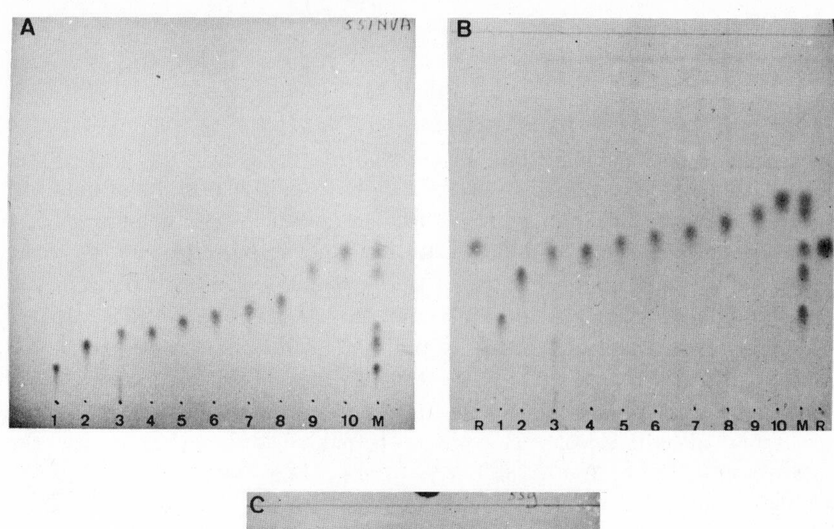

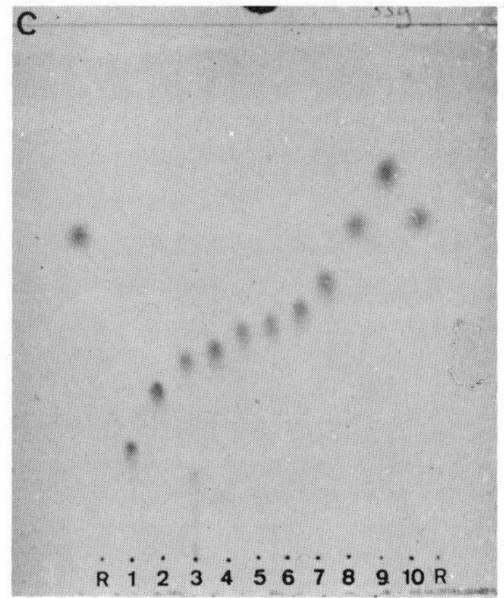

Fig. 21. Improved separation of hypnotics on silica gel in unsaturated chambers as compared to separations in saturated chambers. (*A*) Solvent chloroform-acetone (90 + 10), saturated chamber; (*B*) Solvent chloroform-acetone (80 + 20), saturated chamber; (*C*) Solvent chloroform-acetone (90 + 10), unsaturated chamber. Photography in UV light of 254 nm after activation with ammonia vapor. 1 = heptobarbital; 2 = phenobarbital; 3 = cyclobarbital; 4 = allobarbital; 5 = butobarbital; 6 = pentobarbital; 7 = butalbital; 8 = secobarbital; 9 = hexobarbital; 10 = methylphenobarbital; M = mixture of 1 × 2 + 4 + 9 + 10; R = reference substance 4-nitroaniline. Reproduced from *Anal. Chem.*, **40,** 915 (1968), copyright 1968 by the American Chemical Society; reproduced with permission of the copyright owner.

Changing the solvent ratio in chloroform-acetone (80 + 20) shows an increase of all R_f values, but the spots move over the plate as a zone and the separation is not improved as shown in (*b*). The differences between the R_f values remain almost the same but due to the greater diffusion of the spots the separation becomes worse. In (*c*) the result is shown of the separation in unsaturated chambers with chloroform (90 + 10). The separation and the spread of the spots is much better now. The R_f values of the separations are given in Table IV. For the sake of completeness it should be observed that when the solvent front has reached the 17 cm line more solvent is used in the case of unsaturated chambers because of evaporation of solvent from the plate during the run and, accordingly, solute spots will move higher than in the case of saturated chambers.

TABLE IV

Rf Values of Some Hypnotics Obtained in Saturated and Unsaturated Chambers

	$Rf \times 100$		
	Saturated Chamber, Chloroform-Acetone		Unsaturated Chamber, Chloroform-Acetone
Substance	90 + 10	80 + 20	90 + 10
Heptobarbital	10	24	21
Phenobarbital	17	36	31
Cyclobarbital	20	42	37
Allobarbital	21	42	39
Butobarbital	24	44	42
Pentobarbital	27	45	44
Butalbital	29	46	47
Secobarbital	31	49	52
Hexobarbital	41	52	62
Methylphenobarbital	46	56	73

The vapor adsorption curves of acetone and chloroform from saturated and unsaturated atmospheres are given in Figures 22 and 23. The curves are obtained as described in Section III.C. For comparison reasons the adsorption curve of ether is also given.

The vapor pressures of acetone, ether, and chloroform vapor at 20°C are 178 mm, 435 mm, and 154 mm, respectively (36). Under these conditions the amount of acetone vapor needed for adsorption is again available in

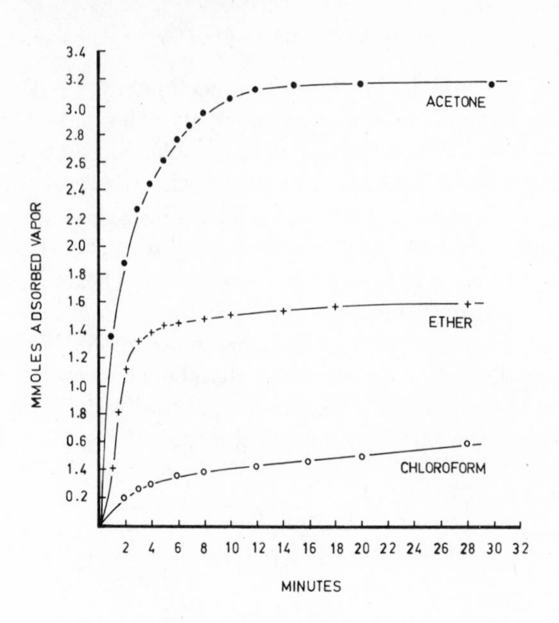

Fig. 22. Vapor adsorption of acetone, ether, and chloroform by a silica gel plate from a saturated atmosphere. Plate size 20 × 20 cm, layer thickness 0.25 mm. Weight measurement with a Mettler K 7 top balance as described in the text. Figures 22 and 23 are reproduced from *Anal. Chem.*, **40,** 915 (1968), copyright 1968 by the American Chemical Society; reproduced with permission of the copyright owner.

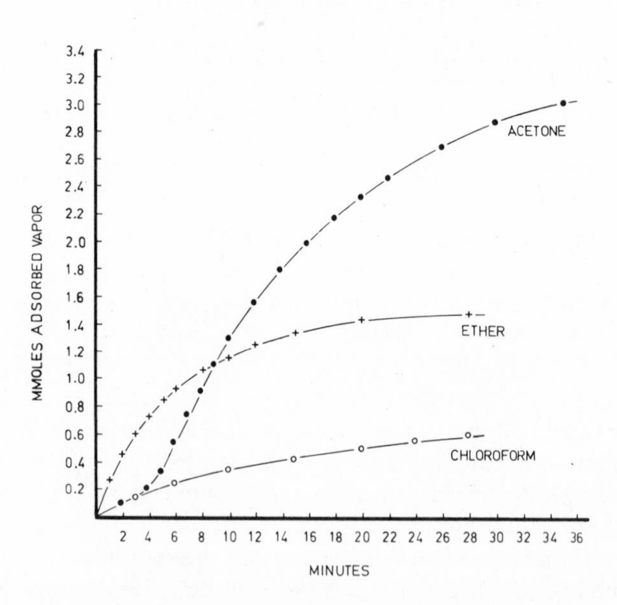

Fig. 23. Vapor adsorption of acetone, ether, and chloroform by a silica gel plate in an unsaturated atmosphere; details as in Figure 22.

less than 1 cm of the vapor present over the plate. From the adsorption curves it can be concluded that, when using chloroform-acetone solvent systems in saturated chambers, the adsorbate will mainly consist of acetone. The uptake of acetone is rapid and in a few minutes the entire plate is covered with a fixed amount of acetone. The ascending solvent then covers the vapor-impregnated areas and the migration of the solutes is determined by the amount of adsorbed acetone.

In unsaturated chambers development is started immediately after the introduction of the solvent. In this case, however, the acetone vapor adsorption is limited in the beginning of development because evaporation of sufficient amounts of acetone takes a certain time. The solvent ascends at once and gradually covers more and more of the dry adsorbent thus preventing further direct vapor adsorption in these areas. Hence, it becomes obvious that because of the ascent of the solvent and the increasing amounts of available acetone vapor, the upper part of the plate is able to take up more acetone vapor than the lower part and, accordingly, a concentration gradient of adsorbed vapor exists before the solvent covers the impregnated areas. Such a gradient is very suitable for separation purposes because the faster moving spots gradually meet areas where more of the polar vapor components have been adsorbed and this results in an acceleration of the migration of these spots. But the slower spots only meet areas with a relatively small amount of polar vapor and their migration rate is affected to a much smaller extent. Thus, better separations can be obtained with multicomponent solvents in unsaturated chambers because a gradient is formed with an increasing polarity when going up the plate. Therefore with multicomponent solvents unsaturated chambers should be preferred to saturated chambers. With single-component solvents no improvement can be expected, but the saving of time, because no saturation period is needed, might be valuable.

It should be noted that the gradient in unsaturated chambers is basically different from what is called gradient elution. In this technique the gradient is a pushing one, coming from the solvent reservoir and pushing up the lower spots which reduces the distances between the various spots. Gradient elution is, therefore, suitable for chromatography of substances belonging to different polarity classes on the same plate. The gradient in unsaturated chambers on the other hand is a pulling one, accelerating the migration of the higher spots and enlarging the distances between the various substances. This will be very useful in separations of chemically closely related compounds.

The rate of separation improvement in unsaturated chambers is, of course, dependent upon the steepness of the vapor gradient. This is influenced by several factors, such as polarity of the vapor components, volatility, chamber volume, chamber geometry, diffusion velocity, temperature, activity of the adsorbent, and composition of the solvent. For example, when using a mixture of hexane and benzene as solvent, the adsorption of benzene shows a very flat adsorption curve and, accordingly, the difference between saturated and unsaturated development is hardly visible. On the other hand, components such as ether, acetone, dioxane, ethylacetate, and lower alcohols will provide much stronger gradients and with solvents containing these components significant improvements can be expected in unsaturated chambers.

Finally, it should be observed that the extra solvent transport across the plate in unsaturated chambers, due to solvent evaporation, tends to diminish the separation improvements. The lower spots are influenced to a stronger extent than the higher spots because the extra solvent supply originates from the solvent reservoir, which causes a pushing up of the lower spots. However, the appearance of many significantly improved separations in unsaturated chambers makes it clear that the adverse effect of the extra solvent transport is, in general, very small in comparison to the positive effect of the vapor gradient.

C. The Use of Troughs for Improving the Vapor Gradient

When searching for optimal separations it often occurs that the composition of the solvent includes some restrictions. An example is the solvent chloroform-isopropanol-25% ammonia (45 + 45 + 10) which, after being introduced by Deiniger (8), was often used for the separation of hypnotics in both paper and thin-layer chromatography. In unsaturated chambers this solvent provided the separation shown in Figure 24 with a number of hypnotics. The same result could be obtained with chloroform-isopropanol (50 + 50) as solvent in unsaturated chambers but in the presence of a trough with 10 ml 25% ammonia at the bottom of the chamber. This once again demonstrates the influence of the vapor. The obtained separation is sufficient, but not optimal, because the more polar substances, heptobarbital and phenobarbital, run too high. It will be clear that a decrease of the migration of these spots can be obtained by decreasing the proportion of isopropanol in the solvent, the isopropanol being the more polar component. Decreasing the amount of ammonia was also considered, but it does not have any effect. So, presumably, the ammonia is to be considered as the pH controlling agent in this case. However, with lower concentrations of isopropanol, demixing of the solvent occurs when adding the 10 parts of ammonia. Thus, it becomes evident that the high amounts of isopropanol are present to prevent demixing. However, we can easily get around the occurrence of demixing by replacing the ammonia in the solvent by a trough with ammonia. Subsequently, the separation as illustrated in

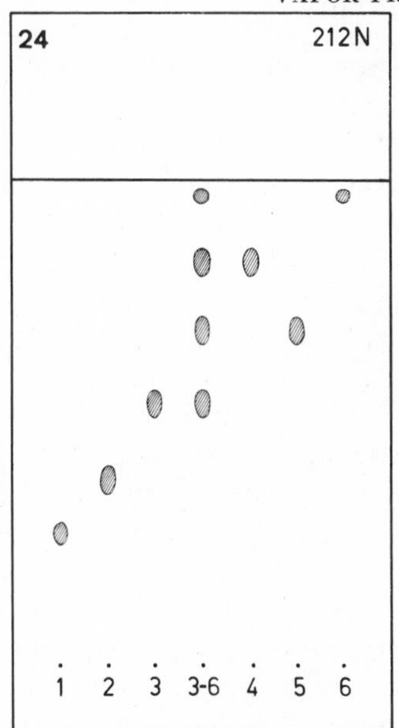

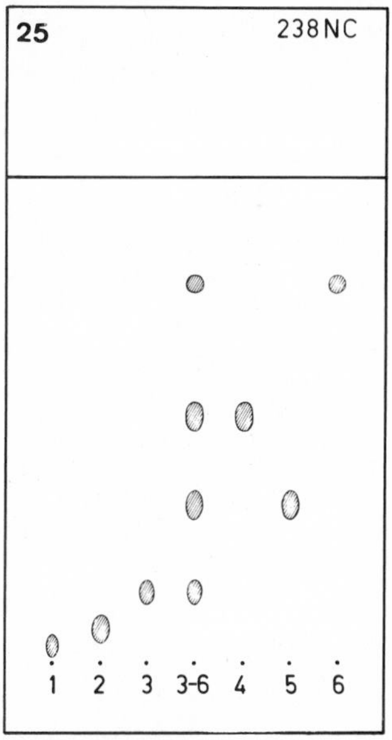

Fig. 24. Separation of hypnotics with chloroform-isopropanol-25% ammonia (45 + 45 + 10) in an unsaturated chamber. 1 = heptobarbital; 2 = phenobarbital; 3 = allobarbital; 4 = hexobarbital; 5 = methylphenobarbital; 6 = bromisoval; 3–6 = mixture of 3 + 4 + 5 + 6.

25
238NC
1 2 3 3-6 4 5 6

Fig. 25. Separation of hypnotics with chloroform-isopropanol (95 + 5) in the presence of a trough with 10 ml 25% ammonia; unsaturated chamber; numbering as in Figure 24.

Figure 25 was obtained with chloroform-isopropanol (95 + 5) in the presence of a trough with 10 ml 25% ammonia. The two lower spots now have the desired small migration rates but the other four less polar substances—allobarbital, hexobarbital, methylphenobarbital, and bromisoval—show decreased R_f values as well, thus having an adverse effect on the spreading. So, obviously, the concentration of isopropanol is too low here, but an increase to 10% in the solvent again resulted in a too high migration of the lower spots, so that optimal spreading could not be obtained by changes in the solvent composition only.

This is caused by the fact that in this separation the evaporation velocity and diffusion rate of isopropanol is too low with the vapor gradient not being steep enough. A higher proportion of isopropanol in the solvent yields more vapor, resulting in higher R_f values of the less polar substances but the more polar substances are then too much influenced by the higher concentration of isopropanol in the ascending solvent.

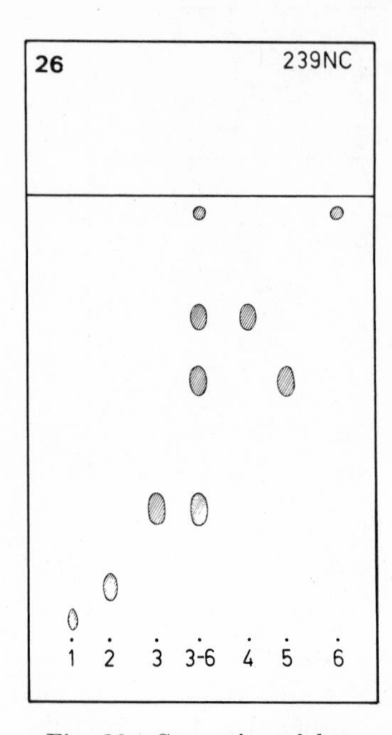

Fig. 26. Separation of hypnotics with chloroform-isopropanol (95 + 5) in the presence of a trough with 20 ml isopropanol-25% ammonia (67 + 33); unsaturated chamber; numbering as in Figure 24.

The latter effect can be prevented, however, by using isopropanol in the trough, together with the ammonia, in combination with chloroform-isopropanol (95 + 5) as running solvent. The extra isopropanol in the trough increases the total amount of vapor and the upper parts of the plate are now able to adsorb these higher quantities, resulting in a further migration of the upper spots. Due to the slow evaporation of isopropanol the lower parts of the plate cannot adsorb extra isopropanol vapor, because of the solvent ascending, so the migration of the lower spots will be scarcely affected. The resulting separation is shown in Figure 26, with the spots being spread across the entire plate. The composition of the liquid in the trough was 20 ml isopropanol–25% ammonia (67 + 33).

The use of troughs also permits the testing of various new solvent systems. Chloroform-methanol-ammonia mixtures have not been used so far in the analysis of hypnotics because too much methanol was required to prevent demixing of the ammonia. These high concentrations of methanol made the systems much too polar. However, when using a trough with ammonia then chloroform-methanol solvents can be easily tested with the methanol being capable of giving a steep gradient. In order to obtain an optimal separation here too, it proved to be necessary to use part of the methanol in a trough, as can be seen in Figure 27. In (*a*) the separation is given with chloroform-methanol (95 + 5) in unsaturated chambers, in the presence of a trough with 17.5 ml 25% ammonia. The spots have rather low R_f values. Changing the solvent ratio to chloroform-methanol (80 + 20) resulted in separation (*b*), but due to the increased migration of the lower spots the separation is not improved. Hence, in (*c*) we used chloroform-methanol (95 + 5) as solvent in combination with a trough containing 25 ml

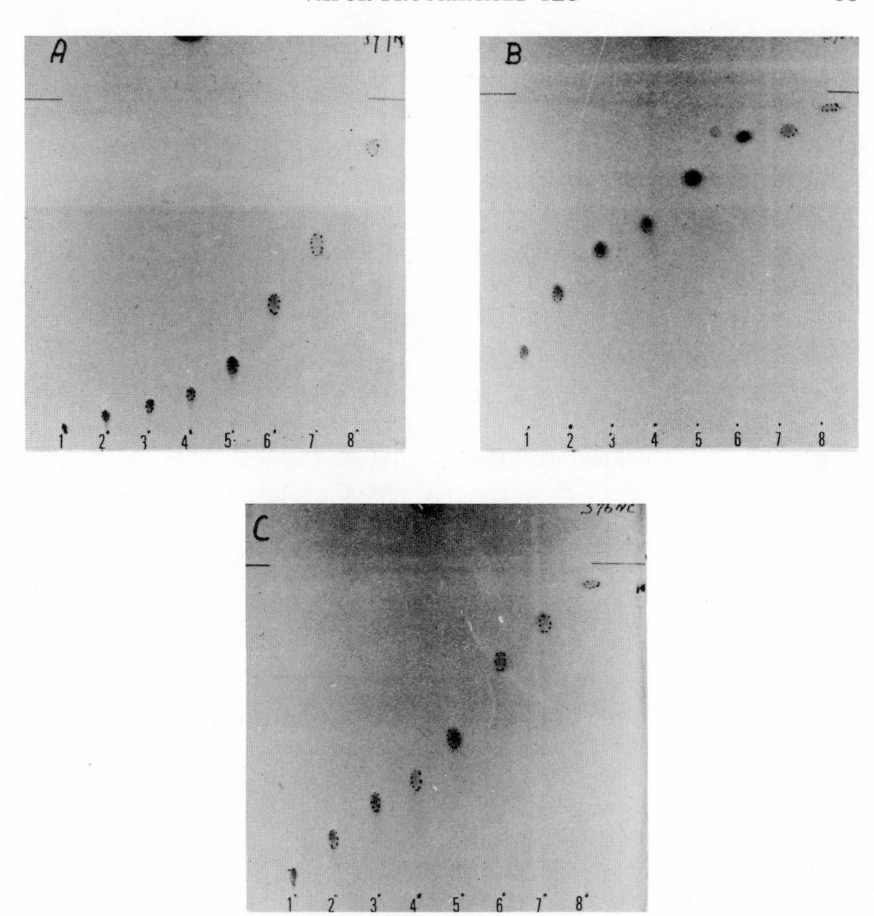

Fig. 27. Optimalization of a separation of hypnotics with the aid of troughs. (*A*) Solvent chloroform-methanol (95 + 5) in the presence of a trough with 17.5 ml 25% ammonia; too small migration of the spots. (*B*) Solvent chloroform-methanol (80 + 20) in the presence of a trough with 17.5 ml 25% ammonia; too high migration of the spots. (*C*) Solvent chloroform-methanol (95 + 5) in the presence of a trough with 25 ml methanol–25% ammonia (30 + 70); optimal separation. Photography in UV light of 254 nm after activation with ammonia vapor. 1 = heptobarbital; 2 = allobarbital; 3 = aprobarbital; 4 = amobarbital; 5 = methylphenobarbital; 6 = hexobarbital; 7 = bromisoval; 8 = carbromal.

methanol–25% ammonia (30 + 70). The stronger methanol gradient thus formed shows a significantly improved separation.

From these experiments it becomes clear that by a suitable change in the solvent ratio, together with the use of troughs the vapor gradient can be influenced so that it can be adapted to a particular separation problem. This is a very valuable factor when searching for optimal separations. Obviously, a thorough knowledge of the properties of the various solvent components and their vapors, as well as of their influence on the migration rate of the solutes, is a primary requisite.

D. Reproducibility

It has been often mentioned that unsaturated chambers provide less reproducible results than saturated chambers (*cf.* 30). Assuming that the other factors affecting the reproducibility can be kept constant when working in unsaturated chambers, then the degree of saturation of the chamber is the deciding factor. This means that reproducibility is dependent on the rate of vapor adsorption. In saturated chambers an attempt is made to achieve maximum vapor uptake and, if this is done properly, fairly reproducible R_f values will result.

In unsaturated chambers vapor adsorption will be different and will take place at a slower rate than in saturated chambers. However, if we can achieve a fixed type of unsaturation and an identical course of vapor adsorption during the run, we can then also expect reproducible R_f values, provided that the other factors are kept constant. This means that the condition in the unsaturated chamber must be standardized. This can be done by starting development of the plate immediately after introduction of the solvent. In that case, no vapor is present at the beginning of the run and, keeping other factors constant, this will lead to reproducible vapor conditions during development, thus giving reproducible R_f values. The results obtained with this technique show the same reproducibility as in saturated chambers as can be seen in Table V. Experiments were done at 22°C and at a relative humidity of 45–51%.

The same principles are valid for the use of troughs. Here too, reproducible vapor conditions are required. Therefore, if the trough procedures are suitably standardized, presaturation of the dry plates is carried out particulary if good results are obtained.

The use of a reference mixture to be spotted on each plate remains valuable. This provides a good control for sufficiently reproducible vapor conditions, while the occurrence of other derangements can also be simply detected.

From this it may be concluded that the observed nonreproducibility of unsaturated chambers in earlier investigations has been caused by an insufficiently uniform unsaturation, due to the fact that one was not aware

TABLE V
Reproducibility of R_f Values in Unsaturated Chambers Obtained in Five Different Experiments

Solvent	Substance	$R_f \times 100$				
Benzene	Indophenol	13	14	15	14	14
	Sudan red G	29	28	31	29	29
	Butter yellow	70	70	73	70	70
Chloroform-ether (75 + 25)	Heptobarbital	25	24	26	25	25
	Phenobarbital	37	38	36	39	37
	Allobarbital	47	50	46	49	48
	Hexobarbital	63	65	61	64	63
	Methylphenobarbital	73	76	72	75	75
Chloroform-isopropanol-25% ammonia (45 + 45 + 10)	Heptobarbital	29	30	31	30	31
	Phenobarbital	37	39	39	38	38
	Allobarbital	53	53	53	53	51
	Methylphenobarbital	69	68	67	67	67
	Hexobarbital	80	80	78	79	77

of the role of solvent vapor on the separation. However, with the use of a standardized unsaturation as described above, the R_f values show good reproducibility.

E. The Appearance of Edge Effects

It has also been stated that the appearance of edge effects represented a second disadvantage of the unsaturated chamber. These edge effects were observed by Demole (9) and Stahl (34) and included a higher migration of the same substances spotted near the edge of the plate in comparison with the substances spotted in the middle of the plate. Figure 28 shows the edge effects with ergot alkaloids as found by Stahl.

Stahl (35) explained the occurrence of edge effects as follows. In unsaturated chambers solvent evaporation from the plate occurs faster at the edges than from the center of the plate. This is caused by the absence of solvent vapor behind the plate so that vapor will travel from the adsorbent side to the back side. The evaporated solvent at the edges is then replenished from the solvent reservoir. Thus, more solvent moves along the edges than in the central area and solute spots also move accordingly.

However, in our investigations we could not establish these edge effects as can be seen in Figure 29 where some results are given that were obtained in unsaturated chambers with single- and multicomponent solvents. Some-

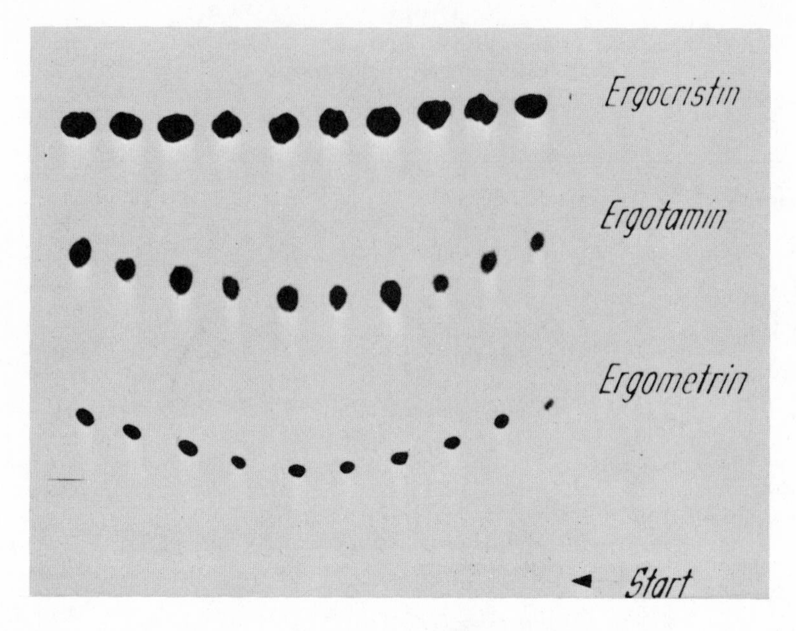

Fig. 28. The apperance of edge effects as found by Stahl, by using unsaturated chambers. Reproduced from *Archiv der Pharmazie*, **292**/64, 411 (1959), with permission of the author and the publisher.

times, a very slight difference could be observed in R_f values of the same solute but this was not found to present any difficulty in identification procedures. Moreover, we could not find any relation between the position of the slightly faster moving spot and the edges. So it might be questioned whether Stahl's explanation of the effects is correct. Solvent evaporation from the solvent reservoir will give vapor at the back of the plate. Furthermore, we could not find any difference between developing a plate in an unsaturated chamber in the normal way and developing a plate with its back placed against the chamber wall. So, apparently the space behind the plate does not play a role. It is also incorrect to assume that the extra supply of solvent, to replenish the loss caused by evaporation, will take place from the solvent reservoir only. Besides that reservoir there will also be a supply from the center to the edges, that is, in a horizontal direction and perpendicular to the direction of development. As a consequence of this horizontal supply the migration of the spots will no longer be vertical but will show a deviation towards the edges. This effect cannot be seen in Figure 28 but

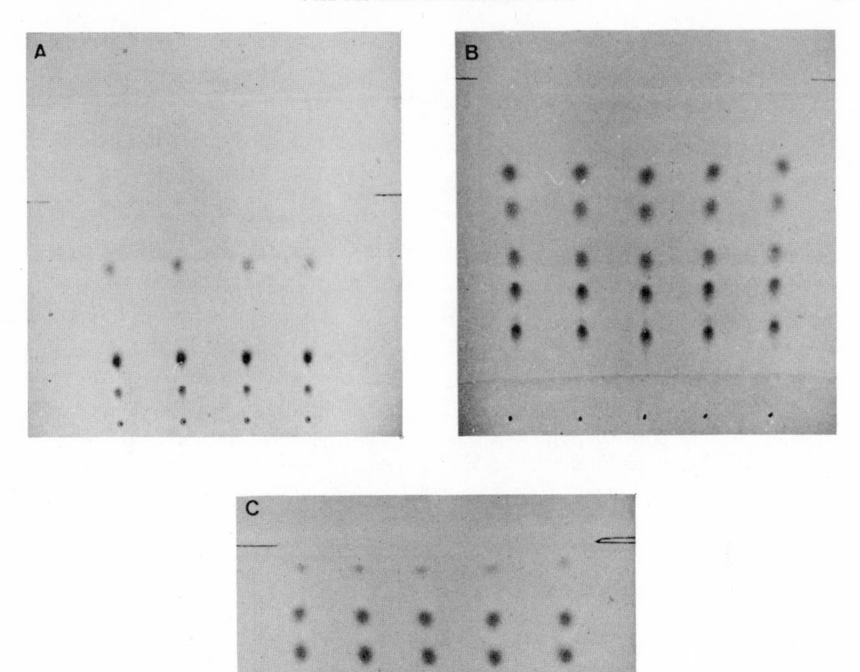

Fig. 29. Absence of edge effects after development in unsaturated chambers. (*A*) Test mixture according to Stahl (Desaga), containing indophenol, sudan red G, and butter yellow, developed with benzene. (*B*) Mixture of hypnotics (heptobarbital, phenobarbital, allobarbital, hexobarbital, methylphenobarbital, bromisoval) developed with chloroform–ether (75 + 25). (*C*) Mixture of hypnotics (see *b*), developed with chloroform-isopropanol-25% ammonia (45 + 45 + 10). Photography in UV light of 254 nm, B and C after activation with ammonia vapor. Reproduced from R. A. de Zeeuw, "Reproducibility of *Rf* values in unsaturated chambers and related development techniques," *J. Chromatog.*, **33,** 222 (1968), with permission of the publisher.

could be observed in our investigations when solvent evaporation at the edges was caused intentionally. This was achieved by development of a plate in a sandwich chamber with the edge seals removed. The result can be seen in Figure 30, which clearly shows the deviation of the spots due to

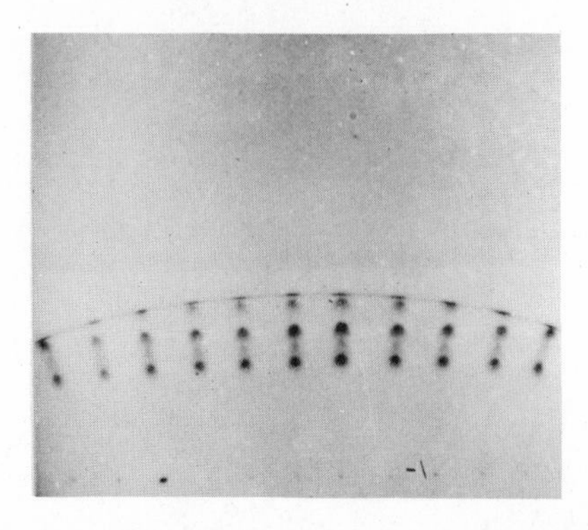

Fig. 30. Appearance of edge effects after intentionally causing solvent evaporation at the side edges. Mixture of hypnotics (heptobarbital, phenobarbital, allobarbital, hexobarbital, methylphenobarbital, bromisoval), developed with chloroform-ether (75 + 25) in an S-chamber, with the side sealing removed.

increased evaporation at the side edges.

So, it is possible that the earlier observations of edge effects were caused by other factors and in our opinion the unsaturated chamber will cause no troubles at all when used properly.

F. Conclusions

The experiments described here have shown that the use of unsaturated chambers offers a variety of possibilities in the separation of closely related substances. The occurrence of a vapor gradient with multicomponent solvents will provide a more efficient separation in comparison to saturated chambers. The separation can be controlled to some extent by using a suitable solvent composition, while the use of troughs provides a second possibility to increase or to decrease the influence of one or more vapor components. The assumed disadvantages of the unsaturated chamber mentioned by earlier investigators can be overcome completely by working with a suitable standard procedure.

It also becomes obvious that the fundamental advantage of TLC is caused by the fact that vapor adsorption on the dry plate can take place. This possibility does not exist in column chromatography or in paper chromatography, due to the very weak adsorption properties of paper. Thus, when searching for optimal separations in TLC the first thing to do is to establish the most suitable solvent vapor influence and it may be expected that further developments in TLC will go in this direction. Therefore, we cannot agree with the opinion that simplification of the usual

TLC procedures is required by using single-component solvents and sandwich chambers (14). For under those conditions one of the most important advantages of TLC would be lost completely.

VI. THE SANDWICH CHAMBER

A. Introduction

Investigations on the phenomena of chamber saturation by Stahl (35) resulted in the introduction of the sandwich chamber. Because of their small inner volume these S-chambers could be used without saturation before development and much less solvent was needed for development. The S-chamber resulted from the desirability to use the smallest possible chamber volume to eliminate any possible difference in the saturation of these volumes. Recently, Kirchner (21) also stressed the use of very small chamber volumes.

Unfortunately, however, very little information can be found in the literature about a comparison of the S-chamber and the N-chamber. Jork (20) obtained better separations of resins in the S-chamber. Jänchen (18) and Kirchner (21) mentioned solvent demixing in S-chambers with binary solvent mixtures. Geiss *et al.* (13) compared both chambers in separations of dyes mainly with single-component solvents. Schweda (31) recently reported a comparative study of glass plates and chromatogram sheets in both S- and N-chambers with the aid of toxicologically important substances and multicomponent solvent mixtures. He concluded that S- and N-chambers were quite different with multicomponent solvents.

No satisfactory explanation of the phenomena observed by these investigators was available and we therefore have tried to obtain a better insight into the different processes taking place in these chambers.

B. Experiments with S- and N-Chambers

We have carried out some separations of hypnotics with a binary and a ternary solvent system in both S- and N-chambers, namely, chloroform-acetone (90 + 10) and chloroform-isopropanol-25% ammonia (45 + 45 + 10). Saturated N-chambers were used and in S-chambers a clean glass plate was used to cover the adsorbent plate. The length of run was 15 cm and the saturation time of the N-chambers was 45 min.

The results of the separations are shown in Table VI. With chloroform-acetone there is a reasonable separation of the substances in both systems, but the R_f values in N-chambers and S-chambers are indeed quite different. Furthermore, solvent demixing occurs in the S-chamber with a second

TABLE VI
R_f Values of Some Hypnotics in Saturated N-Chambers and S-Chambers with Two Different Solvents

	$R_f \times 100$			
	Chloroform-Acetone (90 + 10)		Chloroform-Isopropanol-25% Ammonia (45 + 45 + 10)	
Substance	N-Chamber	S-Chamber	N-Chamber	S-Chamber
Heptobarbital	11	28	19	100
Phenobarbital	20	39	28	100
Cyclobarbital	23	43	45	100
Allobarbital	25	42	39	100
Butobarbital	27	46	52	100
Pentobarbital	29	48	59	100
Secobarbital	33	53	59	100
Hexobarbital	43	65	64	100
Methylphenobarbital	48	72	52	100
Pentothiobarbital	60	78	62	100

front about 3 cm beyond the top front. With chloroform-isopropanol-25% ammonia the nice separation in the N-chamber has disappeared completely in the S-chamber and all hypnotics run with the solvent front. No solvent demixing could be observed here.

C. The Fundamental Differences between S- and N-Chambers

Since we now know the role of solvent vapor, the differences between the chambers can be explained quite easily. Geiss *et al.* (13) have also mentioned the different vapor conditions in their investigations but their explanation was restricted to highly polar vapor components only. The processes involved in TLC with multicomponent solvents using N-chambers can be summarized as follows:

(*a*) Solvent ascends by capillary action and, by a process comparable to frontal analysis, multicomponent solvents are demixed on the plate. If, for example, we use a chloroform-acetone solvent system, a zone of pure chloroform is followed by the binary mixture.

(*b*) In the dry part of the plate, adsorption of solvent vapor will take place with the more polar components being adsorbed more strongly.

(*c*) In the wet part of the plate absorption of vapor as well as solvent evaporation occur.

As a consequence of processes *b* and *c*, the top solvent front will not be pure chloroform in the N-chamber. In the S-chamber, however, the situation is much different. The demixing process will also take place here but because of the small inner volume of the chamber, diffusion of solvent vapor will be limited within the time available. Saturation of the inner volume will be achieved by solvent evaporation from the solvent front and, because this front will consist of pure chloroform only, no acetone will be present in the vapor. Hence, process *b* will give an adsorbate of chloroform vapor only. The remaining top solvent front, when wetting the vapor-impregnated areas, also consists of chloroform only and this explains the occurrence of a second front in the S-chamber. The top front is pure chloroform, the second front is the binary mixture chloroform-acetone. Because the R_f values are dependent upon the amount and the composition of the adsorbed vapor on the dry plate, it will be clear that the separations in the N-chamber and the S-chamber are quite different, because no adsorbed acetone vapor is present on the dry plate in the S-chamber.

When using the chloroform-isopropanol-ammonia system the effects can be explained in the same way. Solvent demixing takes place during development and the highly polar ammonia, giving strong hydrogen bonding with the silanol groups of the adsorbent, is completely removed from the solvent and the remaining ascending liquid contains only chloroform and isopropanol. In the narrow S-chamber little or no diffusion of ammonia vapor can take place and, hence, the hypnotics are separated in their acid form and migrate into the solvent front because of the high proportion of isopropanol in the solvent. This concentration of isopropanol also prevents a further demixing of the solvent.

In the N-chamber the ammonia is also retained from the ascending solvent, but in this case diffusion and adsorption of ammonia vapor can take place over the whole plate. Thus, the influence of ammonia on the separation will only be due to its vapor. This can be confirmed by using a solvent system of chloroform-isopropanol (50 + 50) in the N-chamber together with a trough with 10 ml ammonia at the bottom of the chamber. Because no ammonia is available from the solvent system, any influence of ammonia can only be caused by its vapor. The R_f values in this system are exactly the same as those listed in Table VI, obtained with 10% ammonia in the solvent.

D. Conclusions

From the above explanation it becomes obvious that, due to totally different vapor conditions existent during development, the performance of S-chambers and N-chambers are by no means comparable. Thus, data obtained in one system should not be used as a reference for the other. In fact, S-chambers are closely related to column chromatography because on columns vapor diffusion and vapor impregnation are also limited. Particularly with multicomponent solvent systems this will be true.

Some of the disadvantages of the S-chamber can be overcome to a certain extent by the preadsorption of vapor on a dry plate before development, or by the use of a cover plate with an adsorbent layer which can be impregnated beforehand. For example, a plate can be impregnated with ammonia vapor, followed by development in an S-chamber with chloroform-isopropanol and then the R_f values are similar to those obtained in the N-chamber with chloroform-isopropanol-ammonia systems. With less polar vapors, however, this procedure cannot be used because it turns out to be impossible to preadsorb reproducible quantities of vapor. This is mainly due to the fact that much of the preadsorbed vapor disappears when assembling the S-chamber. The nonuniform amounts which remain adsorbed to the plate then cause nonreproducibility in the separations, thus preventing any identification or comparison with the N-chamber.

Several investigators have used the name "saturation chamber" for the sandwich chamber (7,3) which is, in our opinion, incorrect and which may cause some confusion because in the beginning of development the chamber is unsaturated but during the run vapor adsorption takes place by evaporation of solvent at the solvent front, thus giving some relationship with unsaturated chambers. This can be seen from the fact that when using single-component solvents in the S-chamber the R_f values are more related to those obtained in unsaturated chambers than in saturated chambers. Of course, saturation of the S-chamber is rapidly achieved, but this has little effect on the final separation.

The experiments with the S-chamber make it evident that the advantage of TLC in N-chambers is due to the vapor adsorption on the dry plate. Therefore, when seeking optimal separations it should be remembered that solvent vapor is to be used to affect the best separation. As this is generally not the case for multicomponent systems in sandwich chambers, the use of these chambers should be avoided. However, an exception has to be made for polyzonal TLC with the aid of the BN-chamber (28). This technique has been especially designed to make optimal use of the demixing processes of the solvent and this can be very successful in separations of substances belonging to various polarity classes in one single run.

VII. THE DEVELOPMENT OF THE VAPOR-PROGRAMMING CHAMBER

A. Introduction

It was shown in Section V that the improved separations of TLC with multicomponent solvents in unsaturated chambers are caused by the appearance of a concentration gradient of adsorbed vapor on the adsorbent. The rate of improvement is determined by the polarity and the evaporation velocity of the solvent components and by the character of the adsorbent used. Furthermore, the use of troughs filled with one or more liquid components makes it possible to increase the influence of these components.

Other possibilities to control the extent to which the gradient develops on the plate during development are not available, however, and once the chamber has been closed the only thing to do is wait for the separation result. It will be clear that this involves some difficulties. With highly polar solvent components the gradient may become too steep on the lower part of the plate, the upper parts may show no gradient at all because maximum vapor adsorption will take place. With less polar solvent components the gradient may be too flat or will not reach its maximum due to fast development. Furthermore, with both polar and nonpolar components, differences in evaporation rates between the various components may cause failures in the desired gradient. This can also occur when a separation is required of a number of substances belonging to two different polarity classes. If these groups are subjected to a pulling gradient in an unsaturated chamber, the gap between these groups will be enlarged as well. This effect reduces the possibilities for separation improvements within each group. For such cases one should be able to reduce this gap.

Thus, although unsaturated chambers may well yield improved separations, the conditions are not necessarily optimal for every case, particularly with regard to the influence of vapor. Therefore, we have searched for development techniques providing a much more efficient control of the vapor processes, before and during development, and on every desired area of the plate. This thought led to the apparatus described below, which allows full vapor-programming all over the plate, thus making it possible to affect the migration of each individual spot.

B. The Vapor-Programming Chamber*

The vapor-programming chamber, or the VP-chamber, consists of three parts, all made of chromium-plated brass, which can be seen in Figure 31.

(*1*) The solvent reservoir (A) is a rectangular tank, 20 × 1 × 2 cm, equipped with 2 filling pipes and with an inner volume of about 30 ml.

(*2*) The ground plate (B), 20 × 20 × 1 cm, is fitted with a solvent reservoir holder and a tube for passing warm water (D), insulated from the ground plate by asbestos. The inner part of the ground plate contains a tube system for water circulation, with inlet and outlet visible at F. This

* Patents applied for; the chamber will be obtainable from C. Desaga, Heidelberg.

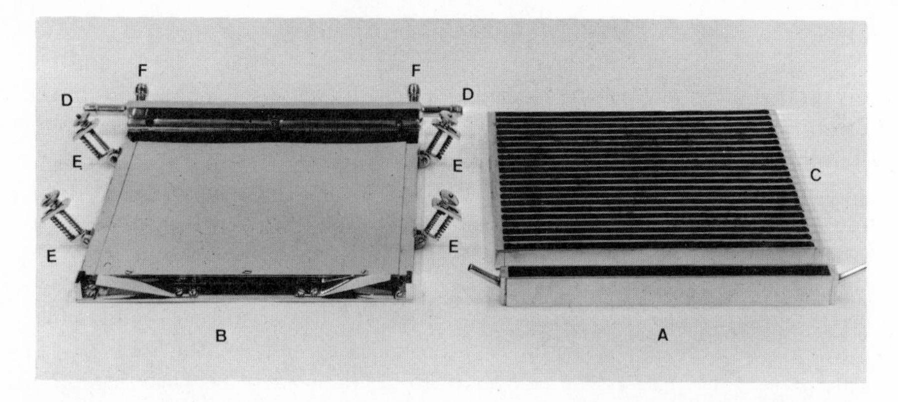

Fig. 31. The vapor-programming chamber: (*A*) solvent reservoir; (*B*) Ground plate; (*C*) Trough chamber. The ground plate is fitted with a tube for passing warm water (*D*) assembly clamps (*E*) and an internal tube system for water-circulating purposes to be connected to a thermostat, the inlet and outlet being visible at *F*.

system is to be used for cooling purposes. Four assembly clamps (E) are located at the sides of the ground plate.

(*3*) The trough chamber (C), containing 21 troughs, 6 mm wide, 12 mm deep with partition walls 2 mm thick. The side walls are 5 mm thick. The chamber is made of welded flat-oval tubes with the tops machined off. The trough chamber is fitted onto the ground plate (B) between the solvent reservoir (A) and the warm water tube assembly (D) as shown in Figure 32. The trough chamber is fixed by assembly clamps.

The assembled vapor-programming chamber for use with 20 × 20 cm plates is shown in Figure 32.

The vapor-programming chamber is used as follows. The samples are spotted on normal 20 × 20 cm TLC plates, 2–2.5 cm from the bottom edge of the plate and at least 1 cm apart. The sides and the bottom are stripped 0.5 cm wide. The troughs are filled with a series of liquids of appropriate compositions to give suitable vapors. Most often, the liquids show increasing polarity from bottom to top. This can be done for example by using mixtures of two or more solvent components of different polarities, with the mixtures having an increased proportion of the more polar component(s). The empty solvent reservoir is placed in the holder and fitted with a folded strip of filter paper, 18.5 × 2 cm, with 1 cm folded over the inner wall. The side walls of the chamber are fitted with small spacers, usually about 0.5 mm thick. The TLC plate is then placed on the chamber with the adsorbent face down, the stripped side edges resting on the

Fig. 32. The assembled vapor-programming chamber.

spacers, and the bottom edge just covering the solvent reservoir. The spacers should prevent the adsorbent from touching the trough walls and the thickness should be of the order of the adsorbent layer. When the plate is properly positioned, the starting spots should lie over the first or second trough, depending on the distance of the spots to the bottom edge of the plate. The solvent reservoir and the filter paper strip are pressed gently against the adsorbent by means of two springy metal strips located beneath the reservoir. The solvent reservoir is then filled with about 25 ml of the appropriate solvent, which is subsequently led to the plate by the filter paper strip and development begins.

Figure 33 shows the vapor-programming chamber in operation. Both water systems are connected by suitable flexible tubing to warm and cold water supplies. The ascending solvent makes the plate transparent so that the underlying troughs and the filter paper strip become visible. At the edges the spacers are visible while the top of the plate protrudes 0.5 cm over the warm water tube.

The use of the VP-chamber, in which development is done horizontally, permits vapor adsorption by the adsorbent from the underlying troughs. Thus, by filling the troughs with suitable liquid mixtures the vapor conditions can be programmed over the entire plate. In this way optimal vapor gradients can be obtained, because every desired polarity can be applied to the various parts of the plate via the vapor phase.

Fig. 33. The vapor-programming chamber at work. The ascending solvent makes the adsorbent transparent, with the underlying troughs and the strip of filter paper becoming visible.

The principle of the VP-chamber is rather simple, but there are some important factors to which special attention must be paid. In the following Section we will discuss these factors in detail.

C. Materials

All parts of the chamber, except the assembly clamps, are made of chromium-plated brass to ensure sufficient thermal conductivity (see Section VII.F) and to allow advantageous processing. The troughs are made from oval tubes having dimensions of 6 × 17 mm with walls 1 mm thick. The tops are machined off so that a trough will result having a depth of about 12 mm. Twenty-one such troughs and side walls are assembled and soldered tight into the integral trough chamber shown in Figure 31. The side walls are made from 5 mm stock. The top surface of the entire trough chamber is machined to a uniformly flat, smooth finish. As brass is not inert to various organic and inorganic liquids, all parts have been chromium plated.

Our VP-chamber has been designed for plates 20 × 20 cm in size but similar chambers can of course be constructed for plates with other dimensions. However, the trough chamber should have a length of at least 15 cm because otherwise it will be very hard to obtain suitable gradients.

The thickness of the TLC plates may vary between 1 and 4 mm. Sagging plates or sheets or insufficiently flat plates cannot be used.

Teflon (Dupont) or other inert, thin, flexible material is suitable for use as side spacers. Teflon is inert to most solvents but is somewhat slippery in use.

Unsuccessful attempts have been made to machine the trough chamber from a solid block of brass. The particular brass stock used was porous and it was impossible to machine thin sections without incurring leaks. Stresses developed due to the machining caused distortions in the chamber.*

Casting of the chamber might prove to be a useful commercial production method for making chambers. This could make available the use of more difficultly processable materials of construction which are more inert to solvents and corrosion. It should be kept in mind, however, that materials having good thermal conductivity are necessary.

D. The Dimensions of the Troughs and the Spacers

The dimensions of the troughs and the thickness of the spacers are very important for the applicability of a certain gradient. It will be clear that due to the differences in liquid compositions in the troughs strong vapor intermixing will occur if the space between the trough walls and the adsorbent layer is too large. The desired gradient then disappears and, moreover, these interdiffusions may well give strong vapor currents, causing local solvent evaporation. Hence, anomalies in the direction of solvent flow will occur, making the separation worthless. Figure 34 illustrates such an anomalous chromatogram. Solvent evaporation has occurred in the center area, followed by an extra supply of solvent from the side edges to the center, thus causing a deviation of the migration direction of the spots. Geiss and Sandroni (15) recently reported similar effects in S-chambers with a diameter of 3–4 mm but were able to prevent the phenomenon by horizontal development.

On the other hand, the spacers cannot be too thin because the adsorbent must not touch the trough walls. If so, development stops immediately and the spots readily diffuse along the walls. It should be observed, however, that most adsorbents swell, more or less, when wetted by the solvent. The rate of swelling is dependent on the polarity of the solvent and the thickness of the strips should be adapted to this phenomenon. In our experiments with layers of 0.25 mm, good results were obtained with spacers of 0.3–0.5

* Note added in proof: Recently, adequate methods have been developed that makes direct machining from a solid block of brass possible.

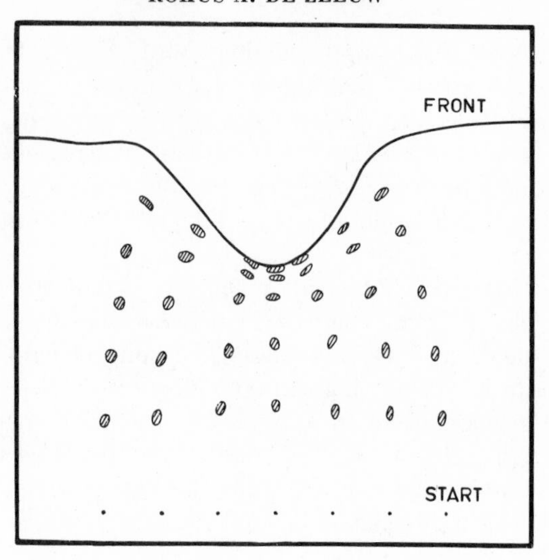

Fig. 34. Irregular development in the vapor-programming chamber, caused by the use of spacers that are too thick.

mm for solvents like hexane, benzene, chloroform, and ether, 0.5–0.8 mm for solvents containing acetone, ethylacetate, and lower alcohols, whereas ammonia-containing solvents needed strips of 1 mm.

When searching for optimal dimensions of the troughs we tested troughs with widths of 2, 6, and 10 mm. In all cases we were obliged to use partition walls at least 2 mm thick, because with thinner walls the chamber could not be milled sufficiently flat and, moreover, wall deformations easily occurred.

Troughs 10 mm wide proved to be too large because only 14 troughs can thus be used for a plate of 20 cm. This does not provide sufficient possibilities to obtain reasonable vapor-programming. When using troughs 2 mm wide, there are ample possibilities to obtain a suitable vapor program but, surprisingly, the results did not show a better efficiency in comparison to those obtained with troughs 6 mm wide. Presumably, the ratio of the trough width to the wall thickness plays a role. Furthermore, the use of 2 mm troughs has the disadvantage of the necessity to fill about 50 troughs before development. This is a rather time-consuming and tedious process, not to mention the difficulties possibly incurred by solvent evaporation from the troughs already filled.

When using troughs 6 mm wide, we obtained very good separations in all experiments. The filling of the 21 troughs could be done in less than 2 min. We, therefore, prefer the 6 mm troughs for the construction of the final model of the chamber. The fact that the oval tubes of this dimension are easily obtainable was another factor in our choice.

Finally, it should be noted that a space of about 1 mm is needed between the trough chamber and the solvent reservoir to prevent disappearance of solvent by capillary action between the walls of these parts.

E. Saturation

To ensure that reproducible quantities of vapor are available for adsorption it is necessary to equilibrate the plate over the filled troughs for 10 min after it is fixed in position. During this period the small volumes over the troughs become almost saturated, resulting in a fixed amount of vapor available for adsorption. Not until after this saturation period is the solvent reservoir filled and development started.

The duration of the saturation period was experimentally established by comparing the reproducibility of the separations. This period is also sufficient for less volatile components such as the lower alcohols. If highly volatile solvent components are used the saturation can be reduced to 5 min.

F. Temperature

The availability of solvent vapor will be determined by the temperature. Thus, it will be clear that constancy of the room temperature within ±1°C is required to achieve reproducible separations. However, in vapor-programmed TLC it was found that at higher room temperatures—over 22°C —poorer separations were obtained than at lower room temperatures, because the migration rates of the spots decreased markedly at higher temperatures. Presumably, this is caused by too much adsorption of vapor, which is then followed by condensation on the plate. Thus, solvent transport from the solvent reservoir will be diminished and, consequently, the migration rate of the spots will decrease.

Cooling of the VP-chamber is therefore necessary. We have equipped the ground plate with an internal tube system for water circulation, to be connected to a thermostatted water supply whose temperature is lower than room temperature. This carefully controlled cooling system provides good control of the vapor adsorption rate. We found, for example, that a separation at a cooling temperature of 16°C can be quite different from a separation obtained with a cooling temperature of 19°C. A thermostat with an accuracy of ±0.1°C is therefore necessary.

The optimum temperature will vary for the different separations, but in our investigations at average room temperatures of 21 ± 1°C optimal separations were obtained with cooling temperatures between 18 and 20°C.

G. Continuous Development

Very careful development is needed to separate closely related substances of almost identical polarity. In these cases a solvent with low polarity and a flat vapor gradient, especially on the lower parts of the plate, should be used, otherwise no useful results will be obtained. However, this implies a slow migration of the solutes, with the spots still migrating at the lower parts of the plate when the solvent front has already reached the top end of the plate. To overcome this difficulty, the VP-chamber can be used for continuous development with the position of the solvent front no longer being the limiting factor and allowing development until optimal separation is obtained all over the plate. The top end of the plate (1 cm) is not acccompanied by a trough underneath but lies 0.5 cm over the tube for passing warm water. The solvent reaching the top will evaporate, thus allowing continuous solvent transport over the plate. With solvents of low volatility the warm water tube should be kept at about 50°C; with highly volatile solvents warming is not necessary because evaporation is fast enough.

When publishing chromatographic data obtained with continuous development, the length of run can no longer be stated. In our opinion, the development time should then be given, or a colored reference substance should be used, stating the required distance to be moved by this reference, for example, 12 cm from the starting points. The optimal development time or the optimal distance to be migrated by the reference compound can be established when searching for optimal vapor gradients. The use of a colored reference provides the advantage of being able to follow the migration throughout development. This in turn permits the rapid detection of any disturbance. In our experiments with drugs we preferred the use of the intense yellow-colored 4–nitroaniline, which has a suitable polarity for the particular solvents involved.

Furthermore, the R_f values for continuous chromatography require a special definition. We prefer to define the R_f values for continuous development as the distance moved by the substance, divided by the distance between the starting point and the top of the plate.

H. The Composition of the Stepwise Gradient and the Use of Decelerators

Until now we have not discussed the composition of the vapor gradients. In vapor-programmed TLC one should use a developing solvent of rather low polarity, giving the substances under investigation a small migration rate. The solvent may be single- or multicomponent. The troughs underneath the starting point (e.g., the first two troughs) are filled with liquids of the same composition as the developing solvent or with liquids of a slightly lower polarity. In trough 3 we then use a liquid composition with a slightly higher polarity than that of the developing solvent, trough 4 is filled with a slightly higher polar liquid than in trough 3, and so forth. This can be done, for example, by using mixtures of two solvents—one polar, one nonpolar—with the mixtures having an increased percentage of the more polar component. For a separation of barbiturates chloroform will be a very suitable developing solvent, giving the barbiturates a slow migration rate with R_f values varying from 0.05 to 0.25. Troughs 1 and 2 are also filled with chloroform, but in trough 3 5% ether is added, having an accelerating effect on the migration rates of the spots. Trough 4 is then filled with chloroform-ether (90 + 10) to give a greater acceleration and this scheme can be continued in the following troughs until pure ether is present. Moreover, the ether can be combined with components of a higher polarity, such as acetone, ethylacetate, or ethanol. Thus, a stepwise vapor gradient will result and, depending upon the desired effect, the polarity in the troughs can be increased to any extent. Decreasing the polarity is also possible. It is obvious that a thorough knowledge is necessary of the properties of the solvent components and of the adsorbent, in particular with regard to their effects on the movements of the substances.

However, one difficulty arises. Solute spots are of finite size and when using the above described gradients, migration of the upper part of a solute spot will be accelerated over that of the lower part each time the spots enter a new area of higher polarity. Since this process is repeated each time the spot enters the next trough, severe tailing results. This phenomenon can be conveniently observed with the aid of dyes.

Fortunately, however, tailing can be avoided completely by interspersing troughs with liquids of low polarity between the troughs containing the polar mixtures. This will have a decelerating effect on the migration of the spots, particularly on the upper part. If, for example, barbiturates should be separated the following vapor-program can be used: solvent, chloroform, troughs 1 and 2, chloroform; trough 3, chloroform-ether (95 + 5); trough 4, chloroform; trough 5, chloroform-ether (70 + 30); troughs 6 and 7, chloroform; trough 8, chloroform-ether (40 + 60); troughs 9 and 10, chloroform; trough 11, chloroform-ether (10 + 90); troughs 12 and 13, chloroform; trough 14, ether-methanol (80 + 20); troughs 15 and 16, chloroform; trough 17, ether-methanol (50 + 50); troughs 18 and 19, chloroform; trough 20, methanol; trough 21, chloroform. Thus, chloroform has a decelerating effect, whereas ether and methanol give an acceleration of the migration rates of the spots. The polarity differences between the decelerating and accelerating troughs increase when going upward over the plate but this is necessary to maintain suitably small spots. In some instances it may even be necessary to use three or more decelerating troughs after one accelerating trough, particularly on the upper parts of the plate.

We do not know how the process of acceleration and deceleration works in reality but the process will produce compact spots without decreasing the improved separation efficiency. We presume that the improvements are not only caused by the acceleration forces but also by a more or less selective deceleration; the less polar substances are decelerated to a higher extent than the more polar substances.

Finally, it should be observed that the accelerating-decelerating stepwise "gradient" obtained in vapor-programmed TLC is basically different from what is called gradient elution. Gradient elution has a pushing effect on the spots, being suitable for separations of substances showing great polarity differences. In vapor-programmed TLC a pulling gradient is obtained which is suitable for separations of closely related substances with small polarity variations.

VIII. SOME APPLICATIONS OF VAPOR-PROGRAMMED TLC

A. Introduction

In this section a variety of examples will be shown of the more efficient separations obtained with vapor-programmed TLC. These examples all include separations of closely related compounds, the selections being mainly chosen from pharmaceutically important drugs. Silica gel and aluminum oxide have been used as adsorbents. A broad variety of liquid components have been used for the compositions of the developing solvents and for the liquids in the troughs, ranging from the nonpolar hexane to the highly polar ammonia.

The results are illustrated by means of photographs so that we can obtain a clearer insight into the separation possibilities, the size of the spots, tailing, and so forth. The separations by means of vapor-programmed TLC are compared with the corresponding separations of the classical saturated chamber generally using two solvent systems. The separations in the vapor-programming chamber are also shown in line diagrams, in which the position of the troughs and the liquid composition can be seen. In these diagrams the position of the first trough has been deleted for simplicity. All starting points are located in the areas over the second trough. The contents of the first trough was always identical to that in the second one.

B. Dyes

Experiments have been carried out with a mixture of indophenol, sudan red G, butter yellow, and 4-nitroaniline. The first three dyes are available as the test mixture for TLC according to Stahl (Desaga) and show a blue, a red, and a yellow color, respectively. The fourth component was the intense yellow-colored 4-nitroaniline. The experiments were done with a mixture of these compounds in benzene. On each plate four or five spots of the mixture were applied.

With benzene as the developing solvent the components show relatively low R_f values and an incomplete separation because sudan red G and 4-nitroaniline partly overlap. Only by means of the color differences in the original chromatogram can it be concluded that two spots are present. From the UV photograph of the separation, which is shown in Figure (35*a*), only one spot can be seen. Changing the solvent to benzene-chloroform (80 + 20), in order to obtain higher R_f values, did not improve the separation. The spots move over the plate as a zone and the mutual distances between the spots are not enlarged (Fig. 35(*b*)). 4-Nitroaniline now coincides with indophenol. The use of unsaturated chambers did not show any advantage. With benzene-chloroform mixtures the vapor gradient is too flat, whereas the use of a more polar component than chloroform results in too strong a gradient, particularly on the lower half of the plate.

However, when using the VP-chamber with benzene as solvent and a simple benzene-chloroform vapor gradient, the separation can be highly improved. The result is illustrated in Figure 35(*c*). The spots are now spread across the entire plate and a complete separation is obtained. Furthermore, it should be noted that the mutual distances between the spots can be changed as desired by the application of slightly different vapor gradients. Thus, if a fifth component should be present, for example, between sudan red G and 4-nitroaniline, the distance between these two spots can be distinctly increased to achieve complete separation between these and the fifth spot. This can be done without disturbing the other separations of the mixture.

C. Barbiturates

The separation of the very closely related barbituric acid derivates is a very difficult problem, as can be concluded from Figure 36(*a*). Twelve barbiturates (see Table VII)—phenobarbital, barbital, cyclobarbital, allobarbital, aprobarbital, butobarbital, amobarbital, pentobarbital, brallobarbital, butalbital, secobarbital, and hexobarbital—have been developed with chloroform-ether (60 + 40) in a saturated chamber. Very little separation is obtained. When using chloroform-ether (75 + 25) as solvent in an unsaturated chamber the separation is improved by the existent ether gradient (Fig. 36(*b*)). The decrease in the ether content of the solvent system prevents the spots from migrating too far.

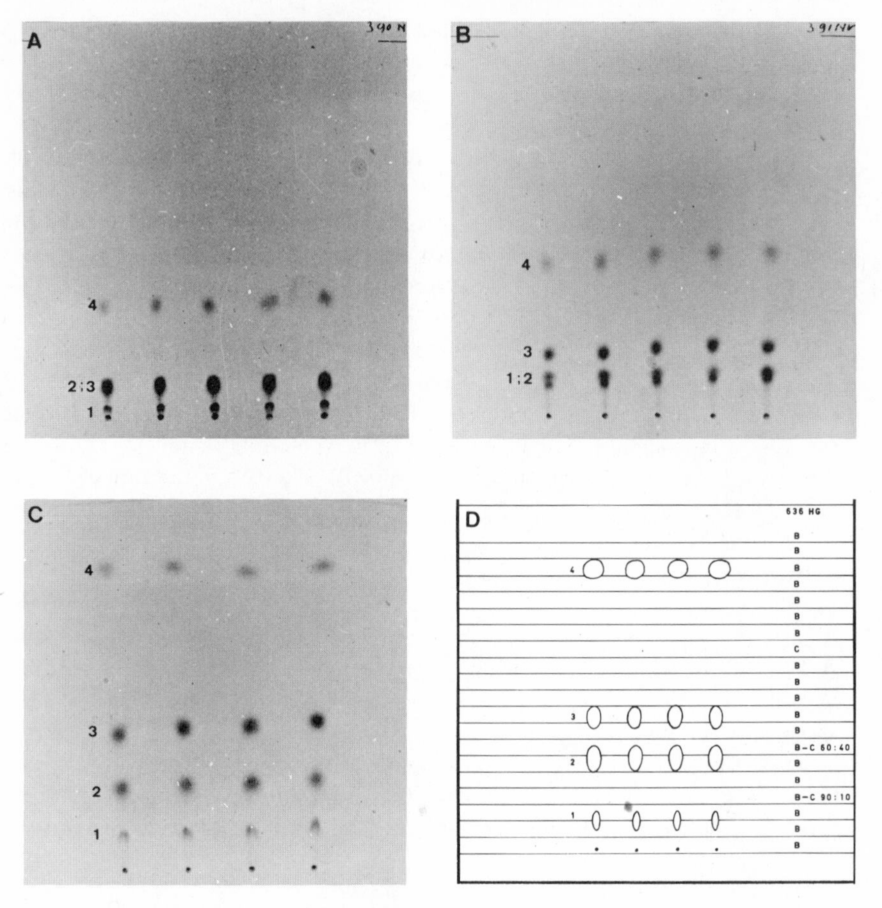

Fig. 35. Improved separation of dyes in the vapor-programming chamber as compared to separations in saturated N-chambers. (*A*) N-chamber, solvent benzene; temp. 21°C, rel. humidity 26%, saturation 45 min, development 36 min. (*B*) N-chamber, solvent benzene-chloroform (80 + 20); temp. 22°C, rel. humidity 30%, saturation 45 min, development 39 min. (*C*) VP-chamber, solvent benzene; temp. 21.7°C, rel. humidity 29%, saturation 10 min, development 110 min, spacers 0.3 mm, cooling 19°C. (*D*) Schematic diagram of the positions of the troughs and the liquid compositions therein during development; B = benzene, C = chloroform. 1 = indophenol; 2 = 4-nitroaniline; 3 = sudan red G; 4 = butter yellow. Adsorbent = silica gel SF 254, load = 20 μg of each substance. Reproduced from *Anal. Chem.*, **40,** (1968) in press, copyright 1968 by the American Chemical Society; reproduced with permission of the copyright owner.

With vapor-programmed TLC, however, a much better separation is obtained, as in Figure 36(*c*), for instance. We used chloroform as the

TABLE VII

Chemical Names of Some Barbiturates

Generic Name	Chemical Names
Heptobarbital	5-methyl-5-phenylbarbituric acid
Phenobarbital	5-ethyl-5-phenylbarbituric acid
Barbital	5,5-diethylbarbituric acid
Cyclobarbital	5-ethyl-5-(1-cyclohexenyl)barbituric acid
Allobarbital	5,5-diallylbarbituric acid
Aprobarbital	5-allyl-5-isopropylbarbituric acid
Butobarbital	5-ethyl-5-butylbarbituric acid
Amobarbital	5-ethyl-5-(3-methylbutyl)barbituric acid
Pentobarbital	5-ethyl-5-(1-methylbutyl)barbituric acid
Brallobarbital	5-allyl-5-(2-bromoallyl)barbituric acid
Butalbital	5-allyl-5-(1-methylpropyl)barbituric acid
Secobarbital	5-allyl-5-(1-methylbutyl)barbituric acid
Hexobarbital	5-(1-cyclohexenyl)-1,5-dimethylbarbituric acid
Methylphenobarbital	5-ethyl-1-methyl-5-phenylbarbituric acid

developing solvent, together with a chloroform-ether-methanol vapor gradient. The spread has been clearly improved with regard to the unsaturated chamber, thus providing more separation potential. Butalbital, secobarbital, and hexobarbital are completely separated and, in addition, butobarbital can be identified from the other substances. It is apparent that all components cannot be separated with only one gradient, especially in view of the close similarity in molecular structures of the compounds. However, when comparing the classical technique of Figure 36(*a*) and the vapor-programmed TLC of Figure 36(*c*), the improvements are significant. In Figure 36(*a*) the spots have been spread over 2.9 cm of the plate, against 8.3 cm in Figure 36(*c*), with the size of the spots not being enlarged. On the other hand, it should be remembered that with other compositions of the gradient new separations become possible.

Chloroform-acetone vapor gradients can also be used for the separation of barbiturates. In classical TLC chloroform-acetone solvent mixtures have often been used in toxicological analysis. Figure 37(*a*) shows the result with chloroform-acetone (80 + 20) in saturated chambers using heptobarbital, phenobarbital, cyclobarbital, allobarbital, butobarbital, pentobarbital, butalbital, secobarbital, hexobarbital, and methylphenobarbital. The separation is far from optimal. In unsaturated chambers with chloroform-acetone (90 + 10) as the solvent a much more efficient separation is

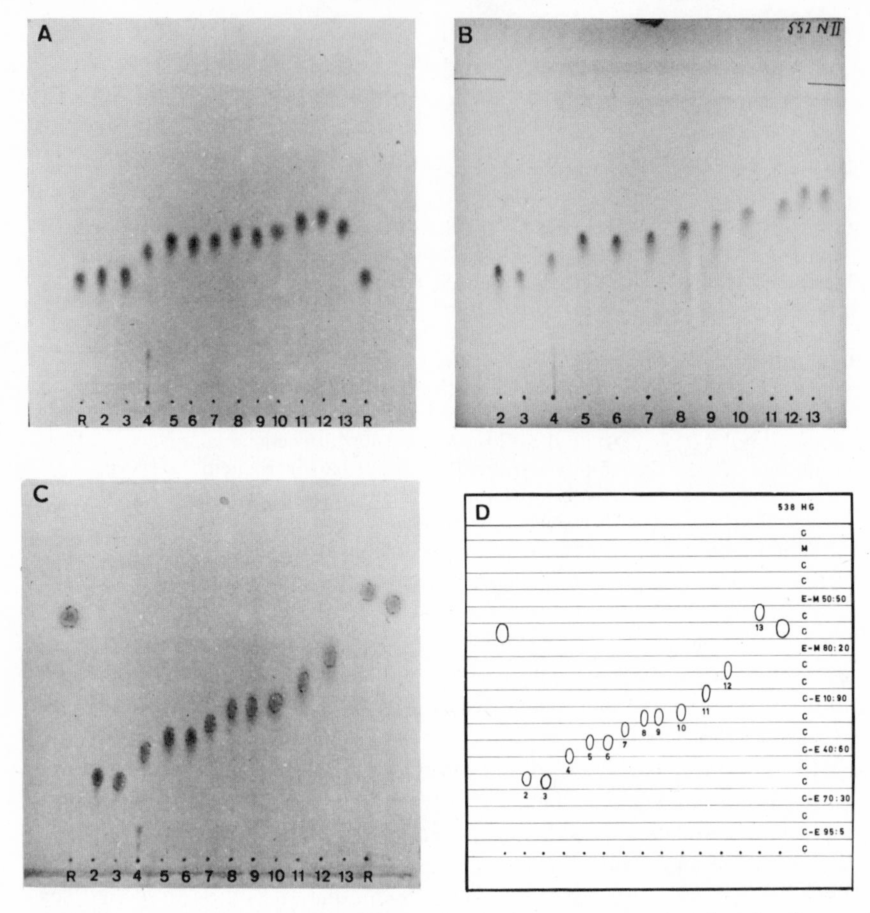

Fig. 36. Improved separation of hypnotics in the vapor-programming chamber as compared to separations in saturated and unsaturated N-chambers. (*a*) Saturated N-chamber, solvent chloroform-ether (60 + 40); temp. 22°C, rel. humidity 34%, saturation 45 min, development 40 min. (*b*) Unsaturated N-chamber, solvent chloroform-ether (75 + 25); temp. 22°C, rel. humidity 30%, development 60 min. (*c*) VP-chamber, solvent chloroform; temp. 22°C, rel. humidity 28%, saturation 10 min, development 75 min, spacers 0.3 mm, cooling 19°C. (*d*) Schematic diagram of the positions of the troughs and the liquid compositions therein during development; C = chloroform, E = ether, M = methanol. 2 = phenobarbital; 3 = barbital; 4 = cyclobarbital; 5 = allobarbital; 6 = aprobarbital; 7 = butobarbital; 8 = amobarbital; 9 = pentobarbital; 10 = brallobarbital; 11 = butalbital; 12 = secobarbital; 13 = hexobarbital; R = reference 4-nitroaniline; adsorbent = silica gel GF 254, load 10 μg.

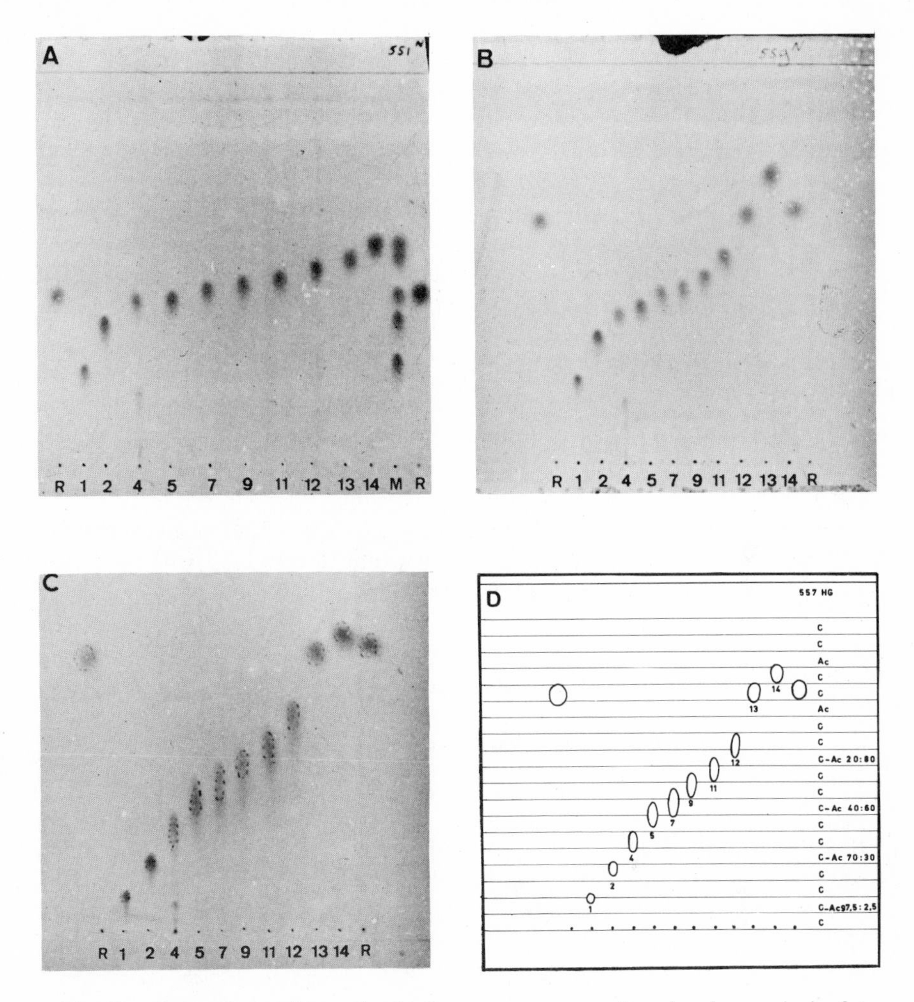

Fig. 37. Separation of hypnotics in the vapor-programming chamber, compared to separations in saturated and unsaturated N-chambers. (*a*) Saturated N-chamber, solvent chloroform-acetone (80 + 20); temp. 22°C, rel. humidity 30%, saturation 45 min, development 36 min. (*b*) Unsaturated N-chamber, solvent chloroform-acetone (90 + 10); temp. 20.5°C, rel. humidity 35%, development 63 min. (*c*) VP-chamber, solvent chloroform; temp. 21°C, rel. humidity 35%, saturation 10 min, development 75 min, spacers 0.5 mm, cooling 19°C. (*d*) Schematic diagram of the positions of the troughs and the liquid compositions therein during development; C = chloroform, Ac = acetone. 1 = heptobarbital; 2 = phenobarbital; 4 = cyclobarbital; 5 = allobarbital; 7 = butobarbital; 9 = pentobarbital; 11 = butalbital; 12 = secobarbital; 13 = hexobarbital; 14 = methylphenobarbital; R = reference 4-nitroaniline; M = mixture of 1 + 2 + 5 + 13 + 14; adsorbent = Silica gel GF 254, load 10 μg of each substance.

obtained due to the formation of the acetone vapor gradient. The nine higher spots which lie so closely together in the saturated chamber are now distinctly pulled apart (Fig. 37(*b*)). The result obtained with the VP-chamber, using chloroform as the solvent and a chloroform-acetone gradient, is shown in Figure 37(*c*). This result is hardly better than the one obtained in the unsaturated chamber; only cyclobarbital and allobarbital are completely separated. Moreover, the spots are tailing a little. Probably, in this case, the continuous gradient in the unsaturated chamber is almost optimal and can hardly be improved by the stepwise gradient in the VP-chamber. The decelerating forces are unable to prevent tailing.

Because of their acid character, the barbiturates can also be developed in alkaline media and then they migrate as anions. Figure 38(*a*) represents the separation of heptobarbital, phenobarbital, barbital, cyclobarbital, allobarbital, aprobarbital, butobarbital, amobarbital, pentobarbital, brallobarbital, butalbital, secobarbital, hexobarbital, and methylphenobarbital with chloroform-isopropanol-25% ammonia (45 + 45 + 10) as solvent in saturated chambers. A reasonable separation is obtained with the spots being spread over 6.5 cm. Figure 38(*b*) shows the result obtained in the VP-chamber, using chloroform-isopropanol (92.5 + 7.5), saturated with 25% ammonia as solvent and a chloroform-isopropanol-methanol-ammonia gradient in the troughs. The separation has become much improved, with the size of the spots not significantly enlarged. The substances are now spread over 15 cm of the plate. This utilizes over 85% of the plate. In spite of the steep gradient the size of the spots is fully controlled. With the aid of slightly different gradients the pairs butobarbital-butalbital and pentobarbital-secobarbital can be separated completely, but this could not be obtained for barbital and allobarbital.

Unsaturated chambers were far less effective than the VP-chamber, because the high polarity of the solvent components prevent the formation of a suitable gradient.

D. Sulfonamides

The investigations with sulfonamides have been started with a selection of substances already studied by Gänshirt (12), namely sulfaguanidine, sulfisomidine, sulfathiazole, sulfacetamide, sulfadimidine, sulfapyridine, sulfisoxazole, and sulfanilamide. The chemical structures of these sulfonamides which are, more precisely, all sulfanilamide derivatives, are given in Table VIII.

With ether-methanol (90 + 10) in saturated chambers a very incomplete separation is obtained, as is shown in Figure 39(*a*). The R_f values are rather low. Increasing the methanol concentration in the solvent to ether-

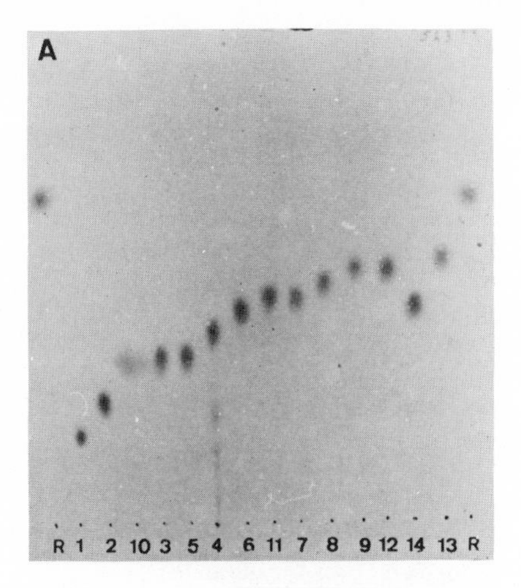

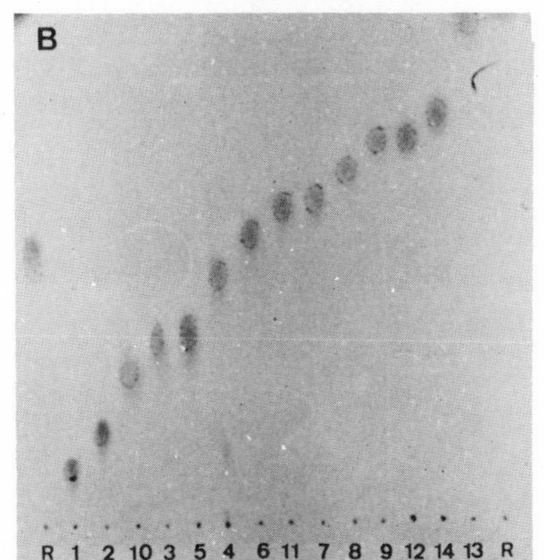

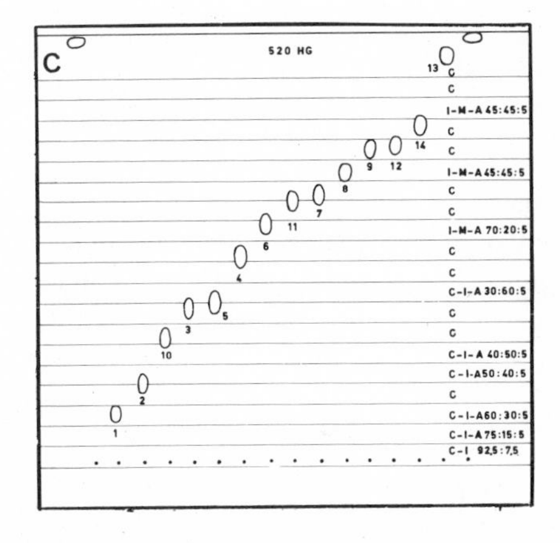

Fig. 38. Improved separation of hypnotics in the vapor-programming chamber, as compared to the separation in the saturated N-chamber. (*a*) N-chamber, solvent chloroform-isopropanol-25% ammonia (45 + 45 + 10); temp. 20.6°C, rel. humidity 39%, saturation 60 min, development 75 min. (*b*) VP-chamber, solvent chloroform-isopropanol (92.5 + 7.5), saturated with 25% ammonia; temp. 21.6°C, rel. humidity 40%, saturation 10 min, development 110 min, spacers 1 mm, cooling 19°C. (*c*) Schematic diagram of the positions of the troughs and the liquid compositions herein during development; C = chloroform; saturated with 25% ammonia; I = isopropanol; M methanol; A = 25% ammonia. 1 = heptobarbital; 2 = phenobarbital; 3 = barbital; 4 = cyclobarbital; 5 = allobarbital; 6 = aprobarbital; 7 = butobarbital; 8 = amobarbital; 9 = pentobarbital; 10 = brallobarbital; 11 = butalbital; 12 = secobarbital; 13 = hexobarbital; 14 = methylphenobarbital; R = reference 4-nitroaniline; adsorbent = Silica gel GF 254; load = 10 μg. Reproduced from *J. Pharm. Pharmacol.*, **20,** 54 S (1968), with permission of the publisher.

TABLE VIII
Chemical Names of Some Sulfonamides

Generic Names	Chemical Names
Sulfaguanidine	N_1-guanylsulfanilamide
Sulfamethizole	N_1-(5-methyl-1,3,4-thiadiazol-2-yl)sulfanilamide
Sulfisomidine	N_1-(2,6-dimethyl-4-pyrimidinyl)sulfanilamide
Sulfathiazole	N_1-2-thiazolylsulfanilamide
Sulfadiazine	N_1-2-pyrimidinylsulfanilamide
Sulfacetamide	N_1-acetylsulfanilamide
Sulfamerazine	N_1-(4-methyl-2-pyrimidinyl)sulfanilamide
Sulfadimidine	N_1-(4,6-dimethyl-2-pyrimidinyl)sulfanilamide
Sulfapyridine	N_1-2-pyridylsulfanilamide
Sulfamethoxypyrimidine	N_1-(5-methoxy-2-pyrimidinyl)sulfanilamide
Sulfamethoxypyridazine	N_1-(6-methoxy-3-pyridazinyl)sulfanilamide
Sulfisoxazole	N_1-(3,4-dimethyl-5-isoxazolyl)sulfanilamide
Sulfadimethoxine	N_1-(2,6-dimethoxy-4-pyrimidinyl)sulfanilamide
Sulfanilamide	4-aminobenzenesulfonamide
Sulfaphenazole	N_1-(1-phenyl-5-pyrazolyl)sulfanilamide

methanol (80 + 20) results in higher R_f values, but the separation becomes worse. Employing the vapor-programming chamber, however, a complete separation of all substances is achieved with ether-methanol (95 + 5) as the solvent and an ether-methanol vapor gradient. Nearly 70% of the plate is now utilized for the spread of the spots. The separation is shown in Figure 39(*c*).

After having obtained this result we wondered if this optimal separation could be maintained if the total number of substances were increased from eight to fifteen by addition of sulfamethizole, sulfadiazine, sulfamerazine, sulfamethoxypyrimidine, sulfamethoxypyridazine, sulfadimethoxine, and sulfaphenazole.

The separations with ether-methanol (90 + 10) and ether-methanol (80 + 20), both in saturated chambers, are given in Figures 40(*a*) and (*b*), respectively. The separations are incomplete and the spreads are far from optimal. In contrast, an efficient separation can be obtained of these many constituents in the vapor-programming chamber, using ether-methanol (95 + 5) as solvent and an ether-methanol vapor gradient (Fig. 40(*c*)). It can be seen that the seven new components can be fitted completely in the separation of the previous eight substances, with only a slight adaption of the vapor gradient and a little stronger deceleration. The latter was obtained by using benzene-ether (50 + 50) instead of ether. Nearly 80% of the plate is effectively used for the spread.

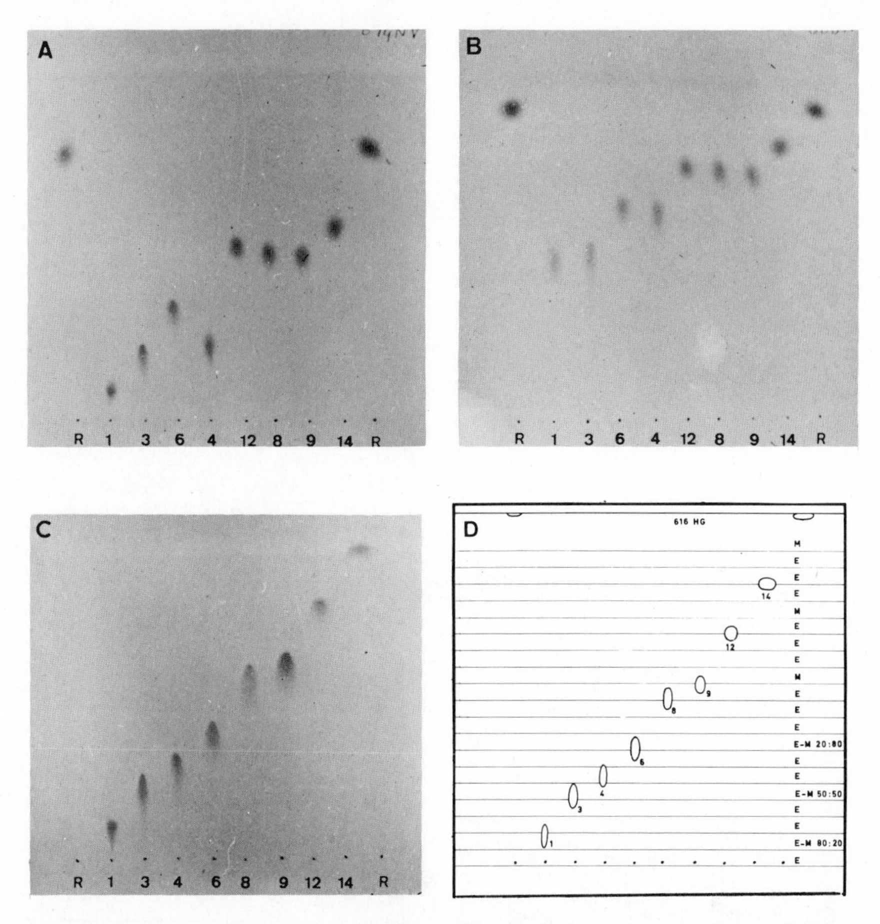

Fig. 39. Improved separation of sulfonamides in the vapor-programming chamber as compared to separations in saturated N-chambers. (*a*) N-chamber, solvent ether-methanol (90 + 10); temp. 21.5°C, rel. humidity 35%, saturation 45 min, development 38 min. (*b*) N-chamber, solvent ether-methanol (80 + 20); temp. 22°C, rel. humidity 33%, saturation 45 min, development 39 min. (*c*) VP-chamber, solvent ether-methanol (95 + 5); temp. 21.5°C, rel. humidity 35%, saturation 10 min, development 63 min, spacers 0.5 mm, cooling 19°C. (*d*) Schematic diagram of the positions of the troughs and the liquid compositions therein during development; E = ether, M = methanol. 1 = sulfaguanidine; 3 = sulfisomidine; 4 = sulfathiazole; 6 = sulfacetamide; 8 = sulfadimidine; 9 = sulfapyridine; 12 = sulfisoxazole; 41 = sulfanilamide; R = reference 4-nitroaniline; adsorbent = Silica gel GF 254; load = 3 μg. Reproduced from *Anal. Chem.*, **40** (1968) in press, copyright 1968 by the American Chemical Society; reproduced with permission of the copyright owner.

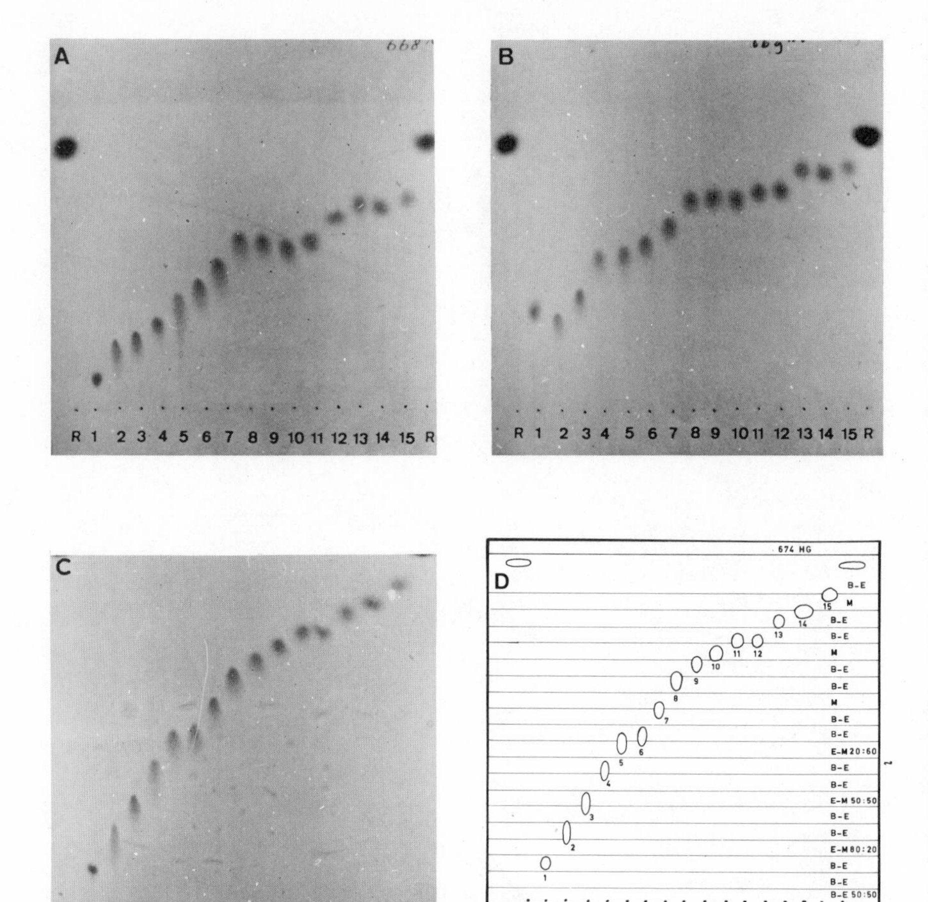

Fig. 40. Improved separation of sulfonamides in the vapor-programming chamber as compared to separations in saturated N-chambers. (*a*) N-chamber, solvent ether-methanol (90 + 10); temp. 21.5°C, rel. humidity 27%, saturation 45 min, development 40 min. (*b*) N-chamber, solvent ether-methanol (80 + 20); temp. 21.5°C, rel. humidity 27%, saturation 45 min, development 40 min. (*c*) VP-chamber, solvent ether-methanol (95 + 5); temp. 21.8°C, rel. humidity 30%, saturation 10 min, development 79 min, spacers 0.5 mm, cooling 19°C. (*d*) Schematic diagram of the positions of the troughs and the liquid compositions therein during development; B-E = benzene-ether (50 + 50); E = ether; M = methanol. 1 = sulfaguanidine; 2 = sulfamethizole; 3 = sulfisomidine; 4 = sulfathiazole; 5 = sulfadiazine; 6 = sulfacetamide; 7 = sulfamerazine; 8 = sulfadimidine; 9 = sulfapyridine; 10 = sulfamethoxypyrimidine; 11 = sulfamethoxypyridazine; 12 = sulfisoxazole; 13 = sulfadimethoxine; 14 = sulfanilamide; 15 = sulfaphenazole; R = reference 4-nitroaniline; adsorbent = silica gel GF 254, load 3 μg.

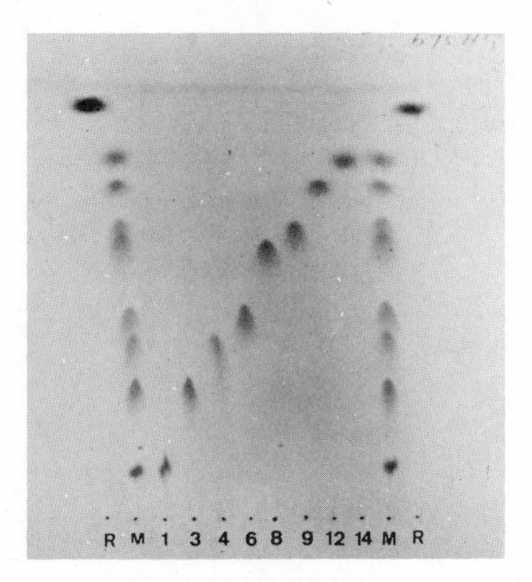

Fig. 41. Separation of a mixture in the vapor-programming chamber. The *Rf* values in the mixture are identical to those of the single components. Solvent ether-methanol (95 + 5), temp. 21.4°C; rel. humidity 31%; saturation 10 min; development 75 min; spacers 0.5 mm; cooling 19°C. Vapor-program and further details as in Figure 40.

It should be observed that the separations of the single components in the VP-chamber remain exactly the same in the separation of a mixture. This is clearly demonstrated in Figure 41, which shows the separation of a mixture of the eight compounds of Figure 39, but now using the gradient of Figure 40(*c*). It is also shown that the conditions perpendicular to the direction of solvent flow are constant across the entire plate: spots of the same substance running in different plate areas show identical R_f values.

In view of the amphoteric character of the sulfanilamides we have also investigated the separation possibilities in alkaline medium. This proved to be very difficult with the classical technique as can be concluded from Figures 42(*a*) and (*b*), using the bottom layer of chloroform-methanol-25% ammonia (70 + 20 + 10) and chloroform-methanol-25% ammonia (50 + 40 + 10) as solvent, respectively. In both cases the separation and the spread remained very poor. However, also in this case significant improvements were obtained with vapor-programmed TLC, as is shown in Figure 42(*c*). This separation was obtained by using the bottom layer of chloroform-methanol-25% ammonia (70 + 20 + 10) as the solvent and a chloroform-acetone-methanol-ammonia vapor gradient. Chloroform, saturated with ammonia, was used as decelerating liquid.

With unsaturated chambers optimal gradients could not be obtained.

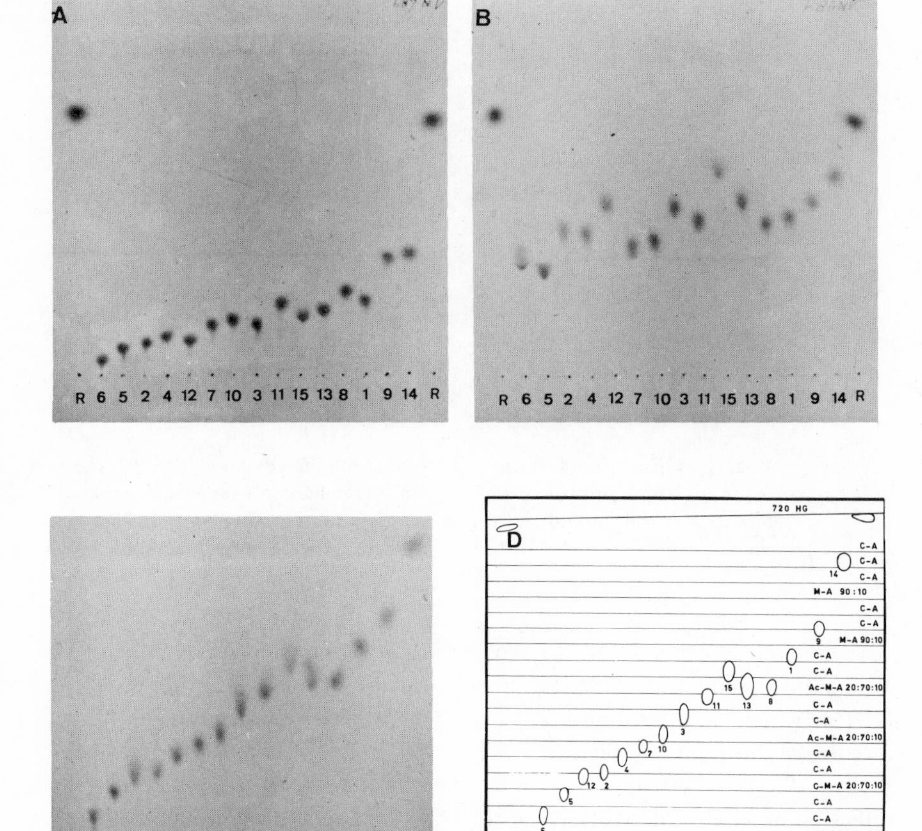

Fig. 42. Improved separation of sulfonamides in the vapor-programming chamber as compared to separation in saturated N-chambers. (*a*) N-chamber, solvent bottom layer of chloroform-methanol-25% ammonia (70 + 20 + 10); temp. 22°C; rel. humidity 41%; saturation 60 min; development 61 min. (*b*) N-chamber, solvent chloroform-methanol-25% ammonia (50 + 40 + 10); temp. 22°C; rel. humidity 41%; saturation 60 min, development 61 min. (*c*) VP-chamber, solvent bottom layer of chloroform-methanol-25% ammonia (70 + 20 + 10); temp. 22°C; rel. humidity 41%; saturation 10 min; development 121 min; spacers 1 mm; cooling 19°C. (*d*) Schematic diagram of the positions of the troughs and the liquid compositions therein during development; C = chloroform; M = methanol; A = 25% ammonia; Ac = acetone; C-A = chloroform saturated with 25% ammonia. 1 = sulfaguanidine; 2 = sulfamethizole; 3 = sulfisomidine; 4 = sulfathiazole; 5 = sulfadiazine; 6 = sulfacetamide; 7 = sulfamerazine; 8 = sulfadimidine; 9 = sulfapyridine; 10 = sulfamethoxypyrimidine; 11 = sulfamethoxypyridazine; 12 = sulfisoxazole; 13 = sulfadimethoxine; 14 = sulfanilamide; 15 = sulfaphenazole; R = reference 4-nitroaniline; adsorbent = silica gel GF 254; load = 3 μg.

E. Local Anaesthetics

As a consequence of the basic character of the majority of the investigated substances—procaine, tutocaine, tetracaine, butacaine, ethylaminobenzoate, butylaminobenzoate and lidocaine—normal chromatography on silica gel is impossible. This adsorbent shows slightly acid properties and, accordingly, basic substances are retained at the starting points and do not show any migration during development. We therefore have used alkaline silica gel plates here by making use of 0.1 N NaOH in the preparation of the plates, instead of water. The chemical structures of the local anaesthetics are listed in Table IX.

TABLE IX
Chemical Names of Some Local Anaesthetics

Generic Names	Chemical Names
Procaine	(2-diethylaminoethyl)-4-aminobenzoate
Tutocaine	(3-dimethylamino-1,2 dimethylpropyl)-4-aminobenzoate
Ethylaminobenzoate	ethyl-4-aminobenzoate
Butylaminobenzoate	butyl-4-aminobenzoate
Butacaine	(3-dibutylaminopropyl)-4-aminobenzoate
Tetracaine	(2-dimethylaminoethyl)-4-butylaminobenzoate
Lidocaine	1-(2-diethylamino-acetamido)-2,6-dimethylbenzene

The separations obtainable in saturated chambers are rather poor as can be seen in Figures 43(*a*) and (*b*). With hexane-chloroform-methanol (60 + 35 + 5) as the solvent all substances, except lidocaine, show low R_f values. With a more polar solvent such as chloroform-methanol (95 + 5) higher R_f values result, but the separation becomes worse.

With the VP-chamber again a complete separation is achieved. With hexane-chloroform-methanol (60 + 35 + 5) as the solvent and a chloroform-acetone-methanol gradient, the separation of Figure 43(*c*) is obtained. In this separation not only the pulling apart of the spots can be seen, but it is also shown that with the VP-chamber mutual distances between two spots can be decreased if necessary. In Figure 43(*a*) there is quite a gap between lidocaine and butylaminobenzoate. When using a normal accelerating gradient in the VP-chamber this gap would be enlarged, with the lidocaine finally running into the solvent front. However, in this case such a gradient is chosen that lidocaine is decelerated on the upper parts of the plate. Thus, without disturbing the further separation the gap becomes smaller and lidocaine does not appear in the front.

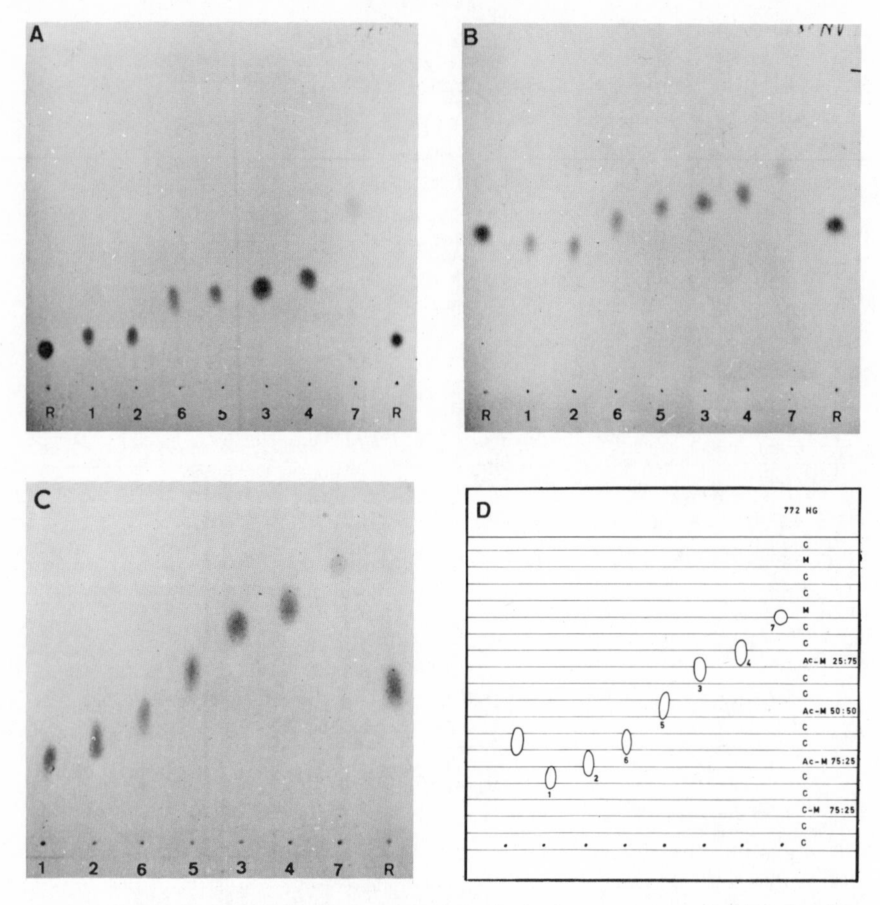

Fig. 43. Improved separation of local anaethetics in the vapor-programming chamber as compared to separations in saturated N-chambers. (*a*) N-chamber, solvent hexane-chloroform-methanol (60 + 35 + 5); temp. 22°C; rel. humidity 40%; saturation 45 min; development 25 min. (*b*) N-chamber, solvent chloroform-methanol (95 + 5); temp. 22°C; rel. humidity 44%; saturation 45 min; development 30 min. (*c*) VP-chamber, solvent hexane-chloroform-methanol (60 + 35 + 5); temp. 22°C; rel. humidity 45%; saturation 10 min; development 32 min; spacers 0.5 mm; cooling 20°C. (*d*) Schematic diagram of the positions of the troughs and the liquid compositions herein during development; H = hexane; C = chloroform; M = methanol; Ac = acetone. 1 = procaine; 2 = tutocaine; 3 = ethylaminobenzoate; 4 = butylaminobenzoate; 5 = butacaine; 6 = tetracaine; 7 = lidocaine; R = reference 4-nitroaniline; adsorbent = silica gel GF 254, prepared with 0.1N NaOH, load = 3 and 4: 10 μg, 1, 2 and 5:15 μg, 6 = 30 μg, 7 = 50 μg. Reproduced from *J. Pharm. Pharmacol.*, **20,** 54 S (1968), with permission of the publisher.

Up to this point in the discussion, silica gel plates have been used in the experiments described above, but vapor-programmed TLC can also be applied on other adsorbents without difficulty. As an example we have used basic aluminum oxide in the next experiments, on which the local anaesthetics can be separated as well. In saturated chambers reasonable results can be obtained, being better than with silica gel, but the separations remain incomplete (Figs. 44(*a*) and (*b*)). The solvents used are hexane-chloroform (50 + 50) and hexane-chloroform-methanol (80 + 15 + 5) with the latter providing a more efficient result. However, with the aid of the VP-chamber the latter separation could still be improved significantly as can be seen in Figure 44(*c*). Hexane-methanol (95 + 5) has been used as the solvent, together with a hexane-chloroform-methanol gradient. All substances are now clearly separated, with about 80% of the plate being utilized for the spread. Again, the steepness of the vapor gradient on the upper parts of the plate has been lowered so as to prevent the faster moving spots from moving into the solvent front. Unsaturated chambers could not provide similar optimal gradients.

F. Conclusions

From these experiments the many advantages and versatility of vapor-programmed TLC are evident. In the VP-chamber it is now possible to govern the chromatographic processes by acceleration and deceleration of the spots at every desired height of the plate and at every rate. Highly improved separations can be achieved, irrespective of the polarity of the substances under investigation or the polarity of the solvent and vapor components used.

Thus it becomes clear that vapor-programmed TLC offers a great many new possibilities in several directions. The first possibility lies in the qualitative analysis of closely related substances, such as homologs, isomers, and derivatives of the same basic structure. With regard to derivatives the new technique can be of great advantage in pharmaceutical chemistry and toxicology, for example in the separation of hormones, sulfonamides, hypnotics, alkaloids, pesticides, analgesics, narcotics, psychopharmaceutics, and anaesthetics, and in the analysis of metabolic products which is highly important in toxicology and biopharmacy.

Second, vapor-programmed TLC can be quite useful in preparative chromatography. It quite often occurs that in normal preparative TLC substances will overlap each other due to the higher loadings. With the aid of the VP-chamber the bands can now be pulled apart completely, thus allowing more efficient isolations. The first results with the VP-chamber in our preparative experiments are very promising. Beside these two major

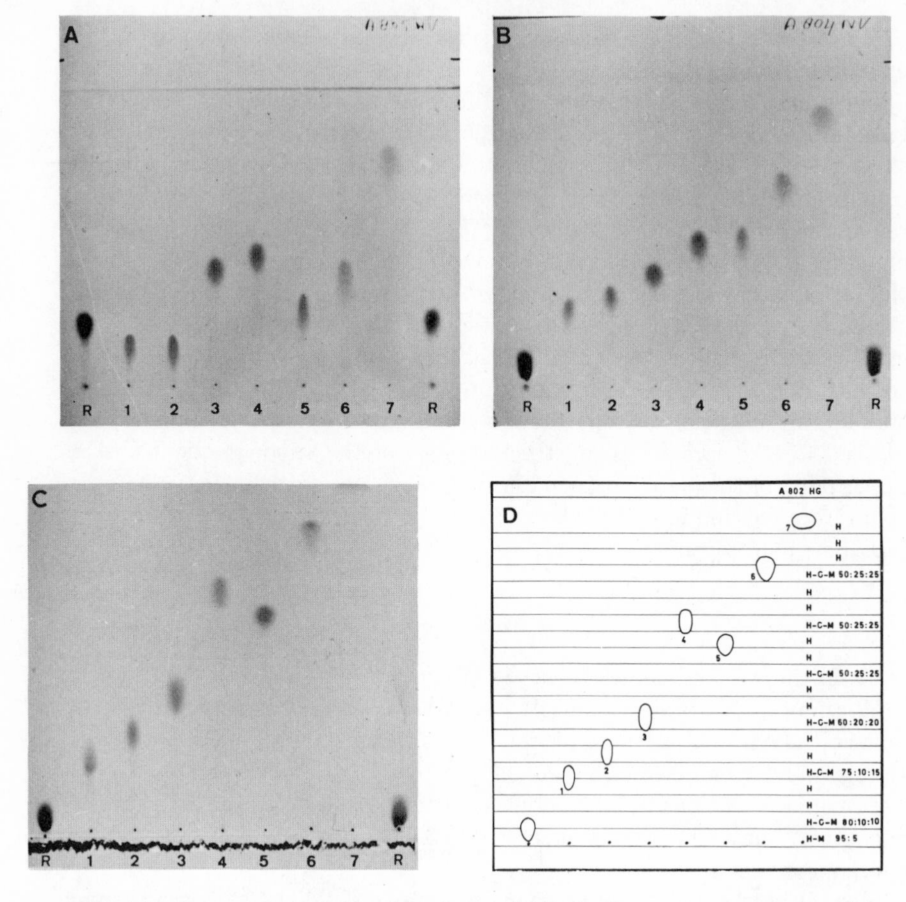

Fig. 44. Improved separation of local anaesthetics in the vapor-programming chamber as compared to separations in saturated N-chambers. (*a*) N-chamber, solvent hexane-chloroform (50 + 50); temp. 21.9° C; rel. humidity 41%; saturation 45 min; development 34 min. (*b*) N-chamber, solvent hexane-chloroform-methanol (80 + 15 + 5); temp. 22°C; rel. humidity 41%; saturation 45 min; development 32 min. (*c*) VP-chamber, solvent hexane-methanol (95 + 5); temp. 21.6°C; rel. humidity 40%; saturation 10 min; development 81 min; spacers 0.5 mm; cooling 20°C. (*d*) Schematic diagram of the positions of the troughs and the liquid compositions herein during development; H = haxane; C = chlorofrom; M = methanol. 1 = procaine; 2 = tutocaine; 3 = ethylaminobenzoate; 4 = butylaminobenzoate; 5 = butacaine; 6 = tetracaine; 7 = lidocaine; R = reference 4-nitroaniline; adsorbent = aluminum oxide GF 254, load = 3 and 4:10 μg, 1, 2 and 5:15 μg, 6:30 μg, 7:50 μg.

fields there are many other application possibilities for vapor-programmed TLC or any other suitable form of vapor-influenced TLC (unsaturated chamber, presaturated plates, troughs, etc.).

Regarding the total time needed for the development procedure it can be concluded that the VP-chamber procedure is of about the same duration as that for the saturated N-chambers. The longer development time in the VP-chamber is compensated by the rather short duration of the saturation period. With unsaturated N-chambers the results are more rapidly obtained, of course, and for those cases in which a suitable continuous gradient is so obtained, the use of the unsaturated N-chamber is recommended. However, useful optimal continuous gradients do not occur very often and in such instances the application of stepwise gradients in VP-chambers should be helpful.

The reproducibility of the separations in the VP-chamber is good. Moreover, the presence of the colored reference substance simplifies the identification procedures. A reference mixture of some substances which are expected to be present can be used as well. The saturation of the small volumes over the troughs takes place very rapidly and does not pose any problem. It will be obvious, however, that the factors involved in the reproducibility, such as room temperature, cooling temperature, relative humidity, saturation, and thickness of the spacers and of the adsorbent layers, must be kept constant. Small variations in these parameters will have great influence on the vapor gradient and, subsequently, on the separation.

Thus, vapor-programmed TLC offers many new possibilities in the analysis of closely related substances which are often inadequately separated by classical TLC techniques, and it may be expected that this technique will contribute to the further progress in thin-layer chromatography.

Acknowledgments

I am grateful to Dr. Egon Stahl and to the publishers of *Archiv der Pharmazie* for their kind permission to reproduce Figure 28. My appreciation is also extended to the publishers of *Pharmaceutisch Weekblad*, *Journal of Chromatography*, *Journal of Pharmacy and Pharmacology*, and *Analytical Chemistry* who kindly gave me permission to reproduce material from my publications.

I am indebted to the following manufacturers who supplied me with samples of the substances used in the investigations.

Abbott, North Chicago, Ill., U.S.A.
U.C.B., Brussels, Belgium
Merck, Sharpe and Dohme, West Point, Pa., U.S.A.

Byk Nederland, Zwanenburg, Netherlands
Labaz, Brussels, Belgium
Hoffmann-La Roche, Basel, Switzerland
Eli Lilly, Indianapolis, Ind., U.S.A.
Boots, Nottingham, England
CIBA, Basel, Switzerland
Benzol Products Company, Newark, N.J., U.S.A.
Siegfried, Zofingen, Switzerland
Asta, Brackwede, Germany
Riedel de Haën, Seelze, Germany
Syntetic, Aarhus, Denmark
Dumese, Copenhagen, Denmark
McNeil, Fort Washington, Pa., U.S.A.
Sterling-Winthrop, Rensselaer, N.Y., U.S.A.
Squibb, New Brunswick, N.J., U.S.A.
May and Baker, Dagenham, England
Novo, Copenhagen, Denmark
Geigy, Basel, Switzerland

References

1. E. von Arx and R. Neher, *J. Chromatog.*, **12,** 329 (1963).
2. H. T. Badings, *J. Chromatog.*, **14,** 265 (1964).
3. L. S. Bark, R. J. T. Graham, and D. McCormick, *Talanta,* **12,** 122 (1965).
4. H. Brockmann and H. Schodder, *Ber. Dtsch. chem. Ges.*, **74,** 73 (1941).
5. M. Brenner, A. Niederwieser, G. Pataki, and R. Weber in *Dünnschicht-Chromatographie,* E. Stahl, Ed., Springer, Berlin, 1962, pp. 115 and 122.
6. H. Dahn and H. Fuchs, *Helv. Chim. Acta,* **45,** 261 (1962).
7. B. H. Davies, *J. Chromatog.*, **10,** 518 (1963).
8. R. Deiniger, *Arzneim.-Forsch.*, **5,** 472 (1955).
9. E. Demole, *J. Chromatog.*, **1,** 24 (1958).
10. J. H. van Dijk and W. J. Mys, *Z. anal. Chem.*, **236,** 419 (1968).
11. R. H. Drost and J. F. Reith, *Pharm. Weekblad,* **102,** 1379 (1967).
12. H. Gänshirt in *Dünnschicht-Chromatographie,* E. Stahl, Ed., Springer, Berlin, 1962, p. 323.
13. F. Geiss, H. Schlitt, and A. Klose, *Z. anal. Chem.*, **213,** 321, 331 (1965).
14. F. Geiss, *J. Chromatog.*, **33,** 232 (1968).
15. F. Geiss and S. Sandroni, *J. Chromatog.*, **33,** 201 (1968).
16. R. B. Hermans and P. E. Kamp, *Pharm. Weekblad,* **102,** 1123 (1967).
17. J. A. Hockey, *Chem. Ind.*, 57 (1965).
18. D. Jänchen, *J. Chromatog.*, **14,** 261 (1964).
19. K. Jones and J. G. Heathcote, *J. Chromatog.*, **24,** 106 (1966).
20. H. Jork, *Deutsch. Apoth. Ztg.*, **102,** 1263 (1962).
21. J. G. Kirchner in *Technique of Organic Chemistry,* E. S. Perry and A. Weissberger, Eds., Vol. XII, "Thin-Layer Chromatography," Interscience, New York, 1967, pp. 99–101.

22. *Ibidem*, p. 8.
23. P. D. Klein, *Anal. Chem.*, **34,** 733 (1962).
24. A. Klinkenberg and G. G. Bayle, *Rec. Trav. Chim.*, **76,** 593 (1957).
25. B. Loev and K. M. Snader, *Chem. Ind.*, 15 (1965).
26. H. K. Mangold, *J. Am. Oil Chemists' Soc.*, **38,** 708 (1961).
27. A. J. P. Martin, *Biochem. Soc. Symposia* (Cambridge), **8,** 4 (1949).
28. A. Niederwieser and M. Brenner, *Experientia*, **21,** 50, 105 (1962).
29. G. Pataki, *J. Chromatog.*, **29,** 126 (1967).
30. K. Randerath, *Dünnschicht-chromatographie*, 2nd ed., Verlag Chemie, Weinheim, 1965, p. 70.
31. P. Schweda, *Anal. Chem.*, **39,** 1019 (1967).
32. W. M. Smit and A. van den Hoek, *Rec. Trav. Chim.*, **76,** 561, 577 (1957).
33. L. R. Snyder in *Chromatography*, E. Heftmann, Ed., 2nd ed., Reinhold, New York, 1967, p. 52.
34. E. Stahl, *Arch. Pharm.*, **292,** 411 (1959).
35. E. Stahl in *Dünnschicht-Chromatographie*, E. Stahl, Ed., Springer, Berlin, 1962, 1967, 1st ed., p. 16, 2nd ed., p. 70.
36. D. R. Stull, *Ind. Eng. Chem.*, **39,** 517 (1947).
37. Tabellenboekje ten dienste van Laboratoria, D. B. Centen, Hilversum, 1962.
38. A. Wehrli, *Can. Pharm. J. Sci. Sect.*, **97,** 208 (1964).
39. R. A. de Zeeuw and M. T. Feitsma, *Pharm. Weekblad*, **101,** 957 (1966).
40. R. A. de Zeeuw, *Pharm. Weekblad*, **102,** 113 (1967).
41. R. A. de Zeeuw, *J. Chromatog.*, **32,** 43 (1968a).
42. R. A. de Zeeuw, *J. Chromatog.*, **33,** 222 (1968b).
43. R. A. de Zeeuw, *J. Chromatog.*, **33,** 227 (1968c).
44. R. A. de Zeeuw, *Anal. Chem.*, **40,** 915 (1968d).
45. R. A. de Zeeuw, *Anal. Chem.*, **40,** 2134 (1968).
46. R. A. de Zeeuw, *J. Pharm. Pharmacol.*, **20,** 54S (1968).
47. D. F. Zinkel and J. W. Rowe, *J. Chromatog.*, **13,** 74 (1964).

"Dry-Column" Chromatography

BERNARD LOEV AND MARJORIE M. GOODMAN

Research and Development Division,
Smith Kline and French Laboratories,
Philadelphia, Pennsylvania

I. INTRODUCTION

"Dry-column" chromatography is an improved chromatographic technique by means of which separations comparable to those obtainable by thin-layer chromatography can be carried out rapidly *in a column* on a preparative scale. This method can be used for the resolution of any mixture that can be separated by thin-layer chromatography (TLC), including those which cannot be separated on the usual "liquid-filled" column. It was found that (*a*) a *deactivated* adsorbent gives separations vastly superior to those obtained with a highly activated absorbent; (*b*) the degree of separation obtained using such an adsorbent in a column is as good as that obtainable by TLC; (*c*) the separation obtained on this type of column is directly related to that of the TLC plates (a valuable relationship that permits the direct transfer of conditions from TLC to the "dry-column");

and (*d*) the dimensions of a column required for a preparative separation can be predicted.

Mixtures containing steroids, alkaloids, lipids, acids, amines, and a large variety of heterocycles have been separated successfully on the "dry-column." It has been used with radioactive substances and for the isolation of metabolites. A "dry-column" impregnated with silver nitrate has been used for the separation of olefins (3a), and a micro method has been used for the preparation of samples for mass spectrometry (3b).

The procedure was first reported in 1965 (1) and improvements were reported in 1967 (2). In this chapter, the pertinent literature is reviewed and complete details of the procedure are given.

II. HISTORICAL

In a paper delivered to the American Philosophical Society in 1897, Day (4), in what appears to be one of the first written descriptions of chromatography, stated

> It may easily be demonstrated that if we saturate a limestone, such as the Trenton limestone, with the oils characteristic of that rock and exert slight pressure upon it so that it may flow upward through finely divided clay, it is easy to change it in its color to oils similar in appearance to the Pennsylvania oils, the oils which first filter through being lighter in color, and the following oils growing darker.

In 1900, Albrecht and Engler (5) independently published similar observations with respect to petroleum analysis. Thus, although Michael Tswett is commonly referred to as the discoverer of chromatography, this is not quite correct. Nevertheless, to him must go credit for the first systematic examination of this technique and its development to its present useful state. In 1903 (6), he reported experiments on the analysis of chlorophyll in which, for the first time, an adsorbate was cleanly separated spatially into several zones on an adsorbent in a vertical tube. In 1906, Tswett labeled this method "chromatography" (6). The importance of Tswett's work went practically unnoticed until 1931 when Kuhn and Lederer (7) used the method for the separation of the polyene pigments. Since that time, column adsorption chromatography has become a standard laboratory procedure. The often tortuous methods that the early workers employed to pack columns with dry adsorbents (see 8, for example), and the difficulties involved in obtaining suitable columns led Winterstein and Stein in 1933 (9) to recommend the use of slurries of adsorbent in a solvent. This became the method of choice for packing and running chromatographic columns.

In 1938, Izmailov and Shraiber (10) described the principles of TLC but used it solely for terpenes; it was not until the late 1950's that TLC became a standard procedure, after Stahl (11) showed the general usefulness of the method. Now TLC has become the indispensable tool for the separation and purification of mixtures. However, the technique has an obvious limitation since only micro- or milligram quantities can normally be separated by this method.

Following the development of TLC, the comparatively poor separations which column chromatography provided and the extensive time and inconvenience involved stimulated many workers to seek ways of attaining the same degree of separation on a preparative scale. "Thick-layer" chromatography was devised, but even with this method several hundred milligrams are the most that can be conveniently separated. Recently, preparative columns consisting of bars made by compressing thin-layer chromatographic adsorbents have been offered commercially. However, they are costly and limited as to the amount of mixture which may be applied.

Dahn and Fuchs (12) employed a dry-packed cellophane column in a horizontal position, but the development of these columns was exceedingly slow (often 20 hr or more) and the capacity was low (maximum of 1 g). They attributed improved separation to the use of small particle size and high adsorbent-to-compound ratio (5000:1) but, as will be shown later, neither of these factors significantly influences the degree of separation.

In 1965, we developed an improved technique for column chromatography which allowed us to obtain the same degree of resolution attainable by TLC. This technique was named "dry-column" chromatography (1,2). All mixtures that can be separated by TLC, including those which are not separable on the usual "liquid-filled" column, can be separated on a preparative scale efficiently and rapidly by this method.

Packing columns dry is by no means a new technique. Indeed, Tswett packed his columns with dry adsorbent in 1903, as did everyone for the next two decades. Unfortunately, the factors required to give separations comparable to those obtained by TLC were not then recognized, and columns prepared along these early historical lines gave separations which were, at best, only *equal* to those obtainable by the now standard "liquid-filled" type columns. Our contribution has been to recognize those factors which convert this method from a hit-or-miss procedure, generally giving poor separations, to one which will rapidly and consistently give separations superior to those of a "liquid-filled" column.

Hall (13) and Stainer and Bonar (14) have recently employed dry-packed columns in which the adsorbent was paraffin-impregnated cellulose, but this represents partition chromatography rather than adsorption chromatography. Bhalla *et al.* (15), in 1967, have extensively investigated the *inverted* (or ascending) "dry-column" procedure which we described in an earlier paper; we have found that the descending procedure, described in this chapter and in our papers (1,2), is far superior.

III. THE "DRY-COLUMN" PROCEDURE

A. General Discussion

Basically, the "dry-column" procedure is carried out by filling an empty column with adsorbent, depositing the mixture to be separated on the top of the column, and developing the chromatogram by allowing the solvent to move down the *dry* column aided by capillary action and the pull of gravity. No liquid flow is observed so there is no channeling, and zone separation is sharp and straight. By the time the solvent reaches the bottom, usually after 15 to 30 min, separation is complete. The separated fractions can then be removed and isolated by one of the techniques described later.

The amounts that can be separated by "dry-column" are limited only by the practical physical limitations of column size. Columns as long as 6 ft have been used, and mixtures of over 50 g have been separated successfully. On a normal laboratory scale, separations up to 15 g are carried out routinely and rapidly by the "dry-column" method. An alumina column $1\frac{1}{2} \times 20$ in. can be used to separate 7 g of a mixture with development taking only about 30 min.

The "dry-column" technique was developed by examining those factors which appeared to be responsible for the superior resolution obtained in TLC and then applying them to a column. The first assumption was that it resulted from a *small* amount of solvent moving over a *dry* adsorbent with only the solvent adsorbed by capillary action being used for development. By contrast, in a "liquid-filled" column, where resolution is much poorer, development occurs on an adsorbent actually suspended in the solvent.

Another, and undoubtedly the most important, factor involved in increasing the resolvability proved to be activity of the adsorbent. In contrast to what might have been expected, it was found necessary to *deactivate* the adsorbent in order to increase its ability to resolve mixtures on a column. This finding was an outgrowth of the observation that most TLC alumina slides or plates, homemade or commercial, proved to have an activity of only II to III on the Brockmann scale (16). Nevertheless, better separations are obtained with these plates than with grade I adsorbent in a column.

Other factors were examined, such as the use of an ascending rather than a descending column. The resolution proved to be the same in both instances but, obviously, the descending method is more rapid and convenient. The particle size of the adsorbent was found to have no bearing on the quality of separation that could be obtained. The use of "TLC grade" adsorbent did not prove to have any advantage over ordinary column-grade sized adsorbents and had the disadvantage of greatly increasing development time. However, by using a suitably deactivated adsorbent and packing the column dry, it was found possible to duplicate thin-layer separations in a column.

A valuable consequence of the relationship between TLC and "dry-column" chromatography is the *direct transferability of the conditions* from the former to the latter. All preliminary studies to determine the optimum conditions for separation of a mixture are done on TLC plates or microscope slides.* Once determined, these conditions can be transferred directly to a column. If, with the column, one uses the same solvent system and the same adsorbent (i.e., adsorbent deactivated to the same activity [II–III] as the TLC plates), a similar degree of separation will be observed, and the compounds will be found on the column close to the positions predicted by the R_f values determined on the TLC plates.

As previously mentioned, Dahn and Fuchs (12) felt that the superior separations obtained by TLC as compared to "liquid-filled" columns were due mainly to the large ratio of weight of adsorbent to compound (5000:1) found in TLC (the ratio in "liquid-filled" column chromatography is normally on the order of 50:1). We have found, however, that in the dry-column procedure, ratios from 70:1 to 300:1 (depending on the difficulty of separation) are all that are required. Thus, it appears that it is not the large adsorbent:compound ratio that makes TLC so effective, but rather the use of a dry, deactivated adsorbent.

The reproducibility of the "dry-column" technique can be seen from the results obtained on chromatography of 0.5 to 15 g of material on different size columns. The results are shown in the following chart:

"Dry Column" Chromatography R_f Values

	Diameter of Column			
Dye	$\frac{1}{2}''$	$1''$	$1\frac{1}{2}''$	$2''$
Blue–I	0.20	0.18	0.17	0.17
Blue–II	0.30	0.33	0.33	0.31
Yellow	0.74	0.75	0.73	0.74

Details of the "dry-column" technique are provided below.

* We prefer to use the TLC microscope slides for our thin-layer work. These 2 × 10 cm plates develop rapidly and, if good quality plates are used, they give as good a separation as may be obtained on the 5 × 20 cm plates commonly in use.

B. Adsorbent

Of the two most commonly used adsorbents for chromatography, silica gel and alumina, we generally prefer alumina because it has a larger capacity (so that smaller columns can be used) and a faster flow rate. Also, the preparation of suitably deactivated adsorbent is more reproducible.

For "dry-column" chromatography, regular column-size adsorbents (100–200 mesh) are employed, and they are suitably *deactivated* to correspond to the activity of the adsorbents on the TLC plates which are used. *It is the activity of the adsorbent that is most important in determining the quality of the separation!* The deactivation is done by adding the proper amount of water, thereby giving an adsorbent whose activity then corresponds to that of TLC plates. To prepare alumina of grade II to III (corresponding to the activity of the average alumina TLC plate), the activity of the adsorbent is checked and, depending on its original activity, from 0 to 6% water is added. For silica gel, the addition of up to 15% water is required for deactivation to a grade corresponding to that of good silica gel TLC plates.

It is most important to check the activity of the column adsorbent *before* deactivation. The percentage of water to be added, as shown in Table I, assumes that the adsorbent, as purchased, is of activity grade I. *This is not always the case.* By checking the activity before deactivation, the amount of water to be added can be adjusted depending on the degree of deactivation already present in the adsorbent. This is particularly important with silica

TABLE I

Water Added (%)	R_f of Appropriate Dye	Activity Grade of Adsorbent According to Brockmann Scale
For Alumina		
0	0	I
3	0.12	II
6	0.24	III
8	0.46	IV
10	0.54	V
For Silica Gel		
0	0.15	I
3	0.22	
6	0.33	
9	0.44	
12	0.55	II
15	0.65	III

gel, where the activity varies considerably even in different bottles of the same brand. Occasionally, bottles come which, as purchased, are already deactivated to grade III and therefore require no addition of water. Because silica gel does deactivate so rapidly to grade II to III with just routine handling, it is likely that the good separations "without deactivation" occasionally reported in the literature were due to the fortuitous use of suitably deactivated adsorbent.

Investigation of different brands of 80 to 200 mesh alumina (among which were Metal Hydrides, Inc., neutral alumina No. 507C-1, Woelm neutral alumina, and Fisher No. A540 alumina, available from Fisher Scientific Co., Pittsburgh, Pennsylvania) showed some variation in suitability for this technique. Certain brands packed too compactly, greatly increasing development time; others had lower capacity or were too sensitive to slight changes in water content. Although all brands were suitable, Fisher material gave the most satisfactory and reproducible results. Recently, an alumina specially made for "dry-column" use has become commercially available (17).

The silicas investigated were Baker No. 3405 silica gel (available from J. T. Baker Chemical Co., Phillipsburg, New Jersey), W. R. Grace & Co. grade 923 silica gel 100 to 200 mesh, and Mallinckrodt No. 2847 silicic acid, A. R., 100 mesh. The Mallinckrodt material was very fluffy and unsatisfactory for packing the "dry-columns"; the Baker silica gel was the preferred brand. A special "dry-column" grade of silica gel has recently become commercially available (17). We stress that these are ordinary chromatographic grades of adsorbents and *not* TLC grades.

To deactivate the adsorbent, the appropriate amount of water is poured onto the adsorbent in a glass jar and equilibrated by rotation on ball-mill rollers for about 3 hr. A satisfactory alternative is rotation in a round-bottom flask on a closed rotary evaporator unit for 3 hr.

The standard method of determining the activity grade of an adsorbent depends upon the ability of a given amount of the adsorbent to separate a mixture of five dyes (16). For greater convenience, we employ a simplified method of determining the activity of alumina or silica gel, based on the distance that a *single* dye moves. For alumina, the dye used to determine the activity is *p*-aminoazobenzene (18); for silica gel, the dye used is *p*-dimethylaminoazobenzene (18).

The activity of the adsorbent is determined in this way: a 1.0×75 mm capillary (19) is filled with the adsorbent* and one drop of benzene is placed

* This is readily done by taking the rubber bulb of a small dropper, punching a small hole in the bottom, inserting the open end of the capillary a short way into the bulb from the outside, and then filling the bulb with adsorbent. Slight tapping of the bulb will cause the capillary to fill rapidly.

on the open end. The tube is inverted (the drop of solvent preventing the adsorbent from falling out of the capillary) and the closed end is snapped off. The damp end is then placed into a small vial filled to a depth of a few millimeters with a 0.5% solution of the appropriate dye in benzene, until a drop of dye solution is adsorbed. Then the capillary is transferred to another vial containing a few milliliters of benzene, and the miniature column is allowed to develop.

The R_f is calculated and the corresponding activity is determined from Table I. When the R_f of the dye on the adsorbent to be used in the column is the same as the R_f of the dye on the TLC plates, the adsorbent is suitably deactivated and ready for use (17).

It is important to check the activity of the TLC plates being used for preliminary work to be sure that they are of grade II–III activity. We find that, particularly with alumina, the activity of most commercial plates is often poor and varies from batch to batch. On the other hand, most commercial brands of silica gel TLC plates have suitable and reproducible activity.

C. Preparation of the Column

1. GLASS COLUMNS

A glass or quartz column with suitable support at the bottom, and with the stopcock *open* to prevent the formation of air pockets, is packed *dry* by pouring in the deactivated adsorbent slowly and evenly while tapping the column with a rubber hammer (Fig. 1) or holding a vibrator (20) against the column (Fig. 2). The column is then loaded with the mixture and developed, as discussed later.

After development, it is necessary to remove the adsorbent from the column in a reasonably intact state in order to isolate the separated components of the mixture. This can be done by applying pressure to one end of the column or by inverting and tapping the end against a stopper. Both of these procedures leave much to be desired, however, and a search for a more convenient method of isolation led to the use of plastic columns described below, which make it possible to simply slice up the column after development to obtain the desired segments. The location of the separated components will be described later.

2. NYLON COLUMNS

In order to facilitate the location and isolation of the separated components of a mixture, an appropriate type of plastic column was sought. A material was desired that would be strong, yet easy to cut and heat-seal by hand, inert to organic solvents, and available at relatively low cost. One

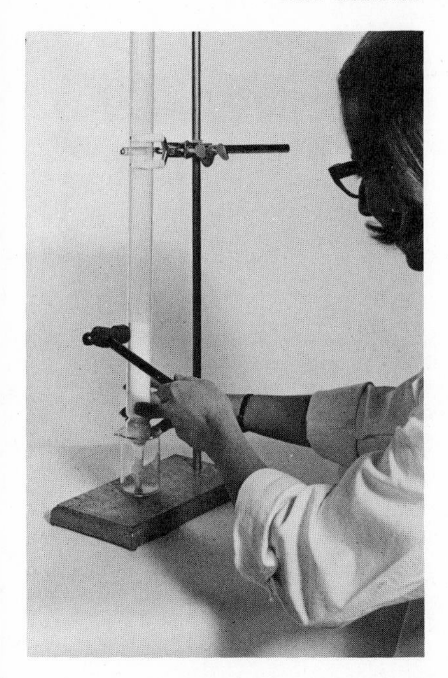

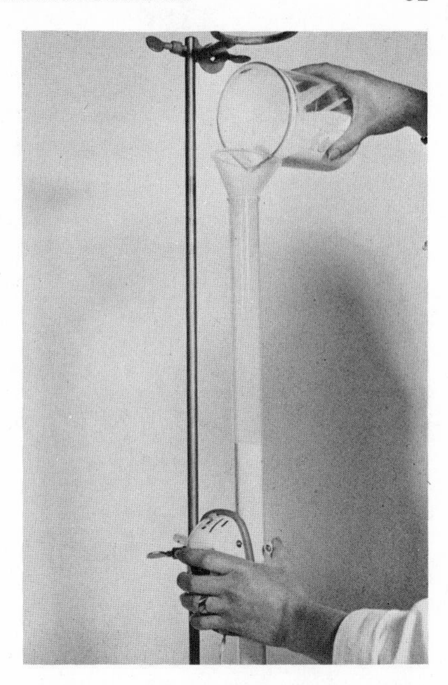

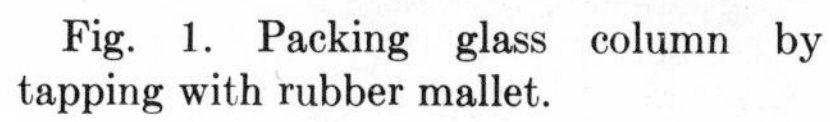

Fig. 1. Packing glass column by tapping with rubber mallet.

Fig. 2. Packing glass column with aid of vibrator.

other property was essential and this was transparency to shortwave ultraviolet light. This property makes possible the location of colorless materials, as described shortly. The requirement of UV transparency limited the available plastics to two—cellophane and nylon. The use of a cellophane column was first described in a little known article by Ochiai and Takeuti (1938) (21), and expanded on by Sabel and Kern (22). However, we found cellophane to be brittle and thus prone to cracking, and not very strong. Nylon, on the other hand, is quite pliable, chemically inert, transparent to ultraviolet light, very strong, and available (23) in rolls of varying diameter and length at relatively low cost. A film thickness of about 1.6 mil was found to be the most suitable for the columns.

The nylon columns are filled and used in the following manner: A column of the appropriate length (discussed later) is sealed at one end (24), a small pad of glass wool is inserted, and two or three holes are made at the bottom (below the glass wool [Fig. 3]) for drainage and to prevent air pockets from forming while the column is being packed. The adsorbent is poured in rapidly (Fig. 4) until the column is one-third filled, at which point the adsorbent is compacted in the column by allowing the bag to drop two or

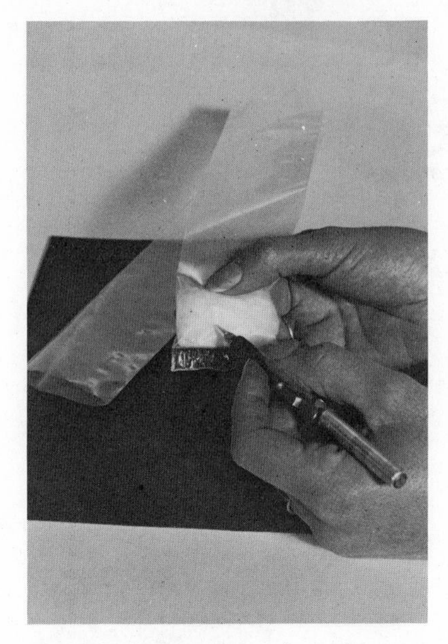

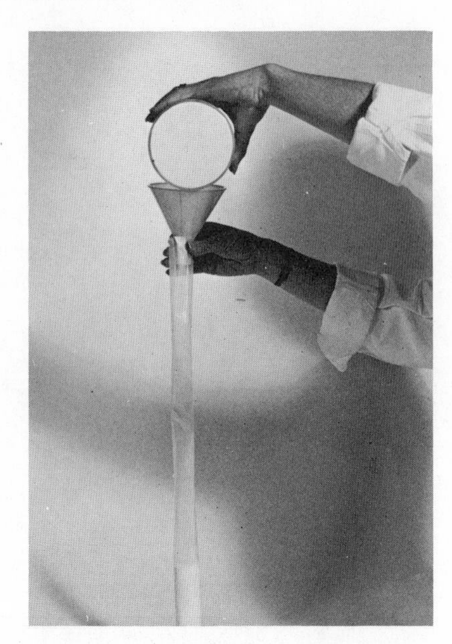

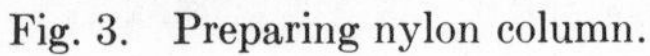

Fig. 3. Preparing nylon column.

Fig. 4. Filling nylon column.

present on the adsorbent from evaporating and changing the activity of three times from a height of about 6 in. onto a hard surface (like packing a melting point capillary, Fig. 5). The bag is then filled another third, again compacted, then filled to the top and compacted once more while being held tightly at the top (Fig. 6). When filled, the column is quite sturdy and can be supported by a clamp (Fig. 7).

D. Loading the Column

Liquids may be poured directly onto the column, or dissolved in a minimum volume of the solvent to be used for development and then distributed evenly onto the top of the column. The solution is allowed to soak completely into the column before development is begun. Solids may be loaded on the column in the same way—that is, by dissolving them in a minimum volume of eluting solvent and pouring the solution on the column or, when very small quantities are being chromatographed, by distributing the powdered solid evenly on the top of the column.

Our preferred method is to *deposit* the solid or liquid on the adsorbent. This is done by dissolving the solid in a low-boiling solvent (ether or methylene chloride), adding about five times its weight of deactivated adsorbent and evaporating the mixture to dryness in a rotary evaporator at 30–40°; the low temperature is used in order to prevent the water the adsorbent. The compound-adsorbent mixture is then distributed

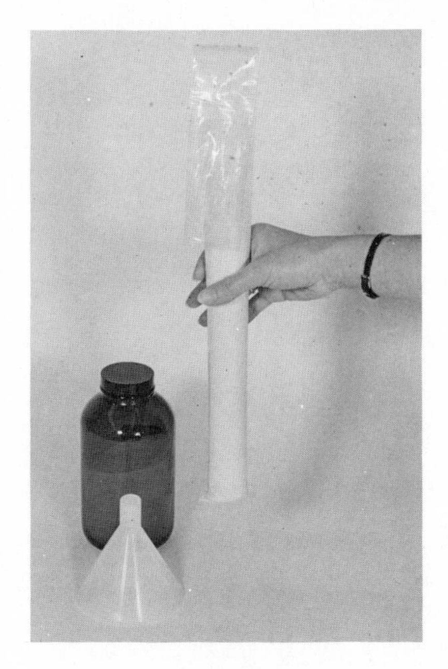

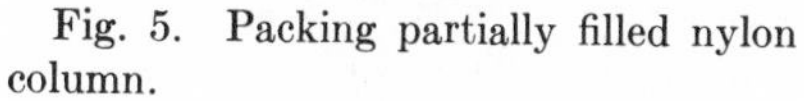

Fig. 5. Packing partially filled nylon column.

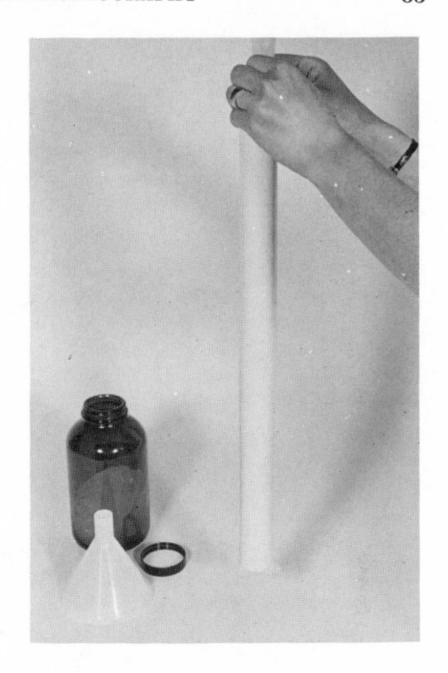

Fig. 6. Final "compacting" of filled nylon column.

evenly on the top of the column and covered with a small layer (3 mm) of sand or glass beads (Fig. 8).

E. Solvent System

The appropriate solvent for use on the "dry-column" is the one which has been found effective for the separation of the mixture on TLC plates. If at all possible (and it usually is if one uses a little patience), a single-solvent system should be used. Lists of solvents in increasing order of elution power can be found in most texts. Based on extensive experience, we have found the following modified and expanded list to be most useful.

1. Petroleum ether
2. Hexane
3. Carbon disulfide
4. Carbon tetrachloride
5. Trichloroethylene
6. Toluene
7. Benzene
8. Isopropyl ether
9. Methylene chloride
10. Chloroform
11. Ether
12. Ethyl acetate
13. Methyl acetate
14. Acetone
15. Propanol
16. Ethanol
17. Methanol

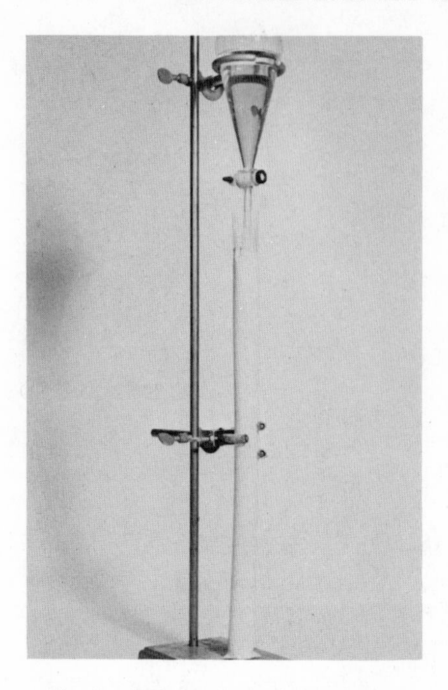

Fig. 7. Filled nylon column ready for use.

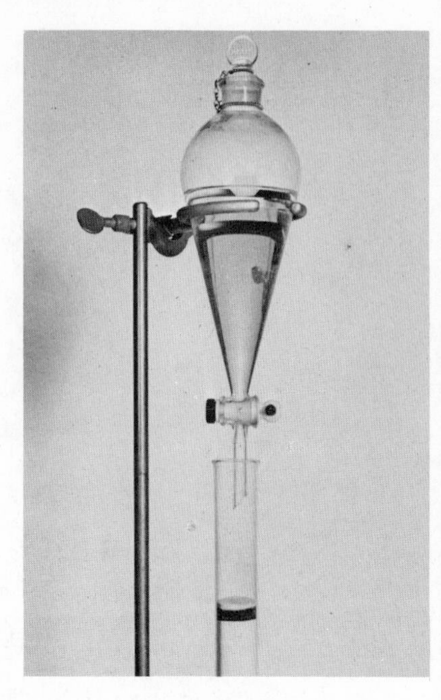

Fig. 8. Glass column showing dye mixture deposited on adsorbent and ready for development.

If a mixed-solvent system *must* be used, a special pretreatment of the adsorbent is necessary or very poor separations will occur. The adsorbent must be additionally equilibrated with some of the solvent mixture that is to be used for development (2). This is done by adding some of the solvent mixture to the adsorbent and reequilibrating in the manner described earlier for deactivation. Preliminary experiments indicate that the amount of solvent mixture to be added should be about 10% by weight of adsorbent (25), but this may vary with the particular solvent mixture used. Once the adsorbent has been equilibrated with the mixed solvent, the "dry-column" is carried out in the usual manner. The use of a mixed-solvent system is discussed in some detail by Casey (25) who utilized it for the separation of a complex lipid mixture.

Chloroform and methylene chloride, *when used with a silica gel column, appear* to soften the nylon column as development occurs, and the column sags. These solvents *may be used,* it is just necessary to be aware of this occurrence; an alternative is to use a glass or quartz column with these specific solvents. These two solvents also have the disadvantage of making

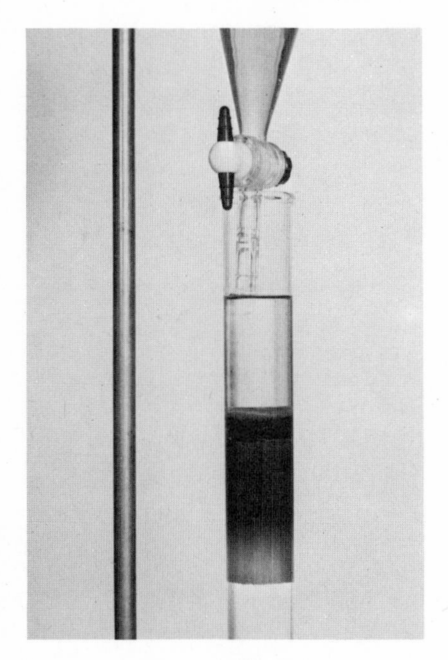

Fig. 9. Dry-column chromatogram (in glass column) of dye mixture shortly after start of development.

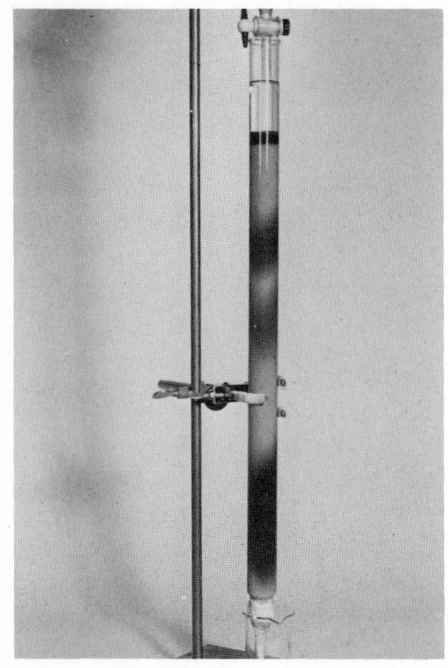

Fig. 10. Completely developed chromatogram (in glass column) showing the three separated components (top to bottom: blue dye I, blue dye II, yellow dye).

the silica gel translucent, so separations are sometimes somewhat difficult to observe. It should be noted that with *alumina* as adsorbent no softening occurs and development is very smooth indeed.

Acetic acid tends to attack nylon, so if it is being used "neat" as solvent, glass or quartz columns should be used. Acetone is a poor solvent to use on alumina as it tends to dimerize, giving diacetone alcohol—the presence of which may lead to confusion and interfere with the collection of the desired materials.

F. Development of the Column

The developing solvent is added under a constant liquid head of 3–5 cm (Fig. 9—a stoppered separatory funnel is convenient for this purpose). The time for development of a 20-in. alumina column varies from 15 to 30 min, depending on the diameter of the column. *When the solvent reaches the bottom of the column, development is complete* (Figs. 10 and 11 show glass and nylon columns, respectively).

Fig. 11. Completely developed chromatogram (in nylon column) showing the three separated components.

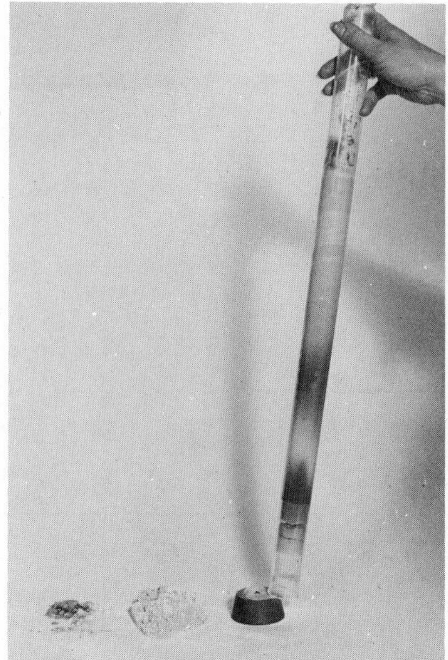

Fig. 12. Extrusion of glass column by tapping on a rubber stopper.

In the case of a glass or quartz column, the column is allowed to run dry, inverted, and tamped down on a cork or rubber stopper (Fig. 12). The packing slides down slowly and is sliced in the usual manner. If a nylon column is used the column may be sliced up, like a sausage, with a sharp knife (Figs. 13 and 14) and the desired segments placed in sintered glass funnels and extracted with methanol or ether.

When the amount of solvent used is limited to that required to develop a column to the bottom, the chromatogram may be left untended since it will stop automatically. The diffusion of zones is extremely slow (several days). The approximate quantity of solvent necessary for this purpose in alumina columns is indicated below:

Column Size (in.)	Volume of Solvent (ml)
$20 \times \frac{1}{2}$	20
20×1	90
$20 \times 1\frac{1}{2}$	300
20×2	500

Fig. 13. Preparing to slice up nylon column.

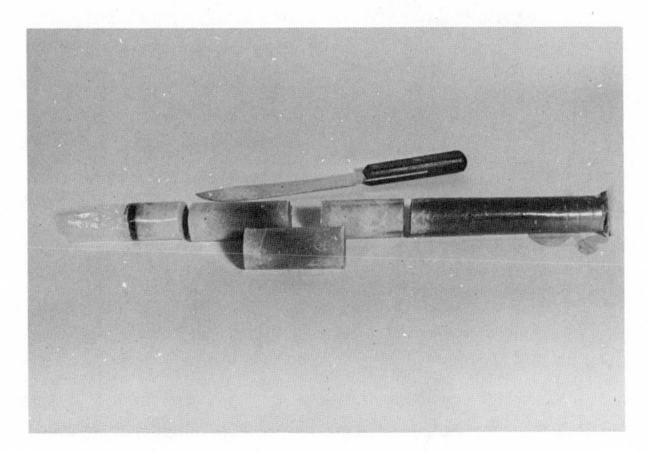

Fig. 14. Sliced nylon column.

These figures also point out the very small amount of solvent required for a "dry-column" compared to a standard type of column.

When working with large diameter columns, it has been found that a compound apparently moves slightly faster at the edges than in the center of the column, and a dome-shaped adsorption band occurs. Although it may not be well known, the same effect occurs in liquid columns (26). In any event, this is not a serious problem in laboratory-scale separations.

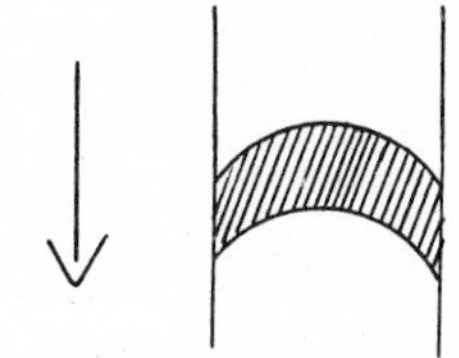

If we continue to add solvent to the column the bands will slowly continue to move down the column, often separating further as development continues. This is equivalent to doubling the length of a TLC plate, or developing the same TLC plate *twice*.* Thus, in instances where a still better separation is desired, we can continue to pass solvent through as long as there is room for separation on the column, or we can continue and elute the separated components off the column. It is usually much more rapid and convenient, however, to stop the development when the solvent reaches the bottom of the column and to then slice the column, since this greatly minimizes the number of fractions that must be worked up.

G. Detection of Compounds on the Column

In the case of colored mixtures, the separation of compounds is easily followed in either glass or nylon columns. The latter makes isolation easier, but glass columns may be used and extruded in the standard manner to obtain the pure components. With colorless mixtures, the development can sometimes be followed by examining the column under long wavelength ultraviolet light in a dark room. Most colorless organic compounds, however, do not show significant fluorescence under long wavelength light. When quartz or nylon columns are used, many more compounds can be located as a result of their fluorescence when examined under *short* wavelength ultraviolet light. However, the fluorescence is often weak and difficult to see, and most compounds still cannot be located this way. The problem of locating substances on a column was solved by utilizing the technique now so commonly used in TLC, that is, by using an *adsorbent* that fluoresces under short wavelength light, and locating compounds by the *quenching* of the fluorescence. Most compounds can be located readily by this relatively little known method, which has been employed in "liquid-flow" columns by Brockmann and Volpers (27) and by Sease (28).

Although fluorescent TLC-grade adsorbents are available, until very recently it was not possible to purchase a fluorescent *column*-grade adsorbent (17). A suitable fluorescent adsorbent is prepared by the addition of

* This addition of further solvent to the column after the front has reached the bottom does not now correspond to developing a "liquid-filled" column. This was demonstrated by first passing solvent through a "dry-column" and *then* putting the mixture to be separated on the column, and developing the column with the same solvent. The separation was the same as was obtained in a standard "dry-column," and far superior to that obtained in a "liquid-filled" column.

In this connection, it is of interest to note that when grade II–III adsorbent is used in a standard "liquid-filled" column, the quality of separation is significantly better than that customarily observed in a "liquid-filled" column with grade I adsorbent, although it *still* does not approach that obtained in a "dry-column."

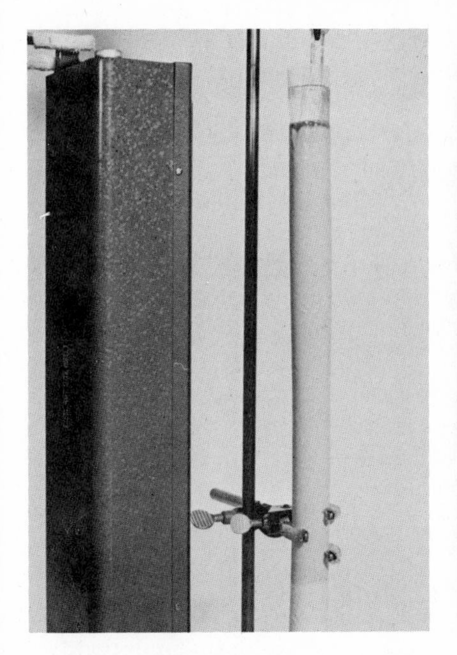

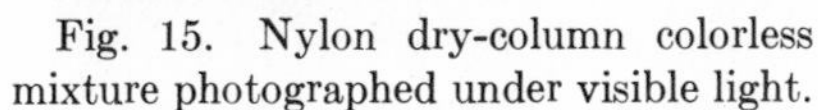

Fig. 15. Nylon dry-column colorless mixture photographed under visible light.

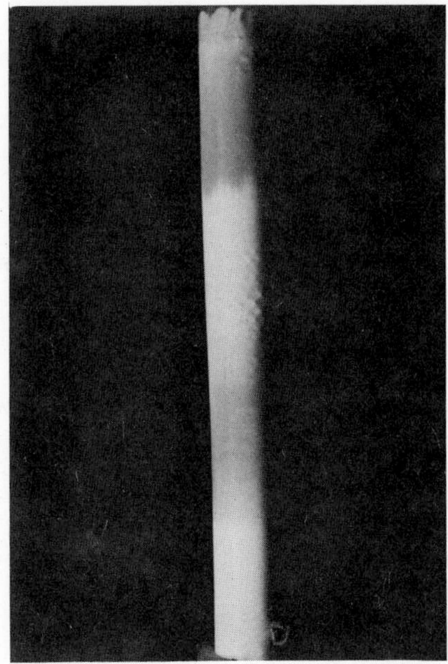

Fig. 16. Column of Fig. 15 photographed under short-wave length light and showing separated components.

0.5% of an inorganic fluorescent powder (29) to the adsorbent at the same time that water deactivation is being carried out, as described in Section III.B. During the equilibration period, the fluorescent material becomes evenly distributed throughout the system.

Now, by using a fluorescent adsorbent, a solvent transparent to ultraviolet light (this excludes benzene and toluene), and a quartz (very expensive and fragile) or nylon column, the separation of a colorless mixture can easily be followed in a dark corner of the room by examining the column under the light of a short wavelength ultraviolet lamp (Figs. 15 and 16) (30). The column is marked under ultraviolet light and then sliced.

Other methods for locating colorless substances are also possible, but they are not as convenient as the above. One method is to lay the developed nylon column down on a table, carefully cut the nylon lengthwise, and then apply a thin streak of a visualization reagent down the length of a column with a brush. Here again, the use of the nylon column permits the easy application of this technique, which was proposed by Zechmeister *et al.* (31). This method, however, does contaminate the desired compounds slightly with traces of the detecting reagents and is somewhat difficult to carry out.

Another rather clever method employs the addition of a colored marker dye, which runs between or with the spots to be separated, thus indicating where the desired compounds are in relation to the marker dye (32,33). Any number of dyes can be used. However, this method obviously necessitates locating a dye with an R_f between the components to be separated, or else separating the marker dye from the compounds; this may prove time-consuming or difficult.

If none of the above methods proves satisfactory, we can use the R_f determined on the TLC plate as an approximate guide to the location of the desired fraction. In this case, the activity of the plate and column material must be very similar.

When all else fails, a very simple and satisfactory solution is to slice the column arbitrarily into sections of a given length, elute the fractions, and run TLCs on these solutions to determine which fractions hold the desired compounds. Fractions that are the same are then combined and concentrated in the standard manner.

H. Choice of Column Size

The selection of the appropriate size column to give optimum separation of a given quantity of mixture is based on the difficulty with which the compounds are separated by TLC. Clearly, the greater the difficulty with which two compounds are separated, the larger the column that will be required. There was found to be a direct relationship between the column dimensions (or weight of adsorbent) and the quantity of mixture being separated; these results are charted for alumina and silica gel columns on Graphs I and II. Mixtures were arbitrarily divided into two classes: (*a*) those which are fairly readily separated on TLC slides (R_f differences of approximately 0.4 and greater), and (*b*) those which are very difficult to separate (R_f differences on the order of 0.1).

In this work, and for the preparation of the graphs, standardized dye mixtures were used. The dye mixture representing a system that can be separated with "average" difficulty was made up of one part of *p*-dimethylaminoazobenzene (18) and one part of *technical* grade *N*-(*p*-dimethylaminophenyl)-1,4-naphthoquinoneimine (18). This same mixture served conveniently as the standard mixture for "very difficult" separations since the technical grade of *N*-(*p*-dimethylaminophenyl)-1,4-naphthoquinoneimine contains, as an impurity, another blue dye which can be separated from the major component only with considerable difficulty. Using dry benzene as the solvent and grade II–III alumina or silica gel, the dye mixture was readily separated into its three components by "dry-column" chromatography. This separation could not be accomplished on the usual

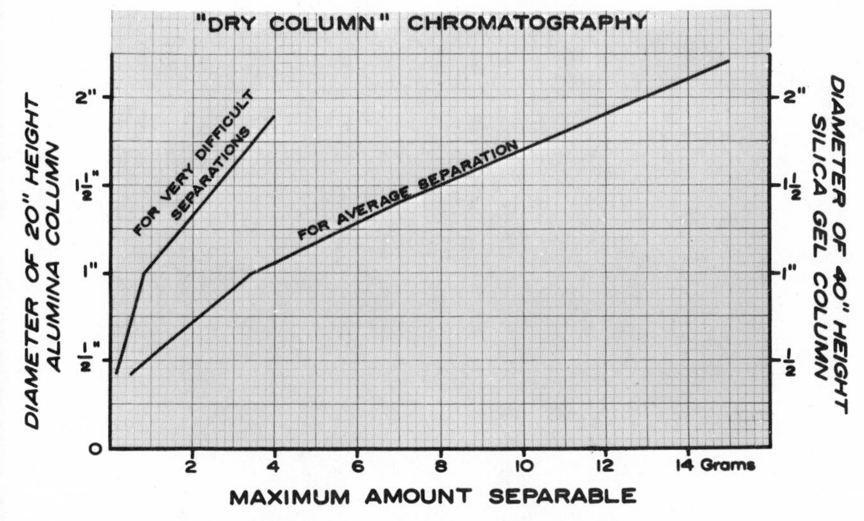

Graph I. Plot showing the column diameter (of a 20″ high column) necessary to separate a given weight of mixture into its components.

"liquid-filled "column, nor on a "dry-column" in which the adsorbents had not been properly deactivated. The colorless standard dye mixture which was used for the study under ultraviolet light was composed of *trans*-stilbene (two parts), benzophenone (one part), and *N*-methyl-6-trifluoromethyl-carbostyril (34) (one part), and was developed with carbon tetrachloride.

Columns of various practical dimensions were examined with respect to the maximum amount of material which could be separated on the column. The experimentally determined curves have been smoothed out, but the actual results proved to fall essentially on straight lines. The results have been plotted in two ways. Since it is often convenient to use columns about 20 in. high, Graph I has been plotted so that one can directly read the diameter of a 20 in. long column to be used for a given weight of a mixture. Thus, in order to separate 2 g of a mixture of "average" separability, a $\frac{3}{4}$-in. diameter alumina column 20 in. long can be used, whereas 2 g of a "difficult" mixture will require a $1\frac{1}{4}$-in. diameter alumina column, 20 in. long.

When plastic columns are used, it becomes very easy to vary column height; consequently, the results have also been plotted in such a way as to take advantage of this fact. From Graph II it can be seen that 2 g of an "average" mixture can be chromatographed on a 2-in. alumina column 8 in. long, or a $1\frac{1}{2}$-in. column 9 in. long, or a 1-in. column 13 in. long. Any one of these columns could be used, the choice being determined only by convenience. The same procedure would hold for a mixture of materials that

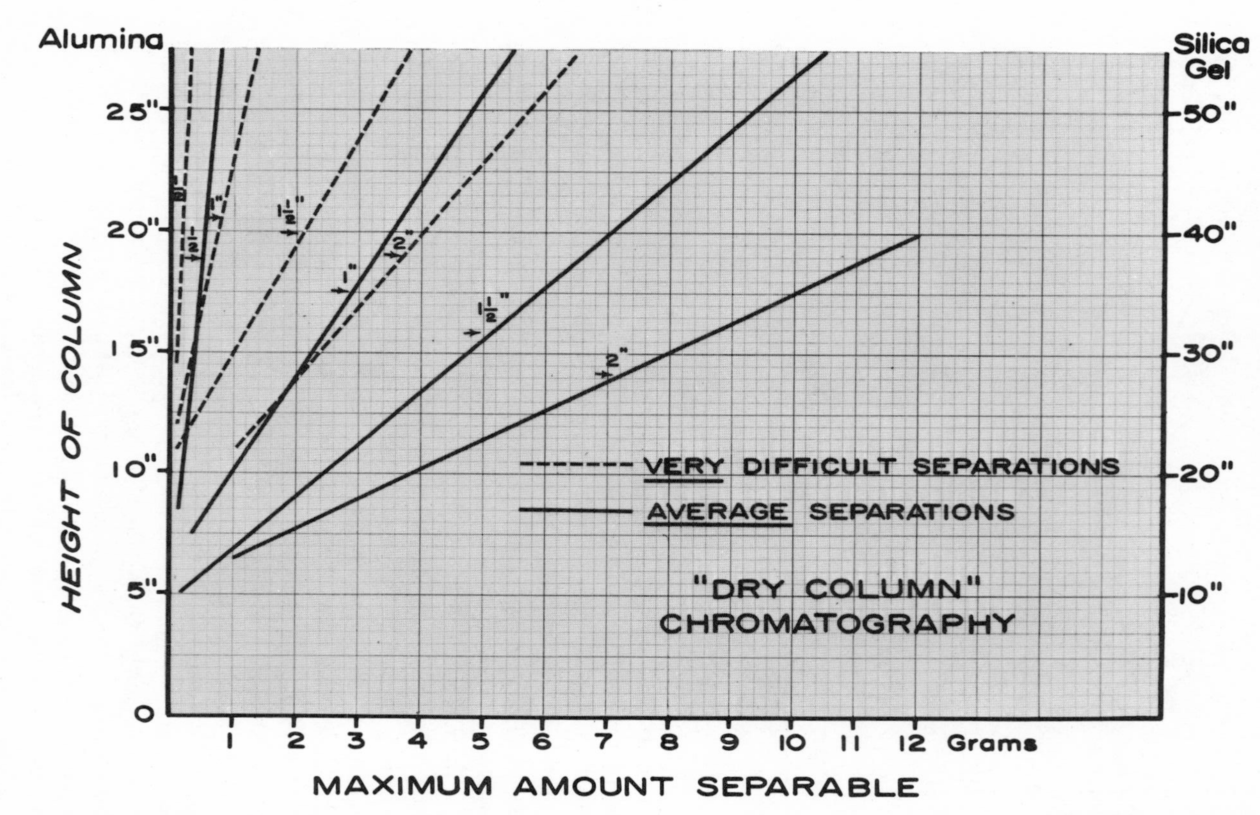

Graph II. Graph used for determining the diameter and height of column required to separate a given quantity of "average" or "difficulty separable mixture by "dry-column" chromatography.

is difficult to separate. If a mixture is difficult to classify, then it should be considered as a "difficult "separation. It should be noted that these graphs represent the *maximum* amounts of material that can be separated on a given size column. For optimum separation, it is preferable to operate below the level of overloading (usually at 50 to 70% of capacity), and preferably with columns at least 20 to 30 in. long. Consequently, in the above example, our choice would be an alumina column 1-in. in diameter and 20 in. long for an average separation, and a column $1\frac{1}{2} \times 25$ in., or 2×20 in. for a difficult separation.

For a given size column the weight of adsorbent used varies somewhat with the particular brand and the particular size. The following are the average amounts needed for a 20-in. high column:

Column Diameter (in.)	Weight (g)	
	Alumina	Silica Gel
$\frac{1}{2}$	40	17
1	240	100
$1\frac{1}{2}$	540	225
2	960	400

Based on the amount of mixture that can be separated and the amount of adsorbent required for the separation, it is found that an average system requires approximately 70 g of adsorbent per gram of mixture to be resolved. A mixture that is very difficult to separate requires approximately 300 g of adsorbent per gram of mixture.

It will be noted from the graph that the column size necessary when using silica gel adsorbent is approximately twice that of the alumina column necessary for separating the same amount of mixture. This difference is due to the density difference between the two adsorbents. The density of the silica gel used is only about half that of the alumina and so a larger column must be used to hold the same weight of adsorbent. However, the *ratio* of weights of compound to adsorbent remains unchanged.

IV. SUMMARY

The "dry-column" method provides an efficient and rapid resolution of mixtures, including those which cannot be separated on a "liquid-filled" column but which can be separated by thin-layer chromatography.

By virtue of the direct transferability of the conditions used with thin-layer plates, all preliminary work can be done quickly on thin-layer microscope slides in the expectation that on a preparative scale, using adsorbent of the same activity and the same solvent system, a similar degree of separation will be observed on the "dry-column" as is obtained on the thin-layer plate. In addition to increasing the degree of separation that can be obtained on a column, this technique has the further advantage that the number of fractions to be processed is determined by the number of cuts made of the column. In general, the maximum will be around 10 although, in practice, often only 4 or 5 are required. Hence, it is no longer necessary to encounter the endless number of fractions usually resulting from "liquid-flow" columns.

Finally, the use of nylon columns and fluorescent adsorbents has simplified detection and isolation procedures to such a degree that "dry-column" chromatography is often quicker, easier, and more reliable than separations done by fractional recrystallization.

References and Notes

1. B. Loev and K. M. Snader, *Chem. Ind.* (London), 15 (1965).
2. B. Loev and M. M. Goodman, *Chem. Ind.* (London), 2026 (1967).
3. (a) R. Teranishi, U.S. Dept. of Agriculture, Western Utilization Research and Development Division, personal communication, 1969; (b) A. J. Bauman and H. G. Boettger, *Microchemical Journal*, **14**, 432 (1969).
4. D. T. Day, *Proc. Amer. Phil. Soc.*, **36**, 112 (1897).
5. E. Albrecht and C. Engler, *Z. Angew. Chem.*, **13**, 1200 (1900).
6. M. Tswett, *Proc. of the Warsaw Soc. of Nat. Sciences*, Biol. Section, XIV, minutes No. 6 (1903); M. Tswett, *Ber. dtsch. bot. Ges.*, **24**, 316, 384 (1906).
7. R. Kuhn and E. Lederer, *Chem. Ber.*, **64**, 1349 (1931).
8. L. Zechmeister and L. Cholnoky, *Principles and Practice of Chromatography*, 2nd ed., Wiley, New York, 1941, p. 67.
9. A. Winterstein and G. Stein, *Z. physiol. chem.*, **220**, 247 (1933).
10. N. A. Izmailov and M. S. Shraiber, *Farmatsiya*, No. **3**, 1 (1938).
11. E. Stahl, *Pharmazie*, **11**, 633 (1956); *Chem. Ztg.*, **82**, 323 (1958); *Angew. Chem.*, **73**, 646 (1961).
12. H. Dahn and H. Fuchs, *Helv. chim. Acta*, **45**, 261 (1962).
13. R. J. Hall, *J. Chromatogr.*, **5**, 93 (1961).
14. E. H. Stainer and A. R. Bonar, *Int. Choc. Rev.* (1965).
15. V. K. Bhalla, V. R. Nayak, and S. Dev, *J. Chromatogr.*, **26**, 54 (1967).
16. H. Brockmann and H. Schodder, *Chem. Ber.*, **74B**, 73 (1941).
17. Properly deactivated, fluorescent grade silica gel and neutral alumina are now commercially available from Waters Associates, Inc., 61 Fountain St., Framingham, Mass., or M. Woelm Co., Eschwege, West Germany. They are sold as "Alumina (or Silica-gel) for 'Dry-Column' Chromatography."
18. *p*-Aminoazobenzene (*p*-phenylazoaniline), item No. 1375 and *N*-(*p*-dimethylaminophenyl)-1,4-naphthoquinoneimine (Tech.), item No. T-478, were purchased from Distillation Products Industries, Rochester, New York. *p*-Dimethylaminoazobenzene was purchased from K and K Laboratories, Plainview, New York, Item No. 4978.

19. Melting point tubes are convenient for this purpose if they are 1.0 mm wide. Such tubes are available as Item No. K-74425, Size A, from Kontes Glass Co., Vineland, New Jersey.
20. An inexpensive hand vibrator for use with glass or quartz columns is available as Item No. 53-C-20550 from Montgomery Ward, Baltimore, Maryland.
21. E. Ochiai and H. Takeuti, *J. Pharm. Soc. Japan*, **58,** 724 (1938).
22. A. Sabel and W. Kern, *Chem. Ztg.*, **81,** 524 (1957).
23. A source of nylon tubing is Walter Coles and Co., Ltd., Plastics Works, 47/49 Tanner St., London, S.E., England; available as COPOL 8 (YK), 160 gauge nylon tubing for "dry-column" chromatography in flat diameters of 1 in. (for about $\frac{3}{4}$ in. diameter column), $1\frac{1}{2}$ in. (1 in. diameter column), 2 in. ($1\frac{1}{2}$ in. diameter column), and 3 in. (2 in. diameter column) in rolls of 100 and 1000 feet, and sealed lengths of 40 in.
24. The nylon can be sealed with a hot iron or with a hand sealer such as item No. S-2020 (used with dial at maximum heat), available from Bel-Art Products, Pequannock, New Jersey. A satisfactory seal can also be made by folding the bottom end over several times and stapling closed.
25. A. C. Casey, *J. Lipid Research*, **10,** 456 (1969).
26. N. Bekesy, *Biochem. Zeit.*, 100 (1942).
27. H. Brockmann and F. Volpers, *Chem. Ber.*, **80,** 77 (1947); **82,** 95 (1949).
28. J. W. Sease, *J. Am. Chem. Soc.*, **69,** 2242 (1947); **70,** 3630 (1948).
29. Fluorescent indicators are available from E. I. duPont, Photo Products Dept., Towanda, Pa., as No. 609 Luminescent Chemical, or from Waters Associates, Framingham, Mass., as Woelm fluorescent green indicator.
30. A portable short wavelength lamp is available as Mineral-Light Model UVS-12 from Arthur H. Thomas Co., Philadelphia, Pa., or as item No. 9815-2 from the Cole-Parmer Co., Chicago, Illinois.
31. L. Zechmeister, L. von Cholnoky, and E. Yjhelyi, *Bull. Soc. Chim. Biol.*, **18,** 1885 (1936).
32. H. Brockmann, *Z. physiol. chem.*, **249,** 176 (1937).
33. B. J. Hunt and W. Rigby, *Chem. Ind.* (London), 1868 (1967).
34. B. Loev, British Patent 1,002,665 (1965).

Ultrafiltration Membranes

Carel J. van Oss

Department of Microbiology, Immunochemistry Laboratory, State University of New York at Buffalo, Buffalo, New York

I. INTRODUCTION

A. Definition

Ultrafiltration is the filtration of a solution under a pressure gradient through thin semipermeable membranes, resulting in at least a partial separation of solute molecules from the solvent or from the solvent and other solute molecules.

Although opinions may differ as to whether viruses should be considered as "molecules," which can be "dissolved," this is largely a matter of semantics; they will be so considered here and their separation with the aid of membranes will be treated as an aspect of ultrafiltration.

The retention of bacteria, cells, and other particulate matter by membranes and other porous media is more a process of fine filtration than of ultrafiltration and will not be treated here, although the influence of the presence of certain cells on some ultrafiltration processes will be discussed.

B. Historical

1. THE FIRST 75 YEARS

The first ultrafiltration experiments were published more than a century ago by Schmidt (1) who demonstrated that a bovine pericardium membrane could partially retain dissolved gum arabic. In 1906 Hertz (2) published a detailed study on protein retentions by various membranes of animal origin. Martin (3) made the first artificial ultrafilter membrane in 1896 by impregnating the porous wall of a Chamberland candle with 10% gelatin (and also with silica), which was completely impermeable to proteins. He applied his ultrafilter to the fractionation of skake-venoms (4). Borrel first proposed the use of collodion (cellulose nitrate) bags for the separation of tetanus and diphtheria toxins from their culture media (Manea (5); see also Duclaux (6), p. 11). This work incited Malfitano to use ultrafiltration as a tool in the study of various colloids (7).

Bechhold introduced the use of flat, reinforced collodion membranes of which the pore size was inversely proportional to the concentration of cellulose nitrate used for their manufacture (8). He was also the first to use the term "ultrafiltration" (9). Asheshov (10) demonstrated that the addition of amyl alcohol or acetone to cellulose nitrate solutions would result in collodion membranes of varied porosity.

The next important advance was made by Elford (11) who, by applying Asheshov's observations (10), established a standard technique for preparing a series of membranes of graded porosity (Gradocol membranes (11). These membranes enabled him to determine the approximate sizes of a large number of viruses and even of some proteins (12). Ferry (13) gives a complete review of all developments in ultrafiltration up to 1936.

2. LACK OF PROGRESS IN THE FOLLOWING 25 YEARS

At this point it becomes necessary to examine more closely why a method, which had been developing quite favorably for three-quarters of a century into a most useful research tool, suffered an almost complete arrest of growth for the next 25 years, during the very era that saw the most stupendous developments in analytical and preparative methods.

The explanation of this paradox seems fairly simple: the very abundance of newer, better, and more precise methods (e.g., analytical ultracentrifugation, light scattering, and electron microscopy) was instrumental in superseding ultrafiltration as an *analytical* method, while the sluggish fluxes of the hitherto known membranes, their doubtful reproducibility, and their proneness to clogging made *preparative* ultrafiltration also much less attractive than the newer techniques (such as chromatography, gel filtration, preparative ultracentrifugation, electrophoresis, and electrodialysis). Thus, for a long time, ultrafiltration as a method was used and studied in very few laboratories indeed.

3. THE MOST RECENT DEVELOPMENTS

It is as a preparative method that ultrafiltration is now beginning to arise from its long hibernation. Thanks to new developments in anisotropic membranes (15–17) and owing to slowly evolved but significant developments in anticlogging devices (18–22; see also 15, 16), ultrafiltration seems at last to emerge as an important and increasingly indispensable preparative tool. In addition, the more refined analytical methods (in particular analytical ultracentrifugation), which have superseded *analytical* ultrafiltration, are the very tools that now promote the development of ultrafiltration as a *preparative* method (23).

Excellent reviews of some of the more recent developments are given by Friedlander and Rickles (24) and by Rickles (25).

II. MEMBRANES*

A. Homogeneous Membranes

Although some of these older membranes may well have been provided with a denser top skin with smaller pores than the bulk of the membrane (see further below and under anisotropic membranes and measurement of porosities), their authors apparently never suspected this to be the case and thus in all probability used these membranes as often with the skinned side down as with the dense skin towards the side of the highest pressure.

* A number of commercially available membranes will be mentioned here and of those, some membranes that are impermeable to blood serum proteins and that have been tested in the author's laboratory will be specifically discussed. Other membranes, particularly some more porous ones, will be tested at a later date and generally cannot at this time be discussed in detail from a viewpoint of the author's own experience. They will, however, be mentioned under the various headings that are most relevant to their application.

In the first case the skin plays only a negligible role in these membranes' ultrafiltration characteristics and we shall therefore consider such crypto-anisotropic membranes among the homogeneous ones.

1. MEMBRANES MADE OF CELLULOSE DERIVATIVES

a. Cellulose Nitrate Membranes

These membranes, generally called collodion† membranes, were the first artificial ultrafilter membranes used with a reasonable yield and a fairly acceptable reproducibility (5–10). They were made in a variety of graded porosities (see above) by Elford (11) and the conditions of their preparation were further refined by Grabar (26). Grabar classified various solvents with respect to cellulose nitrate as follows.

Perfect solvents, which completely dissolve the polymer and which, if used alone to dissolve the polymer, will furnish completely or almost completely impermeable films upon evaporation.

Swelling solvents which, if added to the polymer dissolved in a "perfect solvent," will furnish a gel (see also Asheshof, 10) upon evaporation, which can be transformed into a semipermeable membrane when dipped into a precipitating solvent.

Precipitating solvents are solvents in which the polymer is completely insoluble; in this and other cases most often cold water.

As a working hypothesis this classification has not, to date, been superseded by any more apposite theory (see anisotropic membranes, below). Grabar (26) also showed the importance of atmospheric conditions (humidity and temperature) on the casting of collodion membranes and on the collodion itself, as well as the importance of optically flat and extremely level casting surfaces. For the preparation of graduated collodion membranes Grabar's monograph is without doubt still the best guide for attaining the maximum reproducibility that is possible with this material. For other work on graded collodion membranes, see Gregor and Sollner (27) and for collodion membranes of pore sizes smaller than 5 mμ, see Carr *et al.* (28). The latter membranes are made in an unusual way: first some quite dense membranes are cast after the fashion of Gregor and Sollner (27), then they are dried, and finally the membranes attain the desired pore size by being swollen for 3 hr in ethanol-water mixtures of various proportions, the highest ethanol content furnishing the most porous membranes.

† When consulting the older French literature on the subject of collodion membranes (see particularly Duclaux (6)), one must not lose sight of the fact that there the word "collodion" is often used quite loosely and at times denotes almost any kind of polymer solution and in particular solutions of cellulose derivatives.

Collodion membranes with pore diameters of 5 μ to 50 mμ can be conserved dry. Membranes with smaller pore sizes have to be conserved in water (generally admixed with 20% ethanol to avoid freezing and spoilage), in order to retain the initial pore size.

Ready-made collodion membranes of various pore sizes have been manufactured by the Membranfilter Gesellschaft, Goettingen, Germany, since 1927, according to the methods given by Zsigmondy (29), and they still are manufactured by this firm. In the United States they are marketed by the C. Schleicher & Schuell Co., Keene, New Hampshire. Their protein-stopping membrane (LSG-60) has a flux of 120 ml H_2O/100 cm^2/hr/50 psi, or about seven times the flux of thin cellophane. When retaining blood serum proteins in a solution of $\approx$1% protein, this flux decreases to $\approx$100 ml/100 cm^2/hr/50 psi, and remains remarkably constant at that rate for considerable lengths of time. This is a protein-stopping flux that compares favorably with that of many anisotropic membranes (see below) endowed with pure water fluxes of an order of magnitude higher than this LSG-60 membrane. An even faster protein-stopping membrane of the same origin is LSG-60T: its flux with water is $\approx$400 ml/100 cm^2/hr/50 psi, which gets reduced to 230 ml/100 cm^2/hr/50 psi when retaining serum proteins in a solution of 1%. Both sides of these membranes stop protein, which normally would indicate a homogeneous membrane, but in the case of these membranes, with their very high fluxes, particularly when stopping protein and with their remarkable resistance to clogging, we wonder if these membranes are perhaps endowed with a denser skin on *both* sides and could possibly be classified as *symmetrically anisotropic*.

b. Cellulose Diacetate Membranes

Fairly porous membranes of this type have been described in 1915 by Bertarelli (30). They were first made into excellent ultrafilter membranes with a wide choice of porosities by Duclaux and Amat (31) in 1938. They made use of the solubility of cellulose diacetate in almost saturated water solutions of $MgClO_4$, discovered by Mrs. Dobry-Duclaux in 1936 (32) (see Duclaux, ref. (6), p. 48, for details).

Very dense cellulose diacetate membranes can be obtained by boiling dense membranes (of the Schleicher and Schuell B 20 type) in water. This B 20 membrane is one of a family (B 17, 18, 19, 20) of definitely crypto-anisotropic membranes. One side of them stops protein, the other side does not, but there is no way of visually detecting the right side of any given membrane of this type. When the skinned side has been found by trial and error and is directed towards the high-pressure side, *all these "ultrafine" membranes* (from 17 to 20, or medium, dense, very dense, and super dense) *stop serum proteins in a 1% solution.* This is notwithstanding the fact that their pore diameters, given by the manufacturer, are, respectively 35–20, 20–10, and 10-5 mμ, while the smallest and most abundant of serum proteins (albumin) is a molecule, about 18 mμ long and 4 mμ thick. Obviously, the unsuspected anisotropy of these membranes has led to erroneous conclusions when the pore sizes were calculated according to classical methods that do not take into account the presence of a much denser skin (see below). The fluxes of these membranes are most respectable (see Table I) but the manufacturer could much enhance the usefulness of these membranes if he would mark which side was up during their manufacture.*

TABLE I
Fluxes of Schleicher and Schuell's Cellulose Diacetate Bac-T-Flex Membranes[a]

Membrane	H_2O Flux[b]	Flux When Stopping Proteins in 1% Solution	Flux When Upside Down and Not Stopping Proteins
B 17	2000	1700	25
B 18	400	300	24
B 19	230	200	120
B 20	120	100	20

[a] The high fluxes when stopping protein and the much decreased fluxes when upside down and *not stopping protein* are typical of anisotropic membranes.
[b] Fluxes are expressed in ml/100 cm²/hr/50 psi.

Reid *et al.* (33) obtained the first membranes which under high pressure successfully desalinated seawater by "reverse osmosis" from cellulose diacetate (see Section II-B-3). Cellulose acetate membranes with a pore diameter of 100 mμ, made by freeze-drying from cellulose acetate solutions in *p*-dioxane, were recently described by Rothbaum (34). Cellulose acerate membranes can be made denser by heating and dense membranes can be made even denser by drying (see Section II-C). Cellulose diacetate ultrafilter membranes can be obtained from Schleicher and Schuell (see preceding paragraph) and from Gelman Instruments, Ann Arbor, Michigan.

c. Cellulose Triacetate Membranes

These membranes are less soluble in various organic solvents (particularly acetone, chloroform, and ethylacetate) than diacetate mem-

* Loeb and Sourirajan, as shown in Section B-2, had noticed that B 20 membranes, after heating, only stopped salt with the "rough" side up. The "rough" side is indeed also the protein-stopping side, but unfortunately the distinction between "rough" and "smooth" can only be made *after the membrane has been dried,* a process which reduces its flux 80% or more.

branes. Their principal advantage is their greater resistance to heat, compared with cellulose diacetate and cellulose nitrate, making them readily autoclavable. For this reason cellulose triacetate is finding increasing use in the manufacture of more porous membranes, used for sterile filtrations. It is also used in intermediate porosities, for virus fractionation, for which particularly GM-9 and GM-10 (100 $m\mu$ and 50 $m\mu$ pore diameter), made by Gelman, are quite suitable [see Cliver (34) and also Black (34a)] but it is rarely made into membranes of the relatively small pore sizes appropriate for ultrafiltration, with the exception of an anisotropic membrane (see Section II-B-4-b), and of very dense cylindrical capillary membranes still under development at Dow Chemical Co., Walnut Creek, California (see also the end of Section B-3 for an analogous development).

d. Mixed Cellulose Esters

These membranes, which generally consist of cellulose diacetate and cellulose nitrate in various proportions, are mainly made into very porous membranes, which usually have pore diameters of 0.4 to 0.2 μ and are used for fine filtration or for sterile filtration. Only the Millipore Corporation (Bedford, Massachusetts) seems to use this material for membranes of the intermediate range of 100, 50, and 10 $m\mu$ pore diameter (VC, VM, and VF), mainly aimed at virus fractionation [see Cliver (34)].

2. CELLULOSE MEMBRANES

A large variety of cellulose membranes have been used from the earliest beginnings of dialysis and ultrafiltration.

a. Parchment-Paper Membranes

These are membranes made of paper which has been treated with H_2SO_4 and they have been used in dialysis for more than a century (35). A major drawback of these membranes used to be their denseness and thickness and, because of this, their slow flux and low porosity. Nevertheless, the thinner varieties of "parchment" paper that are commercially available can be quite useful as ultrafiltration membranes. Three varieties have been described (20b, p. 66–67), of 100, 85, and 78 μ thickness, weighing 60, 50, and 45 gs per square meter respectively and having an approximate pore diameter of 4, 8, and 10 $m\mu$ respectively. They attain fluxes from 25 to 200 ml H_2O/100 cm^2/hr/50 psi. These membranes, which are only obtainable in flat sheets can be formed into cylinders by sealing the ends together with the help of undiluted H_2SO_4, which afterwards must be quickly neutralized with an NH_4OH solution, followed by thorough washing in water. This type of paper, being semitransparent and dimensionally very stable, is often used by draftsmen for making the originals of architects' or engineers' drawings. Parhcment-paper membranes are relatively easily compressible and have their highest porosity at rather low pressures.

b. De-esterified Cellulose Membranes

These have an advantage over the original cellulose ester membranes they were made of because they are much more resistant to most organic solvents. After de-esterification they should never be dried, lest they lose part of their porosity and most of their flux.

Duclaux and Hamelin (36) demonstrated the possibility of transforming cellulose nitrate membranes into cellulose membranes, by treating them with an ammonium sulfide solution (see also ref. 6, p. 9). These denitrated membranes generally have a slightly *decreased* pore size, compared with the original membrane. The Cella and Ultra-Cella membranes of the *Membranfilter Gesellschaft* (Schleicher and Schuell, Keene, New Hampshire) are in all probability of this variety.

Cellulose acetate membranes can be deacetylated by treatment with an NaOH solution. These membranes have a *larger* pore size than the original membrane. This treatment is applied to the very dense cylindrical capillary membranes, under development at Dow Chemical Co. (Walnut Creek, California), to obtain protein-stopping deacetylated cellulose triacetate capillaries which are tested for possible use in hemodialysis and hemoultrafiltration (37).

c. Membranes Made from Regenerated Cellulose

These are not membranes that are chemically changed into cellulose, but membranes made *de novo* from solutions of regenerated cellulose. Cellophane, a gelatinous film regenerated from a solution of cellulose xanthate, is the best known representative of this class. McBain and Stuewer (38) indicated its usefulness as a general colloid-stopping membrane. The porosity of commercial cellophane does not vary much: cellophane stops all proteins from a molecular weight of approximately 10,000 on, if used as an ultrafilter membrane. As its thickness does not contribute to its pore size, it is always useful to use the thinnest available variety, in order to obtain the highest possible flux. The thinnest cellophane commercially available is, in the dry state, about 25 to 30 μ and when swollen with water, about 50 μ thick. Commercially available cellophane is dried after being impregnated with glycerol. If cellophane that never has been dried is obtained from the factory, its flux is a bit higher than after drying, but its porosity is not materially different. Cellophane tubing of a very homogeneous quality (No-Jax) is obtainable from the Visking Corporation (Chicago, Illinois). Its flux is about 16 ml $H_2O/100\ cm^2/hr/50$ psi (for

25–30 μ dry thickness). Special types of cellophane membranes can be obtained from the Union Carbide Corporation. Cellophane, on account of its constant homogeneity and its capacity to withstand high pressures, is notwithstanding its rather slow flux the only isotropic membrane now in use in medium-scale preparative ultrafiltration. Its main actual use is in totally removing proteins from biological extracts destined for intravenous or intramuscular injection (20, 39, 40). When well supported, cellophane can easily be used for considerable lengths of time at pressures from 300 to 500 psi.

On a smaller scale its use in deproteinization of protein concentration is not widespread, mainly owing to the lack of laboratory devices designed for the pressures necessary to overcome this type of membranes' slow flux.

Another type of cellulose membrane, Cuprophane, is made by the J. P. Bemberg Company (Wuppertal, Germany); its flux is slightly higher than that of cellophane. Cuprophane is made from cellulose dissolved in Schweizer's liquor, a copper-ammonium hydroxide.

d. Modified Cellophane Membranes

The almost universal but monotonous pore size (approximately 2 mμ diameter) of cellophane, coupled with its otherwise quite advantageous properties of general availability and reproducibility, have prompted numerous authors to attempt to vary it. Within rather narrow limits these attempts have generally been successful.

Treatment of cellophane membranes (25–30 μ dry thickness) with 64% $ZnCl_2$ at 20°C (the density of 64% $ZnCl_2$ at that temperature is 1.83) for 15 min, immediately followed by abundant washing in cold water, gives a membrane with a six times enhanced flux (100 ml H_2O/100 cm^2/hr/50 psi) and an approximately doubled pore size (from the original ≈ 2 to ≈ 4 mμ) (41,42,20b). After this treatment these membranes increase in wet thickness from 50 to ≈ 125 μ (20b). These membranes can be dried after having been impregnated with a 50% glycerol solution in water. While it is very difficult to convert cellophane for use with nonaqueous liquids, $ZnCl_2$-treated cellophane can quite easily be so converted (20b).

Craig has indicated how the pores in cellophane can be somewhat reduced, by acetylation, through treatment with acetic anhydride in pyridine for about 16 hr (43). Another physical method of modifying the pore size of cellophane has been published by Craig and Konigsberg (44). Stretching cellophane in *one* direction results in long, slitlike pores which in practice have the same effect as a reduced pore diameter. Stretching of cellophane in *two* perpendicular directions enlarges its pores. As Craig (43,44) modified the pore size of cellophane with a view to fractionation of solutes by *dialysis* and tested those pore sizes by *dialysis*, no very exact estimation can be given of what the final effective pore sizes would have been, if the membranes were used in ultrafiltration.

Cellophane carries a negative electric charge, most probably due to the presence of some carboxyl groups, amounting to ≈ 0.02 meq/g, according to Heymann and Rabinov (45). These carboxyl groups can be split off by boiling cellophane in 1% NaOH, which reduces the membrane's titratable acidity to ≈ 0.004 meq and approximately halves its flux. The negative charge of cellophane can be *increased* (to approximately double its initial value) by oxidation with bromine in the presence of 0.5% Na_2CO_3, according to Hirsch (46). This treatment tends to weaken the membrane's mechanical strength considerably. Hirsch (47) has also described the influence of pH on the charge of cellophane, as well as the influence of ionic strength (48). The influence of the negative charge of cellophane on the retention of salt ions in an ultrafilter has also been treated (49).

3. ANIMAL MEMBRANES

Although animal membranes were the first ever used for ultrafiltration (1,2), they never became widely popular, doubtless owing to a real or suspected lack of reproducibility. It would seem that in modern times only the present author has used animal membranes successfully for a number of separations on a small and also on a medium large scale (20b, see also 50).

The most important factor contributing to the successful use of animal membranes in the separation and purification of viruses and proteins was the good fortune the author had in hitting upon what is most likely the only animal membrane which not only had all the requisite properties [(*a*) being available in quite respectable lengths without punctures; (*b*) being endowed with a porosity that is variable at will as a function of (1) pretreatment of the membrane and (2) the pressure under which the ultrafiltration takes place; (*c*) being quite thin and thus potentially capable of high fluxes]; but in addition was of exactly the right size to fit the existing tubular ultrafilter of 1 in. diameter.

That animal membrane consists of cleaned small intestines of pigs, pickled in salt and imported from China, which are freely available in Europe and much in use by the sausage-stuffing industries. European pigs' intestines, although equally readily available, are much weaker and riddled with punctures. The much greater resistance of the intestines of Chinese origin is ascribed to the inferior and much rougher food the Chinese porcine species has to subsist on.

It was found by accident that a piece of intestine that had been slightly dried furnished a membrane of a most useful porosity and excellent reproducibility. Thus, after washing out the salt in which these membranes are packed, the membranes are opened by applying some water pressure to them, after which they are blown up with air and allowed to dry for $1\frac{1}{2}$ hr at room temperature inside a draft-free cupboard. The membrane is then ready for use. A remarkable property of such a membrane is its extreme pressure-sensitivity:

At pressure of <0.1 psi the membrane passes erythrocytes, indicating that its pores then are of the order of 2–5 μ in diameter (considering the ease with which erythrocytes can get through holes that are considerably smaller than themselves).

At a pressure of 4 psi the membrane easily passes foot-and-mouth disease virus, but practically completely stops vaccinia virus, indicating a pore size of the order of 100 mμ.

At a pressure of 28 psi the membrane stops all foot-and-mouth disease virus, but passes all blood serum proteins, indicating a pore size of the order of 25 mμ.

Thus, the higher the pressure the *smaller* the pores. This seems to be the only membrane for which we can preselect a pore size, within a wide range between 5000 and 25 mμ by simply turning on the pressure to the appropriate level. Nevertheless the membrane is quite thin; its actual thickness is rather difficult to measure, because of its compressibility, but at 0 psi pressure it seems to be of the order of 10 μ. An increase in pressure does *not*, as long as a solution containing some colloids is ultrafiltered, increase the flux. At all pressures used the flux was of the order of 30–50 ml/100 cm^2/hr. Only with pure water, when no colloidal molecules or particles can be retained by the membrane, were considerably higher fluxes observed at increased pressures.

The main drawback of the membrane is that it cannot be autoclaved so that other means of sterilization must be used.

4. OTHER MEMBRANES

a. Alginate Membranes

Sodium alginate is soluble, while alginate of plurivalent cations are insoluble in water. To take advantage of this phenomenon alginate membranes can be used to concentrate (e.g.) virus suspensions, after which the membranes can be dissolved by converting it into the sodium salt; see Schyma (51) and Moll (52). Calcium-alginate membranes, which are impermeable to all known viruses, are commecially available (B. Alginate) through Schleicher and Schuell. The membranes can be dissolved in 1 or 2 ml of a sodium citrate solution.

Other membranes can be derived from algae: *agar* membranes, which are practically not used, and *agarose* membranes, see below under anisotropic membranes.

b. Gelatin Membranes

All gelatin membranes hitherto described seem to have been formed inside a porous support, as Martin's membranes (3,4) in Chamberland candles, or in or upon collodion membranes, according to Hitchcock (53), or cast from an admixture with cellulose nitrate as described by Loiseleur *et al.* (54,55). Gelatin membranes are not now widely used for ultrafiltration.

c. Synthetic Membranes

The first synthetic membranes were doubtless those of Pfeffer (56) who, almost a century ago, deposited a continuous layer of copper ferrocyanide inside the porous walls of a ceramic pot. These membranes were very dense and were actually used to measure the osmotic pressure of small molecules, lactose for instance. They were formed by allowing a solution of copper sulfate, outside the pot, to diffuse into the porous ceramic wall toward a solution of potassium ferrocyanide, inside the pot (or vice versa). Upon meeting, a continuous precipitate of copper ferrocyanide was formed inside the porous wall. We now know, however, through the work of Hirsch-Ayalon (57), van Oss and Hirsch-Ayalon (58), and van Oss and Heck (59), that all such precipitate membranes only conserve their integral continuity as long as approximately equivalent amounts of the forming ions remain present on either side of the precipitate. Moreover, these are completely impermeable *only to the ions which formed them* (as long as the above-mentioned conditions prevail). Thus, although membranes of this type are of great theoretical interest (57,58,59a) and have contributed to the fundamental understanding of immunological (60) and immunochemical (59) phenomena, and seem to be quite promising in electrodes used for the quantitative measurement of the activities of specific ions (61), *they are not normally long-lived enough to be of much use in ultrafiltration,* nor is it practical in an ultrafilter to maintain permanently two different solutions on either side of the membrane. More promising synthetic membranes for ultrafiltration purposes are various porous "plastic" membranes.

Gregor and Kantner (62) described membranes consisting of a homogeneous polyvinylbutyral film which had been hydrolyzed in $4N$ H_2SO_4 at 60°C for various lengths of time, the porosity increasing with increased time of hydrolysis. Although mainly tested as dialyzing membranes, nothing seems to oppose their use in an ultrafilter. Their pores are small; they seem to vary only between 10 and 1 $m\mu$ diameter. Judging by the published diffusion rates of various low molecular weight substances, their fluxes are fairly slow, probably of the order of 10 ml/100 cm^2/hr/50 psi. Very dense porous polyethylene films have been used as ultrafilter membranes by Sourirajan (63) for the fractionation of hydrocarbon liquids.

The Millipore Corporation (Bedford, Massachusetts) manufactures "hydrophobic" membranes (Aquapel) with pore sizes between 10 and 100 mμ, which are apparently ordinary membranes made of mixed cellulose esters, the material of which has been coated, or otherwise treated, to give it hydrophobic characteristics.

All other porous plastic membranes are either so porous (pore sizes of 200 mμ or over) that they fall outside the scope of this chapter, or they are anisotropic and are discussed below.

B. Anisotropic Membranes

1. *"SKINNED" MEMBRANES AND THEIR PRECURSORS*

Two circumstances which prevail in the preparation of many classical membranes tend to contribute to the formation of a top skin that is considerably denser than the remaining bulk of the membrane.

(*a*) Evaporation of solvent is strongest at the surface of the viscous solution (or gel), which increases the concentration of polymer at the polymer-air interface. As soon as it is formed, this denser top layer hinders further evaporation of solvent through it, thus leaving the remaining underlying solution (or gel) largely unchanged.

(*b*) Leaching out of solvent, as soon as the viscous solution (or gel) is lowered into the precipitating liquid, also is strongest at the surface. This further densifies the top layer of the incipient membrane, which henceforth impedes the penetration of more precipitating liquid and slows down the coagulation of the bulk of the polymer, thus favoring the formation of a very porous under layer (see in particular pp. 220 and 221 of the first article in Ref. 11).

Anisotropic or "skinned"* membranes, which are the ideal devices for combining very small pore diameters with very high fluxes (64), were made and used for more than 50 years before that property was usefully exploited.

Until the work of Loeb and Sourirajan (14,15,65) the very existence of this type of anisotropy was unsuspected and "skinned" flat membranes were thus usually used upside down. This must have contributed very seriously to irreproducible "pore sizes," erratic fluxes, and inexplicable clogging of some membranes but not of others in the same batch. It also probably was the principal hidden factor which caused most of the discrepancies between the calculated and the actual pore sizes, which pre-

* The reader must be cautioned that "skinned" membranes has come to mean "membranes *provided with* a skin," a connotation which is the opposite of the more current anthropomorphic usage of this participium.

occupied a number of authors (11,26,66,67). It finally also explains the mystery which worried some authors (68) of consistently obtaining membranes with widely varying pore diameters, but of an almost constant and quite high total pore volume.

From the description (6) of their manufacture and use it would seem that the sac-shaped collodion membranes, which were formerly much in use, were always used with the "skinned" side toward the highest pressure, which may perhaps account for their one-time popularity (which is otherwise hard to understand, because they are quite tricky to make and they have to be used without any support). This type of collodion bag (CB) is still manufactured by the *Membranfilter Gesellschaft* and marketed by Schleicher and Schuell (Keene, New Hampshire) in only one pore size ($\approx$8 mμ), aimed at the retention of proteins.

Apart from our own experience with the Schleicher and Schuell cellulose diacetate membranes B 17-20 (see Table I) and the above-mentioned discrepancies (see also 11,26,66,67), there are some further indications that at least some of the older membranes must have been "skinned"; see for instance some of Grabar's photomicrographs of sections of collodion membranes (Ref. 26, plate II, Fig. 3-e, f and g). Duclaux (Ref. 7, p. 59), while discussing a collodion membrane which was deposited on a funnel of wet filter paper [a technique of Ostwald's (69)] and which coupled an extraordinarily high flux to a relatively small pore size, professes to be only capable of understanding this "if the layer of nitrocellulose is extremely thin," which, if true, "does not seem very safe."

If the "skinned" type of anisotropy was never suspected to be a property of the membranes that were in common use in the 1930s and 1940s, other asymmetrical properties of these membranes were speculated upon. Elford in particular (11) liked to believe that his Gradocol collodion membranes contained a preponderance of pores perpendicular to the membrane sufrace. His argument for this belief was that "when a dye suspension was filtered through a membrane supported upon a perforated plate, the circular areas, through the pores of which the dye could pass, became deeply stained through the entire membrane thickness and remained sharply defined after long periods of ultrafiltration." Lateral diffusion "therefore occurs very slowly." Unfortunately, this phenomenon, though doubtless true, only proves that pressure ultrafiltration of certain dyes is faster than the unaided diffusion of these dyes (which is not very astonishing), and does not at all prove that there are relatively more perpendicular than lateral pores.

The only well-founded indication of *some* kind of anisotropy of these membranes was given by Grabar and de Loureiro (70) who discovered a "faint but definite" birefringence in the denser collodion membranes when sectioned perpendicular to their surface, but none in sections parallel to the membrane. Most unfortunately, however, no indication was given in which direction the birefringence of these perpendicular sections manifested itself, so that it is very hard to determine now whether this phenomenon supported Elford's perpendicular pores, or whether it was a first hint at the skinned-ness of these membranes. The very fact that the direction of orientation of this birefringence was not mentioned may well mean that the authors could not readily explain it in connection with the speculations (11,26,66) which were current at that period. But only a renewed experimental verification of the phenomenon can settle this matter once and for all.

2. THE DISCOVERY OF ANISOTROPY IN MEMBRANES

In a paper presented before the Division of Water and Waste Chemistry at the American Chemical Society Meeting in Washington, D.C., on March 27, 1962, Loeb and Sourirajan communicated their observation that the very dense Schleicher and Schuell cellulose diacetate membranes (probably B 20) had a "rough" and a "smooth" side and that the membrane (after heat treatment) retained salt, but only with the "rough" side up. They postulated that the membranes consisted of a fine layer of a dense pore structure at the surface of the rough side and that the remainder of the cross section was a region of spongy, relatively large pores (see also 71).

They subsequently made use of this discovery in developing a salt-retaining cellulose diacetate membrane of excellent selectivity and astonishing flux (65). Their hypothesis has since been amply confirmed (72,73,74) and the anisotropic membrane they developed is still essentially the optimal membrane for salt retention by ultrafiltration and, notwithstanding numerous efforts to improve on it by other authors (73;75–78), it has not, to date, even been equalled. Thus they not only made sea-water desalination by ultrafiltration (or "reverse osmosis," see below) economically feasible (65), but they also, by their discovery of the "skinned" membrane, have vastly widened the scope of the application possibilities of ultrafiltration in general.

The advantages of providing fairly porous membranes with a very thin top skin with much smaller pores (which skin is the actual solute-stopping membrane) are considerable. They become immediately obvious when we consider on which parameters the flux J through a given pore of a membrane depends:

$$J = \frac{\pi r^4 \Delta p}{8 \eta l}, \tag{1}$$

postulating Poiseuille flow of a liquid of a viscosity η, under a pressure gradient Δp, through a cylindrical pore with a radius r and a length l perpendicular to the membrane surface, which, for a fairly small pore in a thin membrane, is not too unrealistic an approximation. Thus, for a given pore diameter, which has to remain constant when a given separation is aimed at, and under a given pressure gradient, *the only thing we can do to increase the flux is to decrease l to its absolute minimum still compatible with sufficient membrane strength.* With homogeneous membranes l represents the whole membrane's thickness. With the densest membranes the minimum feasible thickness is of the order of 25 μ, more porous membranes have to be even thicker. Now, the one part of the membrane that stops solute molecules of a given size is, of course, the top surface which is in direct contact with the solution. Thus, if a membrane contains a top layer with pores of the desired diameter 2 r and an infinitesimally thin thickness l', while the remainder of it has pores much larger than 2 r, that remainder can afford to have a total thickness of many times l', to confer it greater mechanical strength. Larger pores in the bulk of the membrane imply a smaller total number of pores, but if those pores in the bulk of the membrane are a times wider than 2 r, there may be up to a^2 times fewer pores, while the flux J per pore will be a^4 times accelerated, so that an increase in pore size of a times in the main mass of the membrane still increases the flux by at least $a^4/a^2 = a^2$ times.

For other conceptualizations of ultrafiltration than those based on pore sizes (79), for example, "diffusional flow" (64,73), or for hypotheses where pore size is not the only consideration, as in "negative adsorption," (80), solute repulsion by low dielectric constants (81), or by electrostatic forces (49;82–85), (see below), the same arguments for explaining the advantages of a very thin active layer supported by a much thicker quite porous layer still hold true (although perhaps in some cases in a slightly modified fashion).

Two other advantages are inherent in the use of skinned membranes:

(*a*) Clogging of the membrane by solutes of a molecular dimension close to the (homogeneous) membrane's pore size, by lodging inside the pore structure of the membrane is impossible with skinned membranes. The only type of clogging still possible with skinned membranes is the sitting of solute molecules on top of the pores, which is generally much easier prevented or cured by turbulence (64).

(*b*) Skinned membranes are self-repairing. This is best illustrated with an example: When a skinned protein-stopping membrane is inadvertently used upside down, with the skin away from the solution that is being ultrafiltered, the membrane is observed to pass protein (and also to get clogged up quite quickly). When the membrane is subsequently turned over, and now used with the skinned side towards the solution, *it stops passing protein* (and no longer gets clogged up).

3. *REVERSE OSMOSIS*

"Reverse osmosis," a term that was invented to denote desalination of salt water by ultrafiltration, appears upon closer analysis to be quite redundant as a description of "ultrafiltration of a solvent through a membrane which is impermeable to the solute, under a pressure which is higher than the osmotic pressure of that solute at its prevailing concentration." Obviously, all ultrafiltration is "reverse osmosis" and the term is neither more descriptive nor more restrictive than ultrafiltration. Still, it has crept into the literature and we might as well continue to use if for all ultrafiltration separations of molecules and ions of smaller than colloidal dimensions, say <2 mμ, because those small solutes do have, at the concentrations at which they usually occur, a considerable osmotic pressure that has to be overcome.

The use of ultrafiltration for desalinating sea water was first proposed by Reid (86) in 1953 and first proven possible by Reid and Breton in 1959 (33). Dense cellulose diacetate immediately proved the most promising membrane material (33), although other materials were tried (75–78;85,87,88). Still, the fluxes obtained were extremely low (<0.3 ml/100 cm^2/hr/50 psi) and had to wait for Loeb and Sourirajan's work on "skinned" membranes (see Section B-2), to attain values of the order of 30 ml/100 cm^2/hr/50 psi for the most modern anisotropic cellulose diacetate membranes, in actual use in the city of Coalinga, California (22,89).

As already mentioned in Section B-2, unity of opinion does not reign about the exact mechanism of ultrafiltration; a discord which becomes the more accentuated the smaller the molecules and ions are that cannot pass the membrane. The predicament is at its sharpest in the interpretation of the mechanism of salt retention by anisotropic cellulose diacetate membranes in "reverse osmosis" (64,73,74;79–81).

Let us first examine the principal facts pertaining to the phenomenon which appears so perplexing:

(*a*) Monovalent salt ions, which are notoriously small, are retained by the membrane, while water molecules which are hardly smaller freely pass it (divalent cations *and* anions are retained even more strongly than monovalent ions (80)).

(*b*) The higher the pressure, the greater the salt retention, (74,90)(the flux is proportional to the pressure).

(*c*) Up to fairly high limits the salt concentration does not materially influence its degree of retention (80) (although at a higher salt concentration the osmotic pressure is, of course, increased and the effective ultrafiltration pressure and thus the flux, decreased).

Facts *a* and *c* immediately show that electrostatic repulsion, which is the major driving force in the salt retention by cellophane (49,91; also 82–85), is of no particular effect here, for the main manifestations of electrostatic repulsion are: preferential retention of the ions of one sign of charge only and maximal salt retention at the lowest initial salt concentration (49).

The major consensus at the moment seems to tend toward the explanation of the phenomenon as "pressure-induced diffusion" of water molecules (fairly quickly) and salt ions (some how very slowly) through the top skin of the membrane, which ideal behavior is (for a very small part) adulterated by a certain degree of leakage of both water and salt through actual pores or holes (64,73,74,90). Although perfectly irreproachable, this explanation does not materially augment our understanding of the mechanism of ion-retention by these membranes. Still, it is easy to understand why the most obvious explanation, that of the membrane's top skin having pores which are just big enough for water molecules, but too small for salt ions to pass (*a*), is at first sight not very satisfying. One reason is that the sizes of water molecules and salt ions do not really differ enough.

Now we must take into account that salt ions dissolved in water are all to a greater or lesser extent hydrated, that is, they each carry a number of more or less tightly bound water molecules with them, which obviously makes them quite a bit larger than just one single water molecule. Although the exact determination of the degree of hydration of given ions is still very difficult and elusive, some quite plausible estimations have been published (92–95), which leave no doubt at least about the *order* of increasing hydrated size of various mono- and plurivalent anions and cations. When we now study which ions are the ones most strongly retained by reverse osmosis membranes (80,90,96), we see that they are the ones with the biggest hydrated diameter, in the order of strongest retention: Ba^{2+}, Sr^{2+}, Ca^{2+}, Mg^{2+}, K^+, Na^+, Li^+, and SO_4^{2-}, Cl^-, Br^-, I^- (97). Thus the discrimination between water molecules and hydrated ions by actual, though perhaps very small (79), pores in the membrane's top skin appears perfectly possible.

Still, this does not yet explain the increased ion retentions at higher pressures (*b*). But higher pressures give rise to higher fluxes, which, once salts are retained at all, tend to enhance their retention (or, thermodynamically speaking, higher pressures enhance the chemical potential of water more than that of the dissolved salts, under most conditions). We also must not lose sight of the fact that increased pressures do give rise to a significant compression of the membrane, which in its turn must result in a decreased pore size (74, see also Section A-3). All the same, some additional effects are most probably at work: one of them may well be the type of negative adsorption which exists at water-air interfaces and which probably also exists at the cellulose-acetate-water interface, causing the "skimming" effect mentioned by Sourirajan (80). The decreased water-content of cellulose acetate at increased pressures, demonstrated by Banks *et al.* (74), is probably relevant here, while the decreased dielectric constant in very close proximity of interfaces (81,98) also most likely plays a role, if indeed it is not the very cause of the surface skimming effect (81).

Anyway, all authors still agree that the top skin of the salt-stopping membrane, whether it contains actual pores or not, is quite dense, while the remainder of the membrane is rather open-pored. Loeb and Sourirajan obtained this effect by adding a solution of the "membrane salt" $MgClO_4$ in water (which had been first used by Mrs. Dobry-Duclaux (32)) to acetone, as a casting solvent for cellulose acetate (65). The actual role played by this and other "membrane salts" has been extensively speculated upon (73,74,87), but it is now becoming clear that solutions of these salts simply serve the purpose of swelling solvents (26) (see Section A-1). In fact, non-salt containing solvents, like formamide (22,106), have advantageously replaced these salt solutions and furnished desalinating membranes with fluxes that are even higher than those of the earlier Loeb membranes (99). All these membranes are subjected to a heat treatment to make them dense enough to reject salt (Section A-1-b).

As salt-stopping membranes with higher and higher fluxes are developed, the accumulation of increased salt concentrations in the close proximity of the membrane ("concentration polarization") and thus the increase of osmotic pressure and the accompanying decrease of total effective pressure and loss of flux, become a serious problem. On a small scale this difficulty is alleviated by vigorous stirring of the salt solution. On a large scale, in continuous-flow installations, only two possible approaches exist: alleviation by laminar flow and alleviation by turbulent flow (100). In the light of the fluxes that are already attainable, it seems likely that turbulent flows of Reynolds numbers of 10^4 and over (89,99) will provide the best solution. It also leaves the greatest scope for membrane improvement: faster membranes unavoidably entail a further increase in polarization. The decision between turbulent and laminar flow is important because it will at the same time dictate which type of device is to be used. Rejection of the laminar flow solution eliminates devices containing membranes in the form of capillary tubes when filtering from the inside (101), although perhaps not when filtering from the outside, as well as other devices featuring large membrane surfaces with their surfaces very close together (which allows them to be housed in small volumes). Most of the latter are of the "Swiss roll" variety (102–105), in which flows of Reynolds numbers $>10^4$ would cause prohibitive pressure losses. This leaves practically only the tubular membrane variety of devices and even there the possible diameters of useful tubes seem to lie between rather close limits (100): probably between 0.1 and 1 in.; the optimal diameter being close to the larger diameter, for practical and mechanical reasons (22,89, see also 20). Three tubular reverse osmosis systems are now commercially available in laboratory, pilot-plant (or even larger) sizes (see also 25) from Universal Water Corporation, Del Mar, California; Havens Industries, San Diego, California; American Standard, New Brunswick, New Jersey. Two "Swiss roll" type systems are available from Desalination Systems, Inc., Escondido, California and General Atomic Division of General Dynamics Corporation, San Diego, California, which has done much work on the development of reverse osmosis (73,104). One "flat plates" system is available from Aerojet-General Corporation, Azusa, California, which also has done much development work on reverse osmosis (87).

Finally the Du Pont de Nemours Company (Wilmington, Delaware) recently disclosed (106) that they have developed a reverse osmosis system comprising millions of capillary fibers, which are used as ultrafilter membranes with the flux from the exterior to the interior; these fibers are made of nylon.

In the same range of porosity as that of salt-stopping membranes, reverse osmosis has also been used for concentration of sugar solutions (107,108) to obtain syrups, with practically no losses; removal of sulphuric acid and other mineral contaminants from mine water (109); removal of glycerol from water (131); removal of urea from water (132), and it will certainly be used for many more separation, purification, and concentration (110) purposes in the near-future (see also 25). Some other reverse osmosis membranes, Amicon's Diaflo UM 1 and UM 2, stopping molecules of molecular weights of respectively 600 and 6000, are treated in Section B-4-a-2.

Recently a new variety of salt-stopping membranes has been developed by Marcinkowsky (135), Kraus (136), Johnson (137), Shor (138), and their collaborators at Oak Ridge. These are "dynamically formed membranes," made by the deposition of minute colloidal particles on a porous support that itself has a fairly large pore size (from 0.1 to several micra). The colloidal particles used are hydrous zirconium oxide (135), organic polyelectrolytes, and even natural products such as humic acid. The advantage seems to be their high flux (up to 2 l/hr/100 cm^2). The colloidal particles have to be continuously added on the upstream side (in a few p.p.m. only) and turbulence is of the greatest importance to prevent clogging of these membranes (138,139). Their salt rejection is lower than that of the cellulose acetate membranes (137). For a complete and recent review of separation and purification by reverse osmosis, see Lonsdale's chapter in this volume (141).

4. PROTEIN-STOPPING MEMBRANES

With all the membranes described under this heading, vigorous stirring of the protein solution is necessary if the membrane is to continue to stop proteins for any length of time.

For work on a laboratory scale only one type of commercially available ultrafilters is satisfactory; they are the ones made by the Amicon Corporation, Cambridge, Massachusetts. They contain suspended magnetic bars that can rotate in a plane parallel to and 1 or 2 mm above the membrane, when activated by one of the manifold available magnetic stirring devices.* The Amicon 52 cell has a capacity of 65 ml and can accommodate membranes $1\frac{3}{4}$ in. in diameter; the Amicon 402 cell has a capacity of $\approx$400 ml and uses membranes with a diameter of 3 in. Both cells are best used with a cylinder of compressed nitrogen; both the Amicon 52 and the Amicon 402 cells are designed to work under pressures up to 75 psi [see (16) for a description and a diagram of a precursor of these cells]. Both cells have transparent housings, which permit continuous visual inspection of the ultrafiltration in progress, and both are provided with safety valves. All parts in contact with the liquid are either made of plexiglass or other plastic material, or of teflon-coated stainless steel; the magnetic bar is also teflon coated. The cells are marketed for approximately $125 and $375 for the Amicon 52 and 402, respectively.

On a pilot-plant (or even larger) scale all the installations mentioned at the end of Section B-3 can probably, without much change, be used with protein-stopping membranes. The three first mentioned tubular devices are probably preferable where very high fluxes prevail. A fourth tubular device, specially designed for protein concentration work (20,39), is manufactured by the Plaatwellerij Velzen, Velzen, Holland. For a similar but more recent, although independently developed, tubular design which, however, contains no provision for either laminar or turbulent flow, see Ref. 111. Dorr-Oliver (Stamford, Connecticut) has also developed a device with a view to protein and other colloid-separation work, which uses disposable cartridges of flat membranes (112).

a. Membranes That Stop All Proteins

There exists a definite need for a nonprotein denaturing method for concentrating several 100 ml of protein solutions several 100-fold in only a few

* In the days before magnetic stirring devices had become generally available, various rocking (19) and shaking (20b) arrangements were used to agitate the entire ultrafilter with the aim of keepnig the liquid inside in motion with respect to the membrane. For some side effects magnetic stirring can have on salt retentions in an ultrafilter, see ref. 21.

hours. Until a few years ago practically only the homogeneous dense cellophane membranes were available to stop all proteins (roughly *all* proteins and polypeptides of $M > 10{,}000$). But at a flux of 15 ml/100 cm²/hr/50 psi, it takes more than 24 hr to reduce 300 ml of a 1% protein solution to a few ml in a normal laboratory ultrafilter. To make protein concentration by ultrafiltration really useful, fluxes of at least 100 ml/100 cm²/hr/50 psi will have to be attained with a 1% protein solution.

With a number of anisotropic membranes such fluxes can indeed be attained and in many cases surpassed.

(*1*) The first membrane that comes to the mind as likely to have good protein-stopping properties combined with a high flux, is, of course, a somewhat more porous Loeb membrane (14,15). We have studied the protein-stopping properties of the most recent variant of that membrane, also developed by Loeb (22) and his collaborators (99), which is made of a mixture of 25 g of cellulose diacetate (Eastman, low viscosity E-398-3), 27 ml of formamide, and 57 ml of acetone but which, after casting and immersion, has *not* undergone the heat treatment necessary to make it stop salt. The flux of such membranes is approximately 100 ml/100 cm²/hr/50 psi with H_2O, but considerably less when stopping protein from a 1% solution and the behavior of these membranes, without heat treatment, is not very constant.

Upon further investigation and multiple trials with other membranes made of the same ingredients but in different proportions, we found (113) that membranes cast of 25 g of (dry) cellulose diacetate (also E-398-3), 50 ml formamide, and 75 ml acetone show very favorable fluxes (790 ml/100 cm²/hr/50 psi) with water and stop all proteins from a 1% protein solution (at a flux of about 390 ml/100 cm²/hr/50 psi). These 25–50–75 membranes are quite easy to make in the laboratory and cost about 5¢ each (of 3 in. × 3 in.) in materials and approximately another 30¢ in labor. When the cellulose acetate is completely dissolved *at room temperature* (by preference with the help of a pestle and mortar), it is subjected to a vacuum for 10 min to remove the air bubbles. The viscous liquid is then cast on a glass plate and spread out with the help of a glass rod or tube reposing on two metal ridges of 0.15 mm thick (133), still at room temperature. The entire glass plate with the solution is then dipped into ice-cold water and left there for about 1 hr. After this the membranes are ready to be cut to the appropriate sizes and can be used. Care must be taken always to use the membrane with the side that touched the glass plate *away* from the solution to be ultrafiltered, in order to be sure that their "skin" is in contact with that solution. Storage of the powdered cellulose diacetate in a vacuum desiccator before use greatly enhances the reproducibility of these membranes. The ultrafiltrate of a 1% blood serum protein solution obtained with such a membrane contains no trace of protein precipitate with the extremely sensitive tannic acid method (114), which otherwise never fails to precipitate even the low molecular weight proteins that are normally always present in serum (115). The optional pressure to be used with these membranes is 30 psi (133). These membranes must either be stored in water or in 20% ethanol, or they can be dried after having been soaked in a 50% glycerol-50% water mixture, without losing their properties, although their flux then decreases as much as 50%. Recent work (140) indicates that membranes like these change even less after having been dried if a surface-active agent is admixed with the plasticizer before drying.

(*2*) Another class of anisotropic protein-stopping membranes is that of the poly-ion type, consisting of complexes of anionic and cationic water-soluble polyelectrolytes developed by Michaels (17) and manufactured and marketed by the Amicon Corporation (Cambridge, Massachusetts). These Diaflo membranes are made in three porosities, all of them stopping proteins, in the order of decreasing pore size: UM 1, UM 2, UM 3—completely stopping molecules of approximate molecular weights of 10,000, 6,000, and 600, respectively (117). Their fluxes with water (containing 0.85% NaCl) are about 800, 500, and 300 ml/100 cm^2/hr/50 psi, respectively, and when stopping protein from a 1% blood serum protein solution: 140, 125, and 85 ml/100 cm^2/hr/50 psi. A more porous Amicon membrane, XM 50, has a flux of 1,300 ml/100 cm^2/hr/50 psi with water and a flux of approximately 400 ml/100 cm^2/hr/50 psi when stopping serum proteins from a 1% solution, but it gets clogged much quicker than the 25–50–75 membrane (133). For other work on the serum protein-stopping properties of an earlier Amicon membrane, which corresponds closest to the actual UM 3, see Ref. 16. All three Diaflo membranes are manufactured with a filter paper basis. The filter paper side must, of course, always be used away from the high pressure side. These membranes must also be stored in water or in 20% ethanol, although they probably also can be dried after treatment with a 50% glycerol–50% water mixture. They are marketed for $6.00 per 3-in. diameter membrane, when ordered in amounts of at least 12 membranes at a time.

Both our 25–50–75 and Amicon's UM 1 and UM 2 protein-stopping membranes stop practically no salt [less than 1%, while cellophane stops 10–20% of the salt (21,40) when ultrafiltering a 10 mM NaCl solution], so that the tedious work of measuring ion (or small molecules) binding of proteins by ultrafiltration (119) can be eliminated. For other work on the binding of ions and small molecules by proteins, measured by ultrafiltration, see Refs. 120, 121, and also 111, 122, and 123; the authors of the last references used no stirring or other means of agitation, so that their results must be taken as rather approximative.

b. Hemoultrafiltration Membranes and Other Membranes That Stop Blood Serum Proteins

These membranes need only to stop proteins of the molecular weight of serum albumin (M $\approx$ 69,000) and over and thus can, at least theoretically, have a slightly larger pore size (and thus higher fluxes) than the membranes mentioned in the preceding section. In practice, however, when actually stopping proteins, membranes are more likely to become clogged by these very proteins, as their pore sizes approach the molecular size of those proteins (20a,64).

Thus, although a cellulose diacetate membrane can easily be made to stop proteins from M $\approx$ 69,000 on (for instance, cast from 25 g cellulose diacetate, 75 ml formamide, and 75 ml acetone), their fluxes, although 2 L/100 cm^2/hr/50 psi, drop to 350 ml/100 cm^2/hr/50 psi, when stopping proteins from a 1% serum rpotein solution (and passing traces of the smaller proteins). In other words, for stopping blood serum proteins, our cellulose diacetate 25–50–75 is at least as satisfactory as this 25–75–75 membrane. In the Amicon series the UM 1 membrane is the optimal one for stopping blood serum proteins, although the XM 50 membrane also seems promising.

One aspect of the retention of blood serum proteins by membranes under pressure is, of course, its possible use in artificial kidneys. Although all types of artificial kidneys now in use are mainly based on *dialysis* of blood, most of these devices can, and sometimes do, exude excess water by *ultrafiltration* when put under a slight surpressure (116). It might be useful to try and reverse this order of importance of the two mechanisms and devise an artificial kidney that is an *ultrafilter* which, after all, is mainly what the glomeruli are too. The first and most important step towards this aim is to investigate if there are conditions under which membranes can stop all blood serum proteins, but pass a serum ultrafiltrate, when *ultrafiltering whole blood, at reasonable fluxes and without undue hemolysis.*

Although some quite unexpected difficulties have arisen, the results of preliminary tests in our laboratory with ACD anticoagulated whole blood seem to show that hemoultrafiltration might be possible. The main difficulty is a very severe clogging of most membranes, due to the presence of the erythrocytes, which make up some 40% of the volume of whole blood. With many membranes the most vigorous stirring is of little avail for dissuading the red cells from sitting on every pore and clogging them up, even when protein retention is no problem. For instance, when membranes are used with pore diameters just small enough to stop all erythrocytes (1 or 0.8 μ in diameter) no flux can be obtained at all. With slightly smaller pores (0.5 μ diameter) fluxes of only 20 ml/100 cm^2/hr/50 psi are obtained, with a filtrate full of proteins. Much denser protein-stopping membranes like Amicon's UM 1 and UM 3 give fluxes of the order of respectively 130 ml and 60 ml/100 cm^2/hr/50 psi, and their XM 50 (although passing some protein) yielded 130 ml/100 cm^2/hr/50 psi. Fluxes of 130 ml/100 cm^2/hr/50 psi were also obtained with our cellulose diacetate membrane 25–50–75 (see Section B-4a1). In the cases of the UM 1, UM 3, and 25–50–75 membranes no significant amounts of protein were found in the ultrafiltrate and little hemolysis in the retentate (as long as the speed of stirring, in an Amicon 52 cell, did not exceed approximately 500 rpm). At this rate and under 50 psi pressure a device with 1 sq ft of membrane would yield 1 L of ultrafiltrate per hour. This quantity, per hour, if discarded and replaced with appropriate amounts of a pure solution, administered either orally or intravenously, seems to be of an order of magnitude where its purifying influence begins to be considerable, without demanding a device of a prohibitive size.

However, for other applications of ultrafiltration to whole blood (for instance, for the removal of the excess of glycerol added to red cells as a freezing protectant), this flux still appears to be rather too slow to merit serious consideration.

5. PROTEIN-FRACTIONATING MEMBRANES

It is rather curious that ultrafiltration has not been used more frequently for the fractionation of proteins of different sizes, because it would at first sight appear more attractive to push proteins in solution through a gel in the shape of a thin membrane then to let them trickle through a column packed with gel beads and to rely on unaided diffusion for their separation. Nevertheless, this method has gained great popularity under the guise of "gel filtration" chromatography. The most important reason for the relative oblivion into which ultrafiltration as a protein fractionation method has subsided during these last 30 years, is without much doubt the incompleteness of separation achievable with the existing membranes (11,12,13). The "cutoffs" of the older membranes were too diffuse and the fractions that could be obtained overlapped too much for ultrafiltration to offer any advantage over other methods.

Our own experience has shown that here too the use of "skinned" membranes can make all the difference between inextricable imbrication and sharp separation of the desired fractions. When investigating the possibilities of separating serum macroglobulins ($M \approx 860{,}000$) from the other serum proteins ($M \approx 160{,}000$ and $\approx 69{,}000$) by ultrafiltration through a layer of agarose-gel, preliminary experiments indicated that the membrane's macroglobulin retention was more complete when the atmospheric conditions had been drier during the setting of the gel. A certain degree of drying evidently conditioned the outer layer of the gel by forming a skin that determined the semipermeability of the membrane. The exact conditions of "skinning" were then determined for the purpose of preparing membranes which consistently retain all macroglobulins, while passing a maximum amount of the other serum proteins (23).

When the agarose gel (5% in physiological saline) has just set, a surface skin is formed by directing the cold air stream of a hair drier for 10 sec from a distance of 25 mm. As soon as possible after this operation the serum is deposited on the membrane. Under a vacuum of approximately 36–45 cm mercury, 2 ml serum filter completely in about 4 hr. When 2 ml serum are applied, approximately 2 ml of a 2.0% ± 0.2% protein solution are collected as ultrafiltrate, containing the normal proportion of medium-sized serum proteins, but no trace of macroglobulin, as shown by analytical ultracentrifugation, immunoelectrophoresis, and immunodiffusion tests of the original serum and the ultrafiltrate (23). For a much faster anisotropic membrane, made of a mixture of cellulose nitrate and cellulose acetate, with a cut-off of M ≈ 300,000, see (134).

Preliminary tests indicate that a 10% agarose membrane can be used for ultrafiltering albumin (M ≈ 69,000), while retaining 7 S globulins (M ≈ 160,000). Other tests, with a cellulose acetate membrane, made with formamide and acetone, in the proportions 25–100–75, also showed that this separation is possible, provided that total protein concentrations not higher than 0.1% are used (134). These results are sufficiently encouraging also to attempt the separation of light and heavy chains of immune globulins with agarose or cellulose acetate (or even other) membranes, which possibilities are now being investigated. Blatt *et al.* (117a) have described the partial separation of serum albumin (M ≈ 69,000) from bovine α-lactalbumin (M ≈ 15,500), with an experimental Amicon membrane (XM-4A).

These examples show, although they are probably the only ones in existence up to now, that the advent of anisotropic membranes is likely to open fresh and extensive possibilities for ultrafiltration as a preparative tool for protein fractionation according to molecular size. The separation of small and medium-sized proteins and other macromolecules from the smallest viruses by ultrafiltration with anisotropic membranes seems a definite possibility. Whether anisotropic membranes will also play a role in the fractionation of viruses of various sizes is not yet quite certain: when pore diameters attain the order of 100 mμ and more, the advantages of anisotropic membranes become much less obvious and the existing homogeneous membranes of those porosities are probably quite adequate (34) and may be hard to improve upon.

C. Conversion of Hydrophilic to Hydrophobic Membranes

Most hydrophilic membranes can be impregnated with nonaqueous liquids after having been completely dried. But the irreversible effects of desiccation on membranes of low porosity are generally too unfavorable to make this a useful process. It is a general rule that the smaller the pores of a membrane, the more it needs to be kept submerged in the liquid (generally water) in which it was originally coagulated, if its porosity as well as its flux are not to be seriously reduced by shrinkage due to dehydration. (Although freeze-drying would probably obviate this difficulty.) Abrupt substitution of the surrounding water by nonaqueous liquids is almost impossible with the denser membranes: only the application of extremely high pressures will make the interstitial water yield to liquids of lower surface tension.

Both the shrinkage induced by dehydration and the high pressures needed to replace water with most other liquids are phenomena that are enhanced with decreasing pore sizes. They both are explicable by Laplace's law:

$$p = 2\gamma/r \tag{2}$$

which states that the pressure p due to surface tension equals 2 $\times$ the surface tension γ at the liquid-liquid or liquid-gas interface divided by the radius of curvature r of the drop or capillary. On studying this relation it will become clear why pores in a membrane will tend to contract under an increasing inward-directed pressure p just before they dry out, if the pore radius r is sufficiently small. It also becomes obvious why such high pressures are needed to replace water by liquids of low surface tension (causing a high interfacial tension γ) in very small pores.

The most adequate method of replacing the interstitial water by nonaqueous liquids is to effect the change extremely gradually (118) (which serves, among other reasons, to obviate desiccation that might otherwise be caused by large osmotic differences), and, where impregnation with a water-immiscible liquid is sought, to utilize intermediary solvents that are miscible with water as well as with the ultimate liquid. For example, interstitial water in a hydrophilic membrane can be replaced by vegetable oil by submerging the wet membrane for at least 1 hr in successive baths of the following composition: 30% water–70% ethanol, 5% water–45% ethanol–50% butanol, 100% butanol (renewed 3 times), 70% butanol–30% oil, 30% butanol-70% oil, 100% oil (67). Fewer steps can be used if small amounts of detergents are admixed with the various solutions (20b). We noticed that cellulose membranes, when imbibed with various nonaqueous liquids, apparently become less prone to compression than they were in the water-swollen state (20b). Cellophane as such could not be impregnated with any nonaqueous liquid that was tried; but cellophane with pores that were slightly enlarged by a treatment with concentrated $ZnCl_2$ (see Section A-2-d) successfully ultrafiltered, after gradual conversion to hydrophobic liquids, vegetable oil, and other organic liquids. By the inverse process hydrophobic membranes made of natural rubber could be induced to ultrafilter water and milk (20b).

The most useful membranes to convert to hydrophobic liquids are cellulose membranes; cellulose nitrate or cellulose acetate membranes are soluble in too large a variety of organic solvents to be of enough versatility. The more porous cellulose membranes that have been dried can of course be used directly with the desired nonaqueous liquid.

Ultrafiltration membranes that are directly made from hydrophobic material are still very rare; Sourirajan mentions using polyethylene membranes (63), but he does not divulge how he made them. In the same article he also mentions spraying dense cellulose acetate membranes with a polyfluorocarbon preparation, which presumably makes them hydrophobic. It is interesting to note that in his efforts to differentiate between ethanol and xylene and *n*-heptane, all the cellulose acetate membranes that were pretreated with absolute ethanol afterwards seemed to pass ethanol preferentially [see also (118)].

D. Leakage Detection of Membranes

Very little has been written on this subject, although the quest for information on this point, particularly with new and unfamiliar membranes, can create the most puzzling problems.

Very dense membranes, which stop molecules of a molecular weight of a few thousand and smaller, practically always show a more or less pronounced, but always measurable, degree of salt retention. In other words, a prior ultrafiltration of a salt solution (containing for instance ≈ 10 mM NaCl) should show a slightly lower salt content in the ultrafiltrate if no leak is present. A quick and easy way to test such salt concentrations is by using a bridge of Wheatstone for measuring resistivities or conductivities.

There are quick ways for a qualitative check on the absence of proteins in ultrafiltrates: if one drop of ultrafiltrate added to a small test tube containing 1 ml of 20% trichloracetic acid or sulfosalicylic acid leaves no noticeable haze or precipitate, it contains less than 0.01% protein. A test for amino acids and small peptides is done in a similar way, but with flavianic acid (2,4-dinitro-1-haphthol-7-sulfonic acid) and for medium- and large-sized polypeptides with picric acid.

Much smaller concentrations (less than 0.001%) of proteins can be shown up with the tannic acid/caffein method (114). When working with proteins it should not be forgotten that ultrafiltering very concentrated protein solutions will much more easily cause the passage of small amounts of proteins than ultrafiltering more dilute protein solutions (20b).

The proper functioning of more porous membranes, which pass proteins of a given size, while stopping larger proteins can of course only be checked by either specific tests for a given protein (immuno-chemical tests are the most obviously indicated here) or by analysis of the ultrafiltrate with the aid of analytical ultracentrifugation (23).

When the ultrafiltration flux exceeds the expected speed by an order of magnitude or more, there is always a strong case for suspecting a leak and one or more of the above-mentioned tests should be applied.

E. Measurement of Pore Size of Membranes

The measurement of pore sizes of *anisotropic* or "skinned" membranes will be the main consideration in this section; the measurement of mean pore sizes of homogeneous membranes having been quite thoroughly treated by Duclaux (67), Grabar (26), and Elford (11,68). It should not be forgotten that their principal method, which is based on the law of Poiseuille (see Section B-2), is, due to the difficulty of determining the exact thickness of the skin, quite inapplicable to the measurement of pore sizes of membranes endowed with the "skinned" type of anisotropy, which, as we have seen in Section B-1, is a fairly common property, even of the older membranes.

Apart from the pragmatic (and probably the best) method of determining the useful pore size of a membrane by the known size of the molecules it just passes (12) or by direct measurement by microscopy (26) or electron microscopy (72), there remains one other class of methods for measuring pore sizes of membranes endowed with a dense top layer of unknown thickness. That is the class of methods based on pressing a liquid-liquid or air-liquid interface through the pores. These methods have been quite thoroughly discussed by Duclaux (67) and will be only briefly outlined here. They are based on Laplace's law [eq. (2)], which permits the calculation of the *radius* r *of the largest pores of the densest layer of the membrane*, if one knows the surface tension γ at the liquid-gas or liquid-liquid interface, as well as the pressure p necessary to press the first air bubble through a wet membrane, or the pressure p necessary to obtain the first drop of an organic solvent (like iso-butyl alcohol) in the ultrafiltrate through a membrane initially wetted with water. The pore radius r in question will be obtained in cm, if the pressure p is expressed in dynes/cm^2 and the surface tension γ of the system in dynes/cm. The surface tension at the water-air interface ≈ 72 dynes/cm at 25°C. As liquid-liquid interfaces usually have much lower surface tensions than liquid-air interfaces, they will penetrate the membrane at a much lower pressure and their use will thus less easily give rise to deformation of the membrane. (For the first air bubble to pass a protein-stopping membrane of 6 mμ pore diameter a pressure of 500 atms, or 7340 psi is required, while for most water-organic solvent interfaces that pressure would be ≈ 40 times less).

The pore size of the more porous membranes (used, for instance, in virus fractionation) can also be checked with the aid of a number of monodisperse polystyrene latices, which can now be made with remarkably uniform particle diameters, varying from 50 to 800 mμ and more (124,125).

F. Attractive and Repulsive Influence of Membranes

1. *INFLUENCE OF ELECTRIC CHARGE OF MEMBRANES*

Few membranes are completely devoid of electric charge; practically all membranes that have been discussed here have a more or less pronounced negative charge. The influence of that charge has on occasion been evoked to explain the decreased ultrafilterability of proteins at high pH (20b,126), although it remains difficult to dissociate the electrostatic repulsion between membrane and protein from other possible mechanisms (like swelling and/or increased hydration of the proteins at high pH), as an explanation for the phenomenon.

The influence of the electric charge of membranes on the retention of small electrolytes is much more clear-cut. The well-known phenomenon of partial salt retention by ultrafiltration through cellophane membranes (19) can only be satisfactorily explained by taking the electronegativity of the membranes into account (20b,49; see also the Translator's note on page 5 of Ref. (19)). The facts are, briefly, as follows.

(*a*) The higher the pressure, the more pronounced the salt retention (19,20b). (The salt retention increases at increased pressures under experimental conditions (19) where compression of the homogeneous membranes is not likely to be an important consideration, given that the increase in flux is in quite linear proportion to the applied pressure. Furthermore, the increased retention is only proportional to the increase in *pressure* and *not* to the increased *flux*, which generally accompany it, as shown by experiments with membranes of various thicknesses and with several layers of membranes, all under constant pressure (49)).

(*b*) The higher the (negative) charge of the membrane (20b,49), the more pronounced the salt retention.

(*c*) Salts with bivalent anions and monovalent cations are more strongly retained than salts with monovalent anions and monovalent cations, while salts with bivalent cations and monovalent anions are hardly retained at all, as compared to salts with monovalent cations and anions (19,20b).

(*d*) The lower the salt concentration, the higher the relative salt retention (19).

It is clear that this type of salt retention must be due to electrostatic repulsion of the *anions* by the negatively charged membrane, a repulsion which is the more pronounced when the anions carry the highest charge and weakest when the cations carry the highest charge. [Ambard and Trautmann's thesis (19) is that all these phenomena can be explained by differences in hydrated volumes of ions at different concentrations. But (hypothetical) differences in hydration can explain neither the difference in influence of multivalency of anions and cations nor the enhanced salt retention at higher pressures; see also the Translator's note on p. 5 of Ref. (19).] For the retention of salts with mono and bivalent *cations* by positively charged membranes, see Ref. 84.

The increase in salt retention at higher pressures is proportional to the streaming potential caused by the applied pressure (20b,49,84) and, indeed, it has been shown (49) that the streaming potential hypothesis can quantitatively account for all the known aspects of the phenomenon, including the remarkably slight influence of temperature differences (19).

For the influence of an electric field on the salt retention by slightly charged membranes in an insulated ultrafilter and for the influence of rotating magnetic fields (magnetic stirring) on these phenomena, see Ref. 21.

For the avoidance of salt retention by membranes (whether due to electric charge or to sheer smallness of pores), see Section B-4-a.

2. ADSORPTION BY MEMBRANES

Adsorption of solute to membranes has been recognized as a factor that could, and often did, complicate ultrafiltration since the beginning of this century. Even in 1907 Bechhold (8) advised shaking the solution that was to be ultrafiltered with little bits of membrane, in order to ascertain if adsorption of solute to the membrane material was likely to occur. Elford published a more thorough discussion of adsorption problems in 1933 (11) and found:

(*a*) that the amount of ultrafiltrate that could be collected before the maximum amount of solute (a dye) found its way into it depended, not very surprisingly, on the concentration of dye in the initial solution;

(*b*) also not very surprising, that the thicker the membrane, the more solute is adsorbed;

(*c*) that the addition of soaps and other surface-active substances prevents the adsorption of other solutes (see also (127)), which he ascribes to some sort of lubrication, although we now suspect that this is rather a manifestation of preferential adsorption of the more hydrophobic detergent (128);

(*d*) that the adsorption of proteins is maximal at their isoelectric point, while the *flux* of a protein solution is minimal at that point (this surely is due to the minimal hydration of proteins at this point and thus to their maximal hydrophobicity and adsorbability, see Ref. 128).

Ferry (13) gives a number of useful hints on avoiding adsorption: ultrafiltering sufficiently large volumes through a membrane; having a high concentration of solute in the initial solution, using high pressures in order to obtain fast fluxes; using thin membranes and adding surface-active substances.

Galloway and Elford (129) made use of the fact that earlier, or simultaneous, ultrafiltration of "broth" helped to diminish the adsorption of foot and mouth disease virus to the membrane; they also noticed that the thicker the membrane, the more virus it adsorbed.

Cliver (34) diminished the adsorption of enteroviruses to the membrane by the addition of serum and found that the pretreatment of the membranes with serum or with a 2% gelatin solution was also effective.

Finally, van Gilse (130), in the course of attempts to concentrate urinary gonadotropic hormone by ultrafiltration, noticed that almost all of the activity could be recovered by dissolving the collodion membrane itself (and discarding the ultrafiltrate as well as the supernatant), and she finally realized that practically all of the hormone could be fixed by direct adsorption onto little pieces of collodion when stirred in the urine.

Acknowledgment

This work was supported in part by Public Health Service Grant GM 14067/GM 16256.

References

1. A. Schmidt, *Poggendorff Ann.*, **114**, 337 (1861).
2. A. F. Hertz, *Z. Physiol. Chem.*, **48**, 347 (1906).
3. C. J. Martin, *Proc. Roy. Soc. New South Wales* (April 5, 1896).
4. C. J. Martin, *J. Physiol.* **20**, 364 (1896).
5. M. Manea, *C. R. Soc. Biol.*, **56**, 317 (1904).
6. J. Duclaux, *Ultrafiltration*, part I (Partie Expérimentale), Hermann & Cie., Paris, 1945.
7. G. Malfitano, *C. R. Acad. Sci., Paris,* **139**, 1221 (1904); **140**, 1245 (1905); **143**, 172 (1906).
8. H. Bechhold, *Kolloid Z.*, **1**, 107 (1906); *Z. physik. Chem.*, **60**, 257 (1907); **64**, 328 (1908).
9. H. Bechhold, *Kolloid Z.*, **2**, 3 and 33 (1907).
10. I. N. Asheshov, *C. R. Soc. Biol.*, **92**, 362 (1952); *J. Bacter.*, **25**, 323 (1933).
11. W. J. Elford, *Proc. Roy. Soc.*, **B106**, 216 (1930); **B112**, 384 (1933); *J. Path. Bacter.*, **34**, 505 (1931).
12. W. J. Elford, *Trans. Faraday Soc.*, **33**, 1094 (1937).
13. J. D. Ferry, *Chem. Rev.*, **18**, 373 (1936).

14. S. Loeb, S. Sourirajan, and S. T. Uuster, *Chem. Eng. News*, **38,** 64 (4–11–1960).
15. S. Loeb, *UCLA Engineering Report*, 61-42 (1961).
16. W. F. Blatt, M. P. Feinberg, H. B. Hoppenberg, and C. A. Saravis, *Science*, **150,** 225 (1965).
17. A. S. Michaels, *Ind. Eng. Chem.*, **57,** 32 (1965).
18. B. Erschler, *Kolloid Z.*, **68,** 289 (1934).
19. L. Ambard and S. Trautmann, *Ultrafiltration*, Charles C Thomas, Springfield, Ill., 1960.
20. C. J. van Oss, (a) Dutch Patent 74,531 (1954), (b) *l'Ultrafiltration*, Doctoral Dissertation, Paris, 1955.
21. C. J. van Oss and N. Beyrard, *J. Chim. Phys.*, **60,** 451 (1963).
22. S. Loeb, *Desalination*, **1,** 35 (1966).
23. C. J. van Oss, A. Scheinmann, and J. E. Lord, *Nature*, **215,** 639 (1967).
24. H. Z. Friedlander and R. N. Rickles, *Analytical Chem.*, **37,** No. 8, 27A (1965).
25. R. N. Rickles, *Membranes, Technology and Economics*, Noyes Development Corp., Park Ridge, N.J., 1967.
26. P. Grabar, *L'Ultrafiltration Fractionnée*, Hermann & Cie, Paris, 1943.
27. H. P. Gregor and K. Sollner, *J. Phys. Chem.*, **50,** 53 (1946).
28. C. W. Carr, D. Anderson, and I. Miller, *Science*, **125,** 1245 (1957).
29. R. Zsigmondy, *Biochem. Z.*, **171,** 198 (1926); *Z. Angew. Chem.*, **39,** 398 (1926).
30. E. Bertarelli, *Zentralbl. f. Bakt., Abt. I*, **76,** 463 (1915).
31. J. Duclaux and M. Amat, *J. Chim. Phys.*, **35,** 147 (1938).
32. A. Dobry, *Bull. Soc. Chim.*, **3,** 312 (1936).
33. C. E. Reid and E. J. Breton, *J. Appl. Polymer Sci.*, **1,** 133 (1959); C. E. Reid and J. R. Kuppers, *ibid.*, **12,** 264 (1959).
34. H. P. Rothbaum, *Nature*, **214,** 285 (1967).
34. D. O. Cliver, *Applied Microbiol.*, **13,** 1 (1965).
34a. F. L. Black, *Virology*, **5,** 391 (1958).
35. T. Graham, *Phil. Trans.*, **183** (1861).
36. J. Duclaux and A. Hamelin, *Ann. Inst. Pasteur*, **25,** 145 (1911).
37. R. D. Stewart, E. D. Baretta, J. C. Cerny, and H. I. Mahon, *Investigative Urology*, **3,** 614 (1966).
38. J. W. McBain and S. S. Stuewer, *J. Phys. Chem.*, **40,** 1157 (1936).
39. C. J. van Oss, *French Patent* 1,075,417 (1954).
40. M. Stoliaroff, C. J. van Oss, G. Martin, and J. G. Gauduchon, *Revue d'Immunologie*, **22,** 546 (1959).
41. J. W. McBain and R. Stuewer, *J. Phys. Chem.*, **40,** 1157 (1936).
42. W. B. Seymour, *J. Biol. Chem.*, **134,** 701 (1940).
43. L. C. Craig, in *Analytical Methods of Protein Chemistry*, P. Alexander and R. J. Block, Eds., Vol. 1, p. 104, Pergamon Press, London-New York, 1960.
44. L. C. Craig and W. Konigsberg, *J. Phys. Chem.*, **65,** 166 (1961).
45. E. Heymann and G. Rabinov, *J. Phys. Chem.*, **45,** 1162 (1941).
46. P. Hirsch, *Rec. Trav. Chim. Pays-Bas*, **71,** 354 (1952).
47. P. Hirsch, *Rec. Trav. Chim. Pays-Bas*, **70,** 567 (1951).
48. P. Hirsch, *Rec. Trav. Chim. Pays-Bas*, **71,** 525 (1952).
49. C. J. van Oss, *Science*, **139,** 1123 (1963); *J. Chim. Phys.*, **60,** 648 (1963).
50. J. P. Thiéry, C. J. van Oss, L. Salomon, and M. P. Doucet, *C. R. Acad. Sci., Paris*, **239,** 1010; 1096 (1954).
51. D. Schyma, *Zentralbl. Bakt., I Orig.*, **181,** 17 (1961).
52. G. Moll, *Z. Hygiene*, **149,** 297 (1963).

53. D. I. Hitchcock, *J. Gen. Physiol.*, **8,** 61 (1925); **10,** 179 (1926).
54. J. Loiseleur and L. Velluz, *Comptes rendus Acad. Sci., Paris,* **192,** 43, 159 (1931); *Bull. Soc. Chim. Biol.*, **14,** 1210 (1932).
55. G. Florence and J. Loiseleur, *Bull. Soc. Chim. Biol.*, **15,** 395 (1933).
56. W. Pfeffer, *Osmotische Untersuchungen*, Engelmann, Leipzig, 1877.
57. P. Hirsch-Ayalon, *Rec. Trav. Chim. Pays-Bas,* **75,** 1065 (1965); **79,** 382 (1960); **80,** 365 (1961); **80,** 376 (1961); *J. Polymer Sci.*, **23,** 697 (1957); *Electrochim. Acta,* **10,** 773 (1965).
58. C. J. van Oss and P. Hirsch-Ayalon, *Science,* **129,** 1365 (1959).
59. C. J. van Oss and Y. S. Heck, *Z. Immunitätsforschung,* **122,** 44 (1961).
59a. C. J. van Oss, *J. Colloid Interf. Sci.*, **27,** 684 (1968).
60. C. J. van Oss and M. Fontaine, *Z. Immunitätsforschung,* **121,** 45 (1961).
61. *Chem. Eng. News,* **1,** 31 (1966), p. 24; **4,** 25 (1966), p. 22; **5,** 30 (1966), p. 50.
62. H. P. Gregor and E. Kantner, *J. Phys. Chem.*, **61,** 1169 (1957).
63. S. Sourirajan, *Nature,* **203,** 1348 (1964).
64. A. S. Michaels, in *Advances in Separation and Purification,* E. S. Perry, Ed., Wiley, 1967.
65. S. Loeb and S. Sourirajan, *UCLA Dept. of Engineering Report,* 60-60, July 1960; *Advan. Chem.*, Ser. **80,** 117 (1963); *U.S. Patent* 3,133,132 (1964).
66. W. J. Elford, P. Grabar, and J. D. Ferry, *Brit. J. Exper. Pathol.*, **16,** 583 (1935).
67. J. Duclaux, *Ultrafiltration,* part II (Partie Théorique et Applications), Hermann & Cie, Paris, 1946.
68. W. J. Elford and J. D. Ferry, *Brit. J. Exper. Pathol.*, **16,** 1 (1935); (see especially Figure 3).
69. W. Ostwald, *Kolloid Z.*, **22,** 72 and 143 (1928).
70. P. Grabar and J. A. de Loureiro, *J. Chim. Phys.*, **33,** 815 (1936).
71. S. Loeb and F. Millstein, *Dechema Monographien,* **47,** 707 (1962).
72. H. K. Lonsdale *et al, O.S.W. Res. Developm. Progr. Rep.*, **150** (1965).
73. U. Merten *et al., O.S.W. Res. Developm. Progr. Rep.*, **208** (1966).
74. W. Banks *et al., O.S.W. Res. Developm. Progr. Rep.*, **143** (1965).
75. C. W. Saltonstall *et al., O.S.W. Res. Devleopm. Progr. Rep.*, **167** (1966).
76. R. F. Baddour *et al., O.S.W. Res. Developm. Progr. Rep.*, **144** (1965).
77. A. S. Michaels *et al., O.S.W. Res. Developm. Progr. Rep.*, **149** (1965).
78. P. S. Francis *et al., O.S.W. Res. Developm. Progr. Rep.*, **177** (1966).
79. E. Glueckauf, *Nature,* **211,** 1227 (1966).
80. S. Sourirajan, *I and EC Fundamentals,* **2,** 51 (1963).
81. E. Glueckauf, *New Scientist,* p. 531, 11–18–65.
82. L. Dresner and K. A. Kraus, *J. Phys. Chem.*, **67,** 990 (1963).
83. K. A. Kraus, A. E. Marcinkowsky, J. S. Johnson, and A. J. Shor, *Science,* **151** (1966).
84. J. G. McKelvey, K. S. Spiegler, and M. R. J. Wyllie, *Chem. Eng. Progr. Symp. Series 24,* **55,** 199 (1959).
85. C. E. Reid and H. G. Spencer, *J. Appl. Polymer Sci.*, **4,** 354 (1960).
86. "Development of Synthetic Osmotic Membranes for Use in Desalting Saline Waters," Research Proposal for the U.S. Department of the Interior, February 19, 1953.
87. B. Keilin, *O.S.W. Res. Developm. Progr. Rep.*, **84,** (1963); 154 (1965).
88. C. E. Reid and H. G. Spencer, *J. Phys. Chem.*, **64,** 1587 (1960).
89. J. Rosenfield and S. Loeb, *I and EC Process Design Developm.*, **6,** 122 (1967).

90. A. S. Michaels, H. J. Bixler, and R. M. Hodges, *J. Colloid Sci.*, **20,** 1034 (1965).
91. L. Ambard and S. Trautmann, *Ultrafiltration*, Charles C Thomas, Springfield, Ill., 1960.
92. J. D. Bernal and R. H. Fowler, *J. Chem. Phys.*, **1,** 515 (1933).
93. R. H. Stokes and R. A. Robinson, *J. Am. Chem. Soc.*, **70,** 1870 (1948).
94. E. R. Nightingale, *J. Phys. Chem.*, **63,** 1381 (1959).
95. D. S. Allam and W. H. Lee, *J. Chem. Soc. A*, **1,** 5 (1966).
96. D. L. Erickson, J. Glater, and J. W. McCutchan, *I and EC Prod. Res. Developm.*, **5,** 205 (1966).
97. F. Hofmeister, *Archiv. Exptl. Pathol. Pharmakol.*, **25,** 1 (1889); **27,** 395 (1890); **28,** 210 (1891).
98. P. Debye, *Polar Molecules*, Chemical Catalog Co., Reinhold Publ. Co., New York, 1929.
99. S. Manjikian, *I and EC Prod. Res. Developm.*, **6,** 23 (1967).
100. T. K. Sherwood *et al.*, *O.S.W. Res. Developm. Progr. Rep.*, **180** (1966); **141** (1965); *I and EC Fundamentals*, **4,** 113 (1965).
101. H. I. Mahon, *Proc. Desalination Res. Conf., Natl. Acad. Sci. Publ.*, **942,** 345 (1963).
102. J. Bevengut, *French Patent*, 555,471 (1923).
103. A. S. Michaels, *U.S. Patent*, 3,173,867 (1965).
104. U. Merten *et al.*, *O.S.W. Res. Developm. Progr. Rep.*, **165,** (1966).
105. D. T. Bray *et al.*, *O.S.W. Res. Developm. Progr. Rep.*, **176** (1966).
106. *Chem. Eng. News*, Technology Concentrates, **5,** 29 (1967), p. 47; *Chem. Eng. News*, **10,** 23 (1967), p. 66; *Chem. Week, Technology Newsletter*, **5,** 27 (1967), p. 64.
107. K. Popper, W. M. Camirand, W. L. Stanley, and F. S. Nury, *Nature*, **211,** 297 (1966).
108. S. Sourirajan, *I and EC Proc. Design Developm.*, **6,** 154 (1967); **7,** 548 (1968).
109. General Dynamics, *Scientific Am.*, **216,** 5, 22 (1967).
110. A. I. Morgan, E. Lowe, R. L. Merson, and E. L. Durkee, *Food Technology*, **19,** 1790 (1965).
111. F. Albert-Recht and C. P. Stewart, *Protides Biol. Fluids*, **8,** 50 (1961).
112. *Chem. Eng. News*, **6,** 12 (1967), p. 105.
113. C. J. van Oss, C. R. McConnell, R. K. Tomkins, and P. M. Bronson, *Clin. Chem.*, **15,** 699 (1969).
114. T. J. Greenwalt, E. A. Steane, and C. J. van Oss, *Fed. Proc.*, **25,** 2, 612 (1966).
115. T. J. Greenwalt, C. J. van OSS, and E. A. Steane, *Am. J. Clin. Pathol.*, **49,** 472 (1968).
116. J. P. Merrill, *Sci. American*, **205,** 1, 56 (1961).
117. W. F. Blatt, S. M. Robinson, F. M. Robbins, and C. A. Saravis, *Analytical Biochem.*, **18,** 81 (1967); **22,** 161 (1968); *Nature*, **216,** 512 (1967).
118. H. Z. Friedlander *et al.*, *Proc. Symp. Membrane Processes for Industry*, p. 235, Birmingham, Ala., 5–19–66.
119. C. J. van Oss, *Rec. Trav. Chim. Pays-Bas*, **77,** 479 (1958).
120. C. J. van Oss, H. Simonnet, and D. Annicolas, *Rec. Trav. Chim. Pays-Bas*, **78,** 425 (1959); *C. R. Acad. Sci., Paris*, **248,** 460 (1959).
121. H. Sasaki and N. Sata, *Koll. Zschr.*, **199,** 49 (1964).
122. J. V. Bennett and W. M. M. Kirby, *J. Lab. Clin. Med.*, **66,** 721 (1965).
123. S. Salminen, *Studies on the Ultrafiltrability of Serum Sodium and Potassium*, Thesis, Helsinki (1961); *Ann. Med. Exp. Biol. Fenniae*, **39,** Suppl. 4 (1961).

124. J. W. Vanderhoff, E. B. Bradford, M. L. Tarkowski, J. B. Shaffer, and R. M. Wiley, *Adv. Chem. Series*, **24,** 32 (1962).
125. E. B. Bradford and J. W. Vanderhoff, *Abstr.* **151***st Am. Chem. Soc. Mtng., D,* **47,** Pittsburgh 3-23-1966.
126. W. J. Elford and J. D. Ferry, *Biochem. J.*, **28,** 650 (1934); **30,** 84 (1936).
127. R. Brinkman and A. von Szent Györgyi, *Biochem. Z.*, **139,** 261 (1923).
128. C. J. van Oss and J. M. Singer, *J. Reticuloendothelial Soc.*, **3,** 29 (1966).
129. I. A. Galloway and W. J. Elford, *Brit. J. Exp. Pathol.*, **12,** 407 (1931).
130. H. A. van Gilse, *The Determination of Gonadotropic Hormones in Urine,* Thesis, Leiden, 1953.
131. S. Sourirajan and S. Kimura, *Ind. Eng. Chem. Proc. Design Developm.*, **6,** 504 (1967).
132. H. Ohya and S. Sourirajan, *Ind. Eng. Chem. Proc. Design Developm.*, **8,** 131 (1969).
133. C. J. van Oss and P. M. Bronson, *Separation Science,* **5,** 63 (1970).
134. C. J. van Oss and P. M. Bronson, in *Membrane Science and Technology,* J. E. Flinn, Ed., p. 139, Plenum Press, New York, 1970; *Analytical Biochem.*, in press, 1970.
135. A. E. Marcinowsky, K. A. Kraus, H. O. Phillips, J. S. Johnson, and A. J. Shor, *J. Am. Chem. Soc.*, **88,** 5744 (1966).
136. K. A. Kraus, A. J. Shor, and J. S. Johnson, *Desalination,* **2,** 243 (1967).
137. J. S. Johnson, Abstr. 155th *Natl. Mtg. Am. Chem. Soc. U-15,* San Francisco, 1968.
138. A. J. Shor, K. A. Kraus, J. S. Johnson, and W. T. Smith, *Ind. Eng. Chem. Fundamentals,* **7,** 44 (1968).
139. D. G. Thomas and J. S. Watson, *Ind. Eng. Chem. Proc. Design Developm.*, **7,** 397 (Z1968).
140. K. D. Vos and F. O. Burris, *Ind. Eng. Chem. Prod. Res. Developm.*, **8,** 84 (1969).
141. H. K. Lonsdale, *Separation and Participation by Reverse Osmosis,* this volume.

High-Pressure Gas Chromatography

Marcus N. Myers and J. Calvin Giddings

Department of Chemistry,
University of Utah,
Salt Lake City, Utah

ABSTRACT

The application of high-pressure methods to gas chromatography is discussed. The potential advantages are increased separation efficiency and speed and, in particular, the enhancement of the volatility of complex molecules whose low normal volatility precludes the application of conventional gas chromatography. A number of such nonvolatile compounds have been "dissolved" and migrated in a dense gas chromatographic system. These include polymers and various biomolecules.

I. INTRODUCTION

There are a number of parameters in gas chromatography which can be altered in a variety of ways to achieve separation goals. Among them are column size, temperature, pressure, various gradients, and the chemical nature of stationary and mobile phases. Pressure is presently the least explored parameter despite the possibility of large and useful effects.

It is a tradition in gas chromatography to avoid large pressure drops through columns in the belief that the nonuniform velocities due to gas expansion will reduce efficiency and thus resolution. Predictions by Knox (1) and careful theoretical examination by Giddings (2) have not confirmed this view; to the contrary the use of high inlet pressures promises several advantages. Most prominent are the possibilities for high efficiencies, high separation speeds, and enhanced migration of intractable, high molecular weight materials.

A. Column Efficiency and Speed

Increased efficiency can be most simply obtained by increasing column length and concomitantly increasing pressure to maintain the flow. This approach was used in the first attempt to demonstrate high-pressure advantages; a packed column 4000 ft long was operated with inlet pressures up to 2500 psi (3). This column yielded over 10^6 theoretical plates. The plate height was not outstanding (.024 in.), probably due to nonoptimal packing and operation, but the feasibility of this method of obtaining high efficiency columns was confirmed.

The next efforts to improve efficiency were directed at columns with large numbers of plates per unit of length. Following the theoretical deductions, the *high inlet pressure micro column* (HIPMC) system was developed (4). Particles in the "sub-sieve" range ($<30\ \mu$) were packed into columns having inside diameters less than .04 in. High inlet pressures were, of course, necessary to obtain reasonable flow. Initial studies with 13 μ alumina (modified with 40% by weight NaOH) in a 0.02 in. i.d. column revealed a plate height of 0.0031 in. at an inlet pressure of 2500 psi. Later work (5) with a similar material (NaI modified alumina) operated at pressures to 30,000 psi yielded plate heights down to and below 0.0028 in., or over 4000 plates/ft. Ordinarily it is a very good column that will yield 1000 plates/ft. The columns tested were not operated under optimum conditions, but again gave an indication of the potential of this approach to high efficiency.

Studies of high-speed separations were made under turbulent conditions in order to utilize the enhanced mass transfer of turbulent flow (6). Gases were forced through a solid coated capillary at velocities up to 70 ft/sec, well into the turbulent region, by inlet pressures to 2500 psi. Under these conditions, with Reynold's numbers up to 16,000, a total of 8200 plates/sec were obtained with nonsorbing peaks and 2500 plates/sec with pentane. The plate height increased with velocity in its usual fashion until a Reynold's number of slightly over 2000 was reached. At this threshold point a sharp drop in plate height occurred followed by slowly decreasing values of the plate height with further increases in velocity.

Analogous experiments with packed columns at Reynold's numbers to 2800 did not, of course, show the discontinuous drop. Here local pockets of turbulence occur at much lower Reynold's numbers and gradually expand with velocity. A clear maximum in the plate height-velocity curve was apparent, but it was more rounded. The declining portion beyond the maximum is of particular value to high-speed operation. One of the packed columns yielded 3800 plates/sec.

Unfortunately these data do not clearly define the role of turbulence in packed columns since coupling (from flow-diffusion interactions) also reduces plate height at high velocities (7). Both effects, however, are favorable to high efficiency and speed as shown by Knox (7a).

B. Enhanced Migration

The most unique and possibly the most valuable application of high-pressure gas chromatography is the enhancement of migration of low volatility solutes by the resulting dense gases. Solubility phenomena in dense gases were first observed by Hannay and Hogarth (8) in 1879 in their studies of ethanol solutions of cobalt and ferric chloride in the critical region. These studies indicated that the usual ionic condition existed above the critical temperature and that the concentration of salt was much higher than the normal vapor pressure of the salt would predict. Since that time geologists have devoted considerable study to this enhanced volatility as a mechanism for the transport of various minerals and, in particular, for the concentration occurring in certain types of ore bodies. Such material transport has also gained the attention of engineers in power generation due to deposition of amorphous silica, quartz, and other materials on the convex sides of steam turbine blades.

The solubility of solid substances in gases was reviewed by Rowlinson and Richardson (9) and also by Booth and Bidwell (10). The solubility of salts is not discontinuous at the critical point of the solutions, but is highly pressure dependent and hence density dependent in this region. The evidence indicates that a true molecular solution exists for enhanced solubility systems.

With the concept that solubility is highly dependent on density and pressure, the capability to change the solvent properties of a single gas simply by varying pressure, a mechanical parameter, introduces an entirely different dimension into chromatography.

The strong pressure dependence of solubility leads to the concept of high-pressure programmed gas chromatography. The application of a steadily increasing pressure will migrate the various species in turn, from the simplest to those most difficult to migrate. The method is analogous to the now commonly used technique of temperature programming in low-pressure gas chromatography.

The stationary phase may also be altered upon increasing pressure. In the case of a liquid stationary phase, the carrier gas will dissolve increasingly with pressure and alter the liquid properties. As the pressure is increased further, the liquid phase itself may dissolve into the gas phase and migrate (bleed), thus becoming useless for the separation. When gas-solid systems are employed, one must consider the competition of carrier molecules for the available surface sites and the consequent weaker retention of solute molecules. This will probably lead to reduced tailing due to the coverage of high energy sites by solvent molecules.

The use of pressure-enhanced migration in gas chromatographic separations was first demonstrated by Klesper, Corwin, and Turner (11) who found that some porphyrins migrated in dichlorodifluoromethane (Freon 12) and monochlorodifluoromethane at pressure up to 2000 psi. Other gases tried (monochlorotrifluoromethane, trifluoromethane, and nitrogen) did not enhance migration.

In 1966, Sie, van Beersum, and Rijnders (12) reported increased volatility of a number of compounds in CO_2 (the C_7—C_{13} paraffins, benzene, toluene, etc.) and their separation at pressures to 1150 psi. In addition to discussing the calculation of second cross virial coefficients from their data, they pointed out that at lower temperatures solute volatility decreases but is more than compensated for by increasing nonideal gas-phase interactions. They demonstrated the separation of high molecular weight polynuclear aromatics with partition and adsorption techniques (13,14) using pentane and isopropanol as the mobile phases.

More recently, McLaren, Myers, and Giddings (15) have reported enhanced volatility of various biomolecules and polymers in CO_2 and NH_3 at 40°C and 140°C, respectively, at maximum pressures of 30,000 and 5000 psi, respectively. Polymers as high as 400,000 mol. wt. were chromatographed. Later extensions of this work (16) involved over 80 different compounds, all in the intermediate ($\sim$400 mol. wt.) to high molecular weight groups. The results are summarized in Table I. Two other gases, helium and nitrogen, were not effective in the dissolution of nonvolatile molecules.

As can be seen from Table I, dense ammonia is most effective in migrating highly polar compounds such as the amino acids and sugars, while dense carbon dioxide is best as a solvent for nonpolar systems, particularly unsaturated molecules. There is considerable overlap, however, with many quite polar molecules migrating in carbon dioxide, and many nonpolar molecules dissolving in dense ammonia.

The pressure programming suggested above for the separation of dissimilar components was demonstrated in the separation of squalane, dinonyl pthalate, and SE-30 silicone gum rubber solely by pressure change. Column separations were also accomplished for several materials. The separation of glycine and two simple peptides was carried out in ammonia using a Porasil B column at 140°C. The carotenes were separated on a Ucon 50 column in carbon dioxide at 40°C. Other separations (sterols, squalane from squalene) have also been accomplished, but the basic emphasis has been on the study of the migration properties of various molecules.

II. EXPERIMENTAL METHODS

The practical application of high pressure to gas chromatography requires a variety of considerations since the type and capability of the apparatus needed, along with its cost and difficulty of operation and maintenance, depend on the nature of the problems encountered.

Generally equipment for high-pressure gas chromatography requires a source of gas at the desired pressure, flow regulation, sample introduction, and sample detection. Since many gases of potential usefulness have critical temperatures above room temperature (see Table II), it may be necessary to heat all or part of the system. Safety precautions must also be observed, as some gases are toxic (Table II), and a danger of rupture due to component failure exists.

A. Pressure Generation

Our basic system, designed for pressures up to 30,000 psi, is shown in Figure 1.

It should be pointed out that the 2500 psi pressure normally existing in full cylinders of the permanent gases is sufficient to migrate components not excessive in molecular weight. In this case the experimental system is greatly simplified. The compressor, control gauge, and reservoir are eliminated. A pressure regulator such as APCO 1341 × 1016 gives satisfactory pressure (and flow) control over the entire 0–2500 psi range. The remaining portions of the equipment can be constructed using conventional valves and fittings. When pressures in excess of commercial tank pressure are required, a compressor or pump and special high-pressure components are required.

Two methods have been used in this laboratory for supplying gases at high pressures. The first utilizes an air-driven diaphragm compressor (AMINCO 46-14021) which can supply up to about 0.1 cc/sec at pressures to 30,000 psi and will operate at temperatures as high as 100°C. These

TABLE I

Relative Migration of Various Species in Ammonia and Carbon Dioxide

Solute	Migration in NH_3 140°C and 200 atm	Migration in CO_2 40°C	Pressure (psi)
Purines			
Adenine	–[a]	+	19,100
Guanine	–	–	19,100
Caffeine	2+		
Xanthine	–		
Nucleosides and nucleotides			
Adenosine, guanosine, uridine	2+	+	19,850
Adenylic acid	2+		
Cortical steroids			
Cortisone, hydrocortisone	+ (react?)	+	19,100
Sterols			
Cholesterol, ergosterol, lanosterol	2+	2+	7,500
Sugars			
Ribose, arabinose, xylose	2+	2+	19,850
Glucose, fructose, mannose, sorbose	2+	2+	19,850
Galactose	+	–	19,850
Maltose	2+	2+	19,850
Lactose	2+	–	19,850
Terpenes			
α-Carotene, β-carotene, lycopene	(decompose)	2+	2,500
Squalene	2+	2+	7,500
Amino acids			
Glycine	2+	+	19,850
Leucine, tryptophane, tyrosine, arginine, lysine	2+	–	19,850
Glycylglycine, glycyl-L-leucyl-L-tyrosine	2+		
Hippuryl-L-arginine	–		
Proteins			
Trypsin, lysozyme, apoferritin, γ globulin	–		
Bovine albumin, bovine hemoglobin, gelatin	–		
Cytochrome C	+		

(Continued)

TABLE I—Continued

Solute	Migration in NH_3 140°C and 200 atm	Migration in CO_2	
		40°C	Pressure (psi)
Lipids			
Cerebroside		+	
Silicone oils and gum rubber			
DC-200		2+	10,300
OV-1	−		
OV-17	2+		
SE-30	−	2+	11,300[b]
Alkanes			
n-Octadecane, *n*-docosane, *n*-octacosane		2+	2,000
Squalane	2+	2+	1,320[b]
Carbowaxes			
Carbowax 400		2+	1,000[b]
Carbowax 1000		2+	1,690[b]
Carbowax 4000		2+	2,800[b]
Carbowax 20,000	2+	+	
Other solutes			
Quercitin	2+		
Amygdaline	−		
Polysulfone	−		
Polyox	−		
Polystyrene 900	−		
Versamid 900	−		
Cellulose acetate	−		
α Glucose penta acetate	2+		
Polyvinyl chloride		+	25,000
Polyethylene		−	22,000
Dinonyl phthalate		2+	1,320[b]
Apiezon L	−	2+	12,000[b]

[a] Migration is indicated by + or 2+, depending on intensity; nonmigration by −.

[b] Threshold pressure.

TABLE II

Critical, Polarity, and Toxicity Data for Gases of Potential Use in High-Pressure Gas Chromatography

Name	Critical Temperature (°C) (26)	Critical Pressure (psi)(26)	Critical Density (g/cc) (26)	Dipole Moment ($\times 10^{-18}$ esu)(27)	Polarizability ($\times 10^{-25}$ cc)(28)	Toxicity Rating (29)
Argon	−122	706	0.531	0	16.2	0[a]
Methane	−82.1	673	0.162	0	26.0	1
Ethylene	9.2	735	0.227	0		2(5500)[b]
Chlorotrifluoromethane	28.8	573	0.58	0		2
Carbon dioxide	31.1	1070	0.468	0	26.5	1(5000)
Fluoroform	33.	691	0.516	1.60		2
Nitrous oxide	36.5	1050	0.457	0.14	30.0	0
Fluoromethane	44.6	853	0.300	1.81		([c])
Sulfur hexafluoride	45.6	545	0.752	0		0
Hydrogen chloride	51.4	1200	0.42	1.05	26.3	3(5)
Propane	96.8	617	0.220	0	62.9	0
Hydrogen sulfide	100.4	1310	0.349	0.93	36.8	3(20)
Dichlorodifluoroethane (30)	111.5	582	0.555	0.505		1(1000)
1,1-Difluoromethane	113.5	652	0.365	2.24		1
Ammonia	132.3	1640	0.235	1.46	22.6	3(100)
Methyl chloride	143.1	969	0.353	1.86	45.6	3(100)
Chlorine	144.	1120	0.573	0	46.1	3(1)
Sulfur dioxide	157.5	1140	0.524	1.62		3(10)
Acetaldehyde	188.	804	0.262	2.72		3(200)
n-Pentane	196.6	486	0.232	0	99.5	1(1000)

[a] The degree of toxidity is indicated by

0 No harm in any concentration, or toxicity caused by unusual conditions or overwhelming dosage.

1 Produces changes in the human body which are readily reversible upon termination of exposure.

2 May produce irreversible as well as reversible changes in the human body; not serious enough to threaten life or cause permanent physical impairment.

3 Can cause injury severe enough to threaten life or cause permanent physical impairment.

[b] Numbers in parentheses indicate maximum concentration for long-term exposure.

[c] Toxicity unknown.

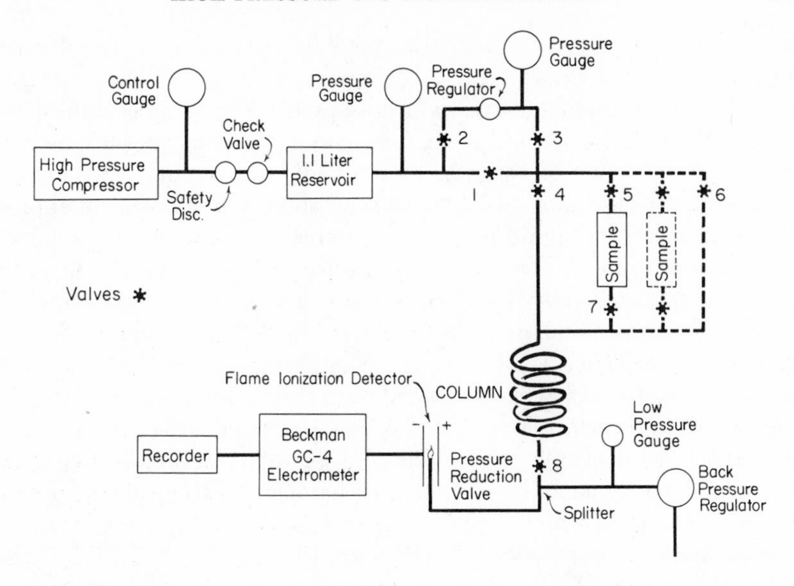

Fig. 1. Schematic diagram of apparatus for high-pressure gas chromatography.

compressors suffer from a difficulty in the air drive transfer system which greatly lessens their reliability.

The second method involves pumping the mobile phase as a liquid at a temperature below critical. The stream is then heated to form the supercritical fluid. This method is particularly advantageous for gases with high critical temperatures since the pump can be operated well below this point. Two pumps have been employed in this fashion, one producing pressures to 5000 psi (Whitey LP-10) and the other to 40,000 psi (AMINCO 46-13715). The latter pump was only recently put into operation.

Since some detectors become flow sensitive when carrier gases such as CO_2 and NH_3 are used, flow stabilization is necessary. A pressure regulator (APCO 1345-4) was used to provide pressure and thus flow control up to 10,000 psi. Above this pressure, the regulator was isolated from the system by closing valves 2 and 3 and opening valve 1 as indicated in Figure 1. The reservoir (AMINCO 41-14735, 1.15 l. vol) then becomes the sole means of "damping out" pressure pulses (which may be several hundred pounds per pump or compressor stroke).

B. Injection

Sample injection at high pressure poses unique problems. The transfer of the small sample from the laboratory environment to the system environment is opposed by the substantial pressure increase. The usual syringe injections become impractical much above atmospheric pressure, while available GC sampling valves will not stand the great mechanical force applied to their interior and will not remain leak-free as pressure magnifies the escape tendency of the gas. Undoubtedly a sampling valve could be designed for these conditions but none are presently available. Our approach has been to let the high-pressure flow source do the work of compression and transfer, using commercial high-pressure valves to properly direct flow.

The sample is usually dissolved in a solvent (preferably one which, like CS_2, does not respond in a flame ionization detector) and loaded on a support material such as Chromosorb W or glass beads, in the same manner as a liquid stationary phase is loaded on a chromatographic packing. This exposes a large surface area and shortens the equilibration time between the carrier gas (solvent) and the sample (solute). A sample is prepared for injection by placing it in the sample chamber at atmospheric pressure, closing the chamber, and raising the gas to the desired pressure. More than one sample can be prepared at the same time by inserting more sample chambers (see Fig. 1). In this case, a bypass is also placed in the system to allow flushing of the tubing between samples.

After the sample chamber has been adjusted to the column pressure, the actual injection into the column is accomplished by closing valve 4 and opening valves 5 and 7 briefly, then reverting to the original. If high flows are used, valve 7 may be left open, and the injection made by closing valve 4 momentarily. This produces a small pressure drop at the tee just before the column and an injection surge from the sample chamber. This procedure has resulted in sample plugs as short in duration as 0.01 sec by using a specially designed pneumatically operated valve manipulator for closing and opening valve 4. Sample plugs of about 0.03 sec duration were routinely obtained for the turbulent flow studies.

C. Detection

Ideally, sample detection would be accomplished at the column outlet pressure. Unfortunately none of the high-sensitivity detectors for gas chromatography are readily adaptable to high-pressure use. One device for detection at high pressure is a UV cell which is being evaluated at the present time.

Alternately, we could cool the gas to a liquid, decompress in that form, then use liquid detection technology. This approach has been used by Sie and Rijnders (13,14) who employed a UV detection method. Unfortunately, liquid detection systems are usually relatively insensitive. In addition, the small volume integral to our system puts severe limitations on detector volume and geometry. This is aggravated by the low molar volume and diffusivity characteristic of the liquid state.

If, following a third route, the dense gas is decompressed at the exit of the column, we could use any one of the gas chromatography detectors now available. However, complex solute molecules reach a state of extreme thermodynamic instability as the pressure is lowered drastically. Immediate molecular association and wall condensation are expected. If the time between decompression and detection is kept short, a certain amount of solute will remain in the vapor phase for detection. We have used this approach in a majority of our studies, utilizing the high sensitivity of flame ionization detectors. These operate at atmospheric pressure following decompression and stream splitting. One part of the split stream goes to the detector and the remainder is vented through a back pressure regulator (APCO 1103-11) to keep the pressure at the splitter constant and, hence, provide constant flow to the flame detector.

The pressure reduction valves used for decompression were conventional high-pressure valves (High Pressure Equipment Co.) with special stems having a 1° taper (fabricated at our laboratory). Considerable difficulty was experienced in maintaining a steady flow; generally the flow tended to decrease with time, apparently due to Joule-Thompson cooling with attendant dimension changes in the valve, and possibly due in part to condensation at the orifice. Heating the valve was somewhat beneficial but did not completely overcome the difficulty. A new valve has been developed in this laboratory which overcomes the major problems (details will appear in a later publication).

When large samples were used, the coalescence of solute molecules, mentioned above, led to a visible fog as shown in Figure 2. Frequent plugging of the tubing and "spiked" peaks (Fig. 3) also occurred. The spikes are perhaps due to large clusters of molecules entering the flame.

One approach to eliminate the condensation problem would entail the use of pyrolysis conditions near the column exit preceding decompression. This method, currently undergoing evaluation, would hopefully cause the large molecules to break into small, volatile species which are stable after pressure reduction. Limited data recently obtained indicate this approach should be useful for many compounds.

D. General Apparatus and Methods

The valves and fittings used for high-pressure gas chromatography instrumentation have been supplied by the High Pressure Equipment Company. The connecting stainless steel tubing was .125 in. o.d. and either 0.02 or 0.04 in. i.d. Many of the columns were also made from this tubing.

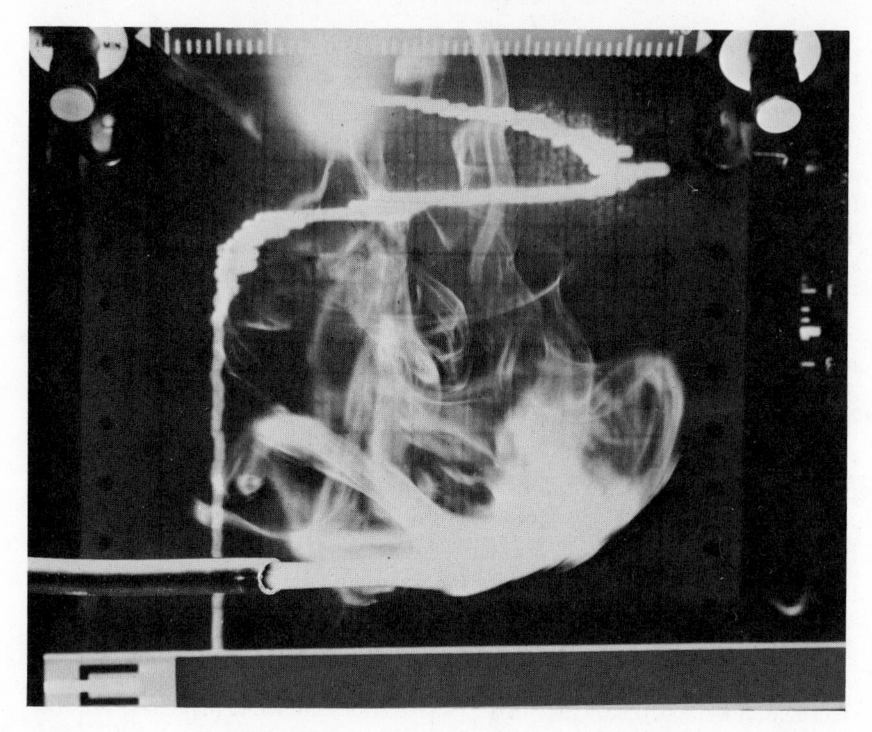

Fig. 2. Fog produced when the pressure on CO_2 containing squalane as a solute is abruptly decompressed to atmospheric pressure from 20,000 psi.

The small i.d. makes special care necessary in packing. The most effective procedure has been to introduce small amounts of packing material into a small funnel attached to the steel tubing, and then to tap the column repeatedly against a solid surface with the column in a vertical position. Columns made from .25 in. o.d. and 0.083 in. i.d. stainless steel were packed by conventional methods.

The choice of liquid stationary phase is limited in that many of the conventional materials migrate or "bleed" at high pressures, as mentioned earlier. However, these may be used at pressures below that at which they first migrate (i.e., their threshold pressure), as was done with SE-30 silicone gum rubber, Ucon 50, and Apiezon L. Of course, solid materials such as the Porasils and alumina give no bleeding problem. A third approach is to bond the liquid phase to the support surface. This has been suggested by Abel *et al.* (17) for high temperature work. These authors used a long-chain alkyl di- or trichloro silane which is reacted with the support, then treated with water vapor to polymerize the excess material into a coherent liquid

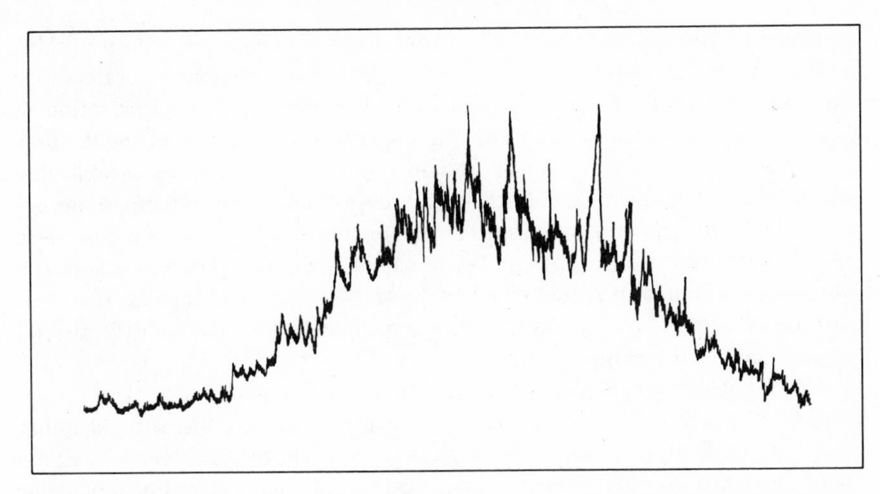

Fig. 3. Typical peak obtained with overly large samples using a flame ionization detector. The spikes are perhaps due to large clusters of solute molecules entering the flame.

phase. Our initial attempts with this particular type of support have not been too successful, but the concept appears promising. Halasz (17a) has recently introduced a similar material where other types of bonds to the silica surface are made. This material is currently being marketed under the trade name Durapak by Waters Associates. Our initial work with this material has been promising.

High-speed studies, such as those made in the turbulent region, must utilize components with minimal time constants. For such work we used a Beckman GC-4 electrometer (time constant 0.004 sec) coupled with a CEC 5-124 galvanometer recorder (capable of 1300 cps).

Pressure programming has received considerable attention in our work because of its potential for handling wide-ranging components in single runs. Both continuous and stepwise programs have been used. The former was achieved by continually raising the column pressure at the rate of 3000 psi per hour by means of a clock drive attached to the control gauge or the high-pressure regulator. In these cases the sample chamber was continually purged with valve 4 closed and valves 5 and 7 open. Stepwise programming was done by changing the pressure by various increments and allowing several minutes at pressure before sampling by the usual injection described earlier; this method was most sensitive. However, continuous examination of a sample mixture was not possible except with the continuous programming method.

Proof of migration is very important in these studies since chromatographic migration under such conditions is a new phenomenon. Recorder response was used here for the first indication of migration. Inspection of the sample was usually made at the conclusion of the experiment. This generally indicated sample loss and in many instances total sample disappearance. Chemical reactions could occasionally be imagined as responsible for such disappearance, but ordinarily the components were unreactive toward one another. We have also used IR confirmation on the condensed effluent in a few cases to show that chemical change does not ordinarily occur. Several exceptions were indicated by the greatly altered appearance of the sample.

Contamination can also be deceptive in high-pressure gas chromatography. Peaks have been traced to fingerprints, thread lubricants, cleaning solvents, and solutes incompletely removed in earlier experiments. Some polymers, such as polystyrene, may contain small amounts of low molecular weight solvents that are difficult to remove. These often yield spurious peaks. In such cases, the polymer was dissolved in carbon disulfide, and then the solvent was boiled off under reduced pressure at the glass point of the polymer.

The solvent power of the dense gas may result in unexpected problems. The choice of "O" rings in valve packing is critical, since neoprene and Viton are severely attacked by dense ammonia at 140°C. One Viton O ring placed in a sample chamber completely disappeared except for a small residue. Ethylene propylene rings resisted attack by ammonia but were affected by the high temperature and required frequent replacement. Teflon O rings have since proved to be much better, except the valves are stiffer in operation.

III. DISCUSSION

The careful consideration of the effects of pressure, temperature, and carrier gas for a particular separation problem may show a moderate pressure (a few hundred atmospheres) to be sufficient. Some of the problems mentioned earlier may thus be eliminated. Where high pressures are necessary, the choice of optimum conditions will yield more effective results.

Guidelines for choosing optimum operating conditions have been only partially formulated (16). The effect of such factors as column dimensions or flow rate is fairly well understood from general chromatographic theory (7), but the role of carrier gas and the pressure needed to exhibit a sufficient solubility for various compounds lacks clear definition at present. As a first approximation, the consideration of solubility in liquids and how it is affected by intermolecular forces may yield indications of the solution behavior of dense gases. Workers in liquid solubility usually consider four modes of interaction between molecules: dispersion (London) forces, dipole–dipole interactions, dipole-induced dipole interactions, and hydrogen bonding. These interactions collectively give rise to the cohesive energy overcome as a liquid becomes a gas. The relationship between this and solubility is represented in approximate form by the Hildebrand (18) solubility parameter

$$\delta = (\Delta E/V)^{\frac{1}{2}} \tag{1}$$

where ΔE is the energy of vaporization of the liquid and V is the volume per mole.

A number of workers have used this parameter as a criterion of polymer solubility, a subject of interest because dense gas chromatography has proven applicable to macromolecules. Solubility is indicated when the solubility parameters for both polymer and solvent are similar (usually within ± 1, with occasional variations as wide as ± 2.5). However, this simple consideration often fails and other characteristics must be examined.

The hydrogen bonding tendency, although roughly indicated in the solubility parameter, has an additional effect for some solutes and solvents. Pimentel and McClelland (19) classify solvents in the following way: proton donors (such as many highly halogenated compounds), proton acceptors (like ketones, olefins), simultaneous proton donors and acceptors (water, alcohols, etc.), and those with no hydrogen bonding. Often solvents are simply grouped as strong, weak, or non-hydrogen bonding. A finer measure is sometimes desirable for accurate formulation, and Gordy (20) has used a spectroscopic method relating the frequency shift of the OD bond of deuterated methanol in various solvents to the hydrogen bonding strength of the solvent. These data, coupled with the solubility parameter, gave greatly improved solubility predictions (21,22).

In some instances dipole moments also needed separate consideration as reported by Crowley, Teague, and Lowe (23), and by Hansen (24) and Gardon (25), although from a slightly different approach.

Unfortunately, the hydrogen bonding ability and solubility parameters of many of the potentially useful gases are unavailable. Solubility parameters can be found only for the liquid form of the gases. Giddings *et al.* (16) have suggested from consideration of van der Waals equation that the solubility parameter of dense gases could be approximated by

$$\delta = 1.25\, P_c^{\frac{1}{2}}[\rho/\rho_{\text{liq}}] \tag{2}$$

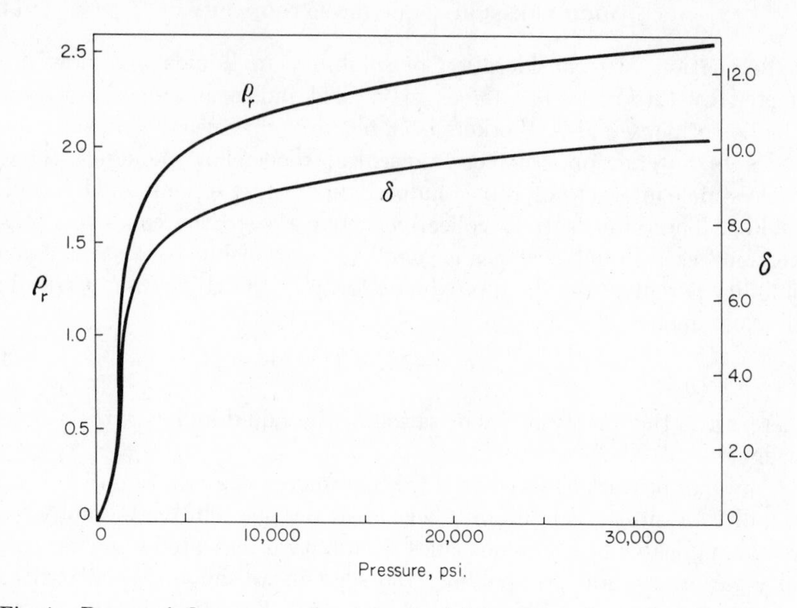

Fig. 4. Pressure influence on reduced density (ρ_r) (see the scale on the left ordinate) and the solubility parameter (δ) (see the scale on the right ordinate) of CO_2 at 40°C [calculated from $\delta = 1.25 P_c^{\frac{1}{2}} (\rho/\rho_{\text{liq}})$].

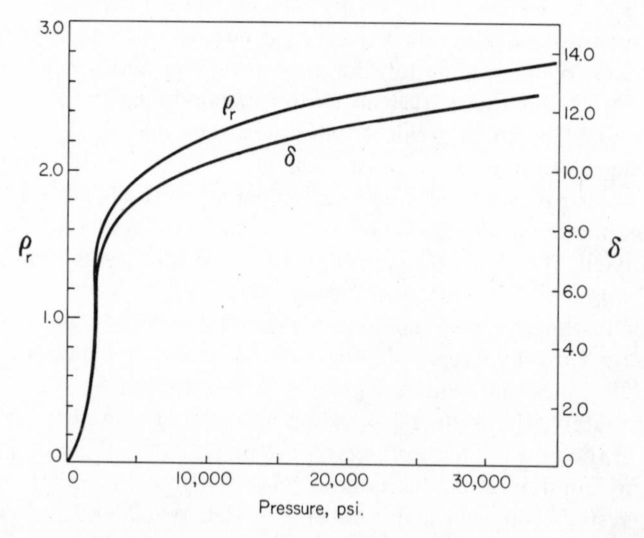

Fig. 5. Pressure influence on reduced density (ρ_r) (see the scale on the left ordinate) and the solubility parameter (δ) (see the scale on the right ordinate) of NH_3 at 140°C [calculated from $\delta = 1.25 P_c^{\frac{1}{2}} (\rho/\rho_{\text{liq}})$].

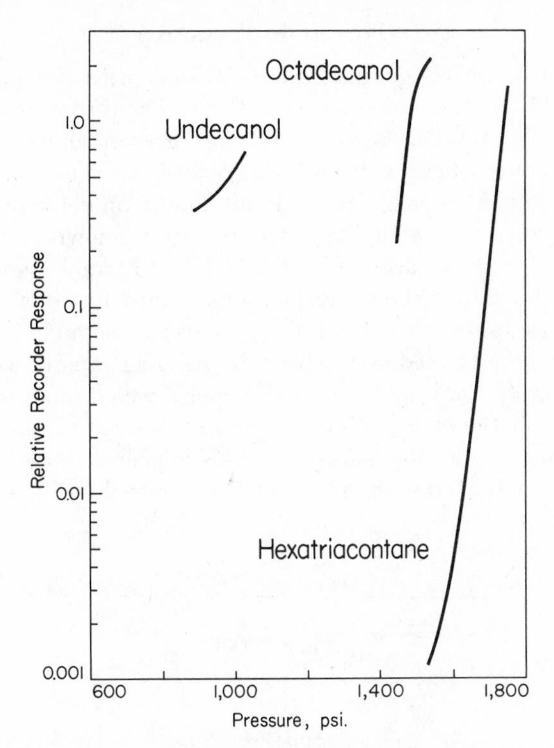

Fig. 6. Solubility increase as indicated by recorder response near the threshold pressure.

where δ is in $(\text{cal/cm}^3)^{\frac{1}{2}}$, P_c is the critical pressure in atmospheres, ρ is the density of the gas, and ρ_{liq} is the density of the parent liquid. The "state" effect (relative compression and temperature) is given by $[\rho/\rho_{\text{liq}}]$ while the remainder—$1.25\ P_c^{\frac{1}{2}}$—is associated with a "chemical" effect, depending on solvent hydrogen bonding tendencies, dipole interactions, and dispersion forces. The pressure-volume terms are not included, nor are solvents capable of hydrogen bonding with a solute but not with themselves considered. The latter may exhibit extreme departures as they do in liquid solubility studies.

The solubility parameters calculated from eq. (2) for CO_2 at 40°C and NH_3 at 140°C are shown in Figures 4 and 5, along with the reduced densities for these gases.

Our experimental studies have revolved mainly around carbon dioxide and ammonia. A number of other gases should be useful in high-pressure separations. Table II is a compilation of critical polarity and toxicity data for a few promising gases with critical temperatures below 200°C.

A. Threshold Pressures

The rapid increase of solubility (or volatility) with pressure has been noted before (15,16). As a consequence of this rather abrupt dependence, we can define a fairly unique *threshold pressure* for each solute-gas pair at a given temperature which indicates the lowest pressure for detectable migration. This parameter will depend only mildly on detector sensitivity.

The rapid increase in solubility with pressure is shown in Figure 6 for three substances: hexatriacontane ($C_{36}H_{74}$), octadecanol, and undecanol. The first two substances show a response increasing by 10 or more with a pressure increase of less than 100 psi. This effect is magnified because it is in the region where the reduced density is changing rapidly with pressure (Fig. 4). In Figure 7 we show response versus δ which more truly reflects the abruptness of the change. The undecanol does not exhibit such rapid increases in response as the other two. This response trend confirms our earlier deduction that the sharpness of the threshold will increase with molecular weight.

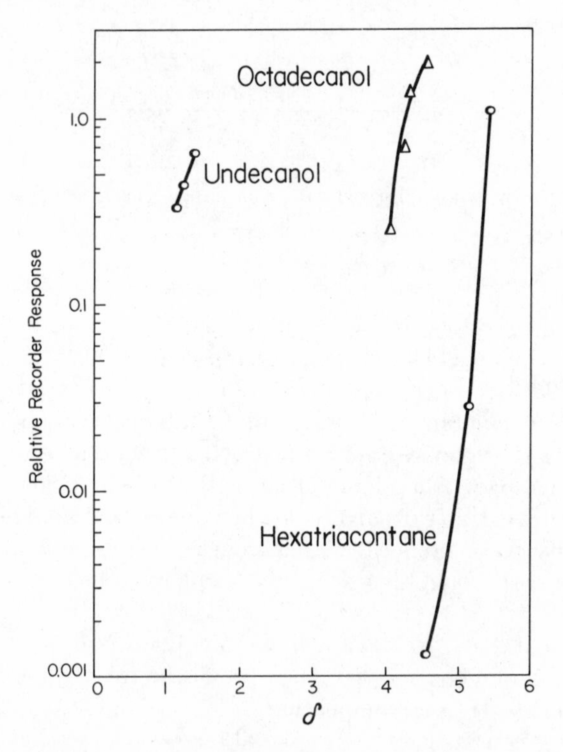

Fig. 7. Solubility increase (as indicated by recorder response) with increasing solubility parameter (δ) in the threshold region.

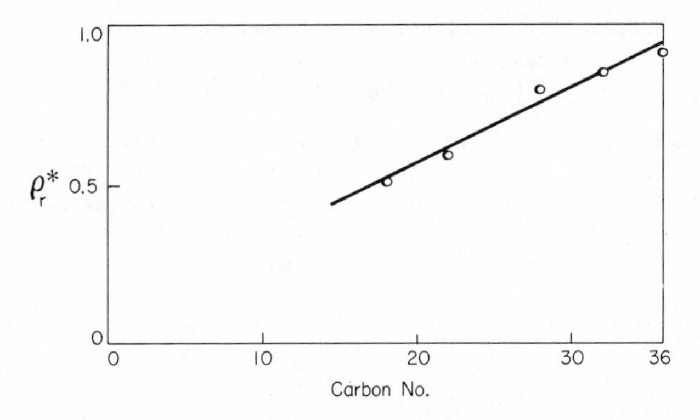

Fig. 8. Reduced densities at which migration begins (ρ_r*) for alkanes in the range C_{18}–C_{36}.

Reduced threshold densities for several *n*-alkanes in the range C_{18}—C_{36} are shown in Figure 8 with carbon dioxide as the compressed carrier gas. Octadecane with a solubility parameter of 8.0 migrates at a CO_2 solubility parameter value of 2.0 as calculated from eq. (2). By contrast the larger hexatriacontane (C_{36}) requires a solubility parameter of 4.5–5.0 for migration. The threshold density appears to show linear relationship to carbon number over the range studied.

Acknowledgment

We thank D. Claude Price for experimental work contributing to Figures 6, 7, and 8. This investigation was supported by Public Health Service Research Grant GM 10851-13 from the National Institute of Health.

References

1. J. H. Knox, *J. Chem. Soc.*, **1961,** p. 433.
2. J. C. Giddings, *Anal. Chem.*, **36,** 741 (1964); also in *Gas Chromatography*, **1964,** A. Goldup, Ed., Elsevier Publishing Co., Amsterdam, 1965.
3. M. N. Myers and J. C. Giddings, *Anal. Chem.*, **37,** 1453 (1965).
4. M. N. Myers and J. C. Giddings, *Anal. Chem.*, **38,** 294 (1966).
5. M. N. Myers and J. C. Giddings, *Separation Sci.*, **1,** 761 (1966).
6. J. C. Giddings, W. A. Manwaring, and M. N. Myers, *Science*, **154,** 146 (1966).
7. J. C. Giddings, *Dynamics of Chromatography, Part 1. Principles and Theory*, Dekker, New York, 1965.

7a. J. H. Knox, *Anal. Chem.*, **38,** 253 (1966).

8. J. B. Hannay and J. Hogarth, *Proc. Roy. Soc.* (London), **29,** 324 (1879).
9. J. S. Rowlinson and M. J. Richardson, *Advances in Chemical Physics*, Vol. II, Interscience, New York, 1959, p. 85.

10. H. S. Booth and R. M. Bidwell, *Chem. Revs.*, **44**, 447 (1949).
11. E. Klesper, A. H. Corwin, and D. A. Turner, *J. Org. Chem.*, **27**, 700 (1962).
12. S. T. Sie, W. van Beersum, and G. W. A. Rijnders, *Separation Sci.*, **1**, 459 (1966).
13. S. T. Sie and G. W. A. Rijnders, *Separation Sci.*, **2**, 729 (1967).
14. S. T. Sie and G. W. A. Rijnders, *Separation Sci.*, **2**, 755 (1967).
15. L. McLaren, M. N. Myers, and J. C. Giddings, *Science*, **159**, 197 (1968).
16. J. C. Giddings, M. N. Myers, L. McLaren, and R. A. Keller, *Science*, **162**, 67 (1968); J. C. Giddings, M. N. Myers, and J. W. King, *J. Chromatography Sci.*, **7**, 276 (1969).
17. E. W. Abel, F. H. Pollard, P. C. Uden, and G. Nickless, *J. Chromatog.*, **22**, 23 (1966).
17a. I. Halasz, Fifth International Symposium on Chromatography, Las Vegas, Nevada, January 1969.
18. J. H. Hildebrand and R. L. Scott, *The Solubility of Nonelectrolytes*, 3rd ed., Reinhold, New York, 1950.
19. G. C. Pimentel, and A. L. McClellan, *The Hydrogen Bond*, W. F. Freeman and Co., San Francisco, Cal. 1960.
20. W. Gordy, *J. Chem. Physics*, **7**, 93–99 (1939); *8*, 170–177 (1940); **9**, 204–214 (1941).
21. H. Burrell, *Offic. Dig. Feder. Soc. Paint Technol.*, **27**, No. 369, 748 (1955); *Interchem. Rev.*, **14**, 3, 31 (1955).
22. E. P. Lieberman, *Offic. Dig. Feder. Soc. Paint Technol.*, **34**, No. 444, 30 (1962).
23. J. C. Cowley, G. S. Teague, Jr., and J. W. Lowe, Jr., *J. Paint Technology*, **38**, 269 (1966).
24. C. M. Hansen, *J. Paint Technology*, **39**, 105 (1967).
25. J. L. Gardon, *J. Paint Technology*, **38**, 43 (1966).
26. R. C. Reed and T. K. Sherwood, *Gases and Liquids*, 2nd ed., McGraw-Hill, New York, 1966.
27. A. L. McClellan, *Tables of Experimental Dipole Moments*, W. H. Freeman and Co., San Francisco, Cal., 1963.
28. H. H. Landold and R. Burnstein, *Zahlenwerte und Funktionen*, Vol. I, pt. 3, Springer Verlag, Berlin, 1952.
29. N. I. Sax, *Dangerous Properties of Industrial Materials*, Reinhold, New York, 1957.
30. *Matheson Gas Data Book*, Herst Litho., Inc., New York, 1966.

Plasticizing Effect of Permeates on Membrane Permeation and Separation

NORMAN N. LI AND ROBERT B. LONG

Corporate Research Laboratories,
Esso Research and Engineering Company,
Linden, New Jersey

Polymeric membranes have long been used to separate hydrocarbon mixtures. The separation, however, cannot be predicted from the permeation data of single compounds, because the permeates plasticize the membrane (1,2,3). Plasticization of the membrane changes the solubility and diffusivity of the permeate. This paper reviews the results of such effect on membrane permeation and separation for both gas and liquid permeations.

I. THEORETICAL BACKGROUND

The passage of gas through a plastic membrane is usually considered to involve three independent physical phenomena (1,4,5) as follows:

1. Solution or sorption of the gas or vapor at one surface of the membrane.

2. Diffusion of the dissolved gas through the membrane.

3. Reevaporation or desorption.

Since the flow process is generally slow, the use of an equilibrium relationship between the concentrations of sorbed gas at the interfaces and the respective partial pressures is permitted. Henry's law can thus be assumed to apply:

$$C = Hp \tag{1}$$

Polymeric films in general can be regarded as interspersed crystalline and amorphous regions. The gas molecules are assumed to be soluble only in the amorphous region. They pass through the membrane by a diffusion process following Fick's law:

$$J = -D(dC/dX) \tag{2}$$

If D, the diffusion constant, is independent of concentration C and when the boundary conditions are:

$$X = 0, \qquad C = C_1$$

$$X = L, \qquad C = C_2,$$

Fick's law can be integrated to give

$$J = (D/L)(C_1 - C_2) \tag{3}$$

From eqs. (1) and (3)

$$J = \frac{D(H_1p_1 - H_2p_2)}{L} \tag{4}$$

If the Henry's law constant is assumed to be a function only of temperature and both membrane surfaces are at the same temperature, then

$$H_1 = H_2 = H$$

and

$$J = (DH/L)(p_1 - p_2) \tag{5}$$

The permeability constant P is defined as

$$P = DH = \frac{JL}{p_1 - p_2} \tag{6}$$

or

$$J = \frac{P(p_1 - p_2)}{L} \tag{7}$$

The fundamental postulation by Graham (68) that Henry's law applies to sorption and desorption, and that Fick's law applies to the diffusion process, has been well established (1,5,6) and is generally used as the basis for studies of permeation processes.

For unsteady-state permeation, the amount of permeates retained per unit volume of film is equal to the rate of change of concentration with time:

$$-dJ/dX = dC/dt \tag{8}$$

Combining eq. (8) with eq. (1) leads to the Fick's second law:

$$dC/dt = (d/dX)[D(dC/dX)] \tag{9}$$

A solution for the case of a finite solid with constant diffusion coefficient can be used to evaluate the total amount of permeates Q passing through the film from $t = 0$ to $t = t$:

$$Q = (DC_1/L)\cdot t - (C_1L/6) \tag{10}$$

Eq. (10) shows that Q increases linearly with t. If we now extrapolate back the linear portion of the Q versus t plot to the time axis, we obtain an intercept $t = \tau$ where $Q = 0$ and

$$D = L^2/6\tau \tag{11}$$

τ is known as the "time lag" and provides an experimental method for the determination of D.

When the plasticizing effect is significant, the integrated forms of Fick's first and second laws [eqs. (3), (7), and (10)] do not hold because the diffusion, solubility, and permeability coefficients all become concentration- or pressure-dependent.

II. PERMEATION OF SINGLE GASES AND VAPORS

A. Permeation of Organic Gases and Vapors

1. EXPONENTIAL RELATIONSHIP BETWEEN PERMEABILITY AND PRESSURE

The permeability of organic gases and vapors shows a complicated dependence on pressure and concentration by virtue of the strong interaction between solute and membrane. Early work on the pressure effect was carried out at low-pressure conditions, usually at the condition of 1 atm on the upstream side of the membrane and vacuum downstream side. The results show that exponential equations can be used to relate the solubility coefficient and the integrated diffusion coefficient to vapor activity (4). The plasticizing effect of hydrocarbons and Freon was investigated at elevated pressures by Li and Henley (7). They found that permeability is a strong function of pressure as illustrated in Figure 1 and expressed by the exponential equation:

$$P = P_o' e^{Ap} \tag{12}$$

The constant A is essentially a plasticizing factor because it shows the magnitude of the effect of permeate concentration on the mobility of the permeate in the plastic film. If A is high, then a small amount of permeate causes a large change in diffusivity. If it is low, a high concentration of permeate is required to obtain a small change in permeability.

It is important to note that the above equation only holds when Henry's law holds. This is because the equation is derived from eq. (15) on the assumption that the solubility coefficient is constant.

Combining the well-established eq. (13) with eq. (14), we obtain eq. (15)

$$D = D_o e^{AC} \tag{13}$$

$$P_o = D_o H_o \tag{14}$$

$$P = (H/H_o)\, P_o e^{aHp} \tag{15}$$

If the solubility coefficient is independent of pressure, then

$$(H/H_o)\, P_o = \text{constant} = P_o' \tag{16}$$

$$aH = \text{constant} = A \tag{17}$$

and eq. (15) is reduced to eq. (12). As discussed in detail later, for some gases, the solubility coefficient is independent of pressure approximately up to the critical pressure of the permeating gas. In this case, eq. (15) is applicable approximately up to the critical pressure of the permeate.

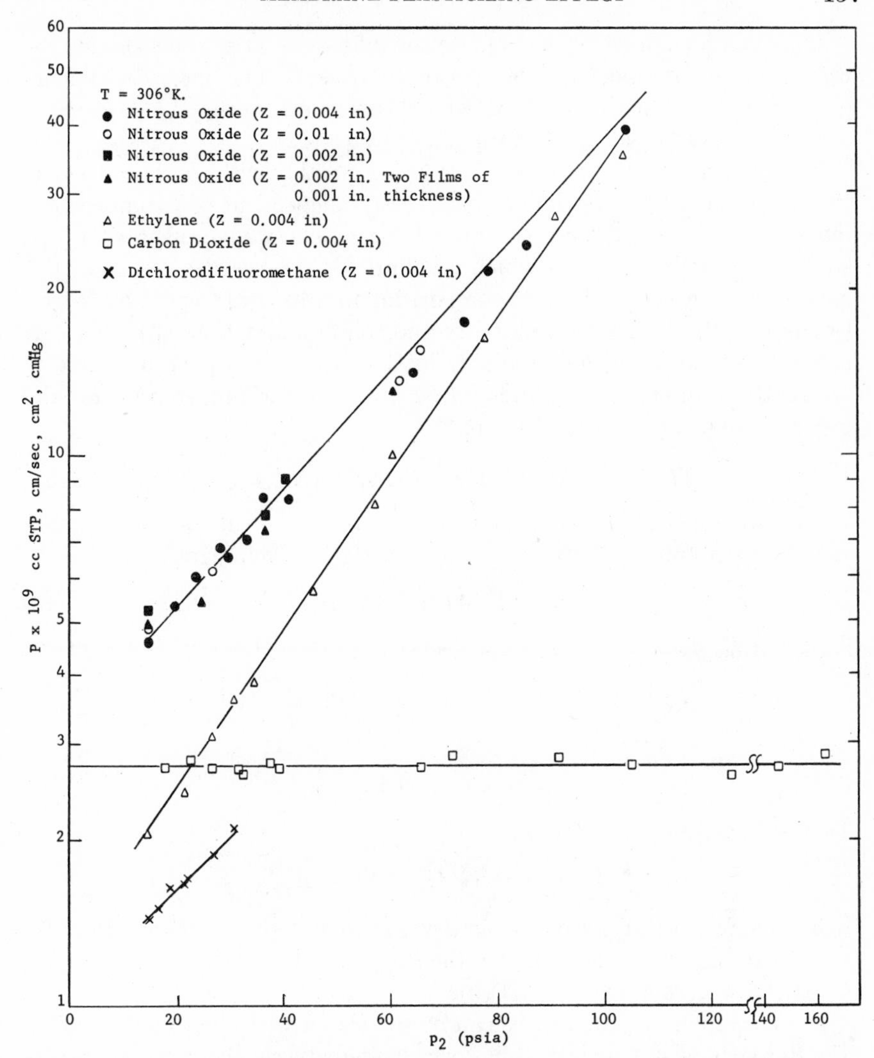

Fig. 1. Permeation constants of ethylene, nitrous oxide, carbon dioxide, and dichlorodiffluoromethane [Li and Henley(7)].

When the permeability coefficient varies with pressure, usually the diffusion coefficient also varies with pressure. This is because diffusivity appears to be more strongly affected by permeate concentration than solubility (1,8,9). However, Kishimoto and Matsumoto (10), as well as Mears (11), observed that the exponential concentration dependence of diffusivity applies only in the region of high concentration. This finding indicates that it is a dangerous practice to evaluate the diffusivity at zero penetrant concentration by extrapolating the plot of log D vs. C. The linear dependence of diffusivity on concentration which sometimes occurs at low vapor activity was discussed by Nishijima and Oster (12).

Since D varies with pressure and thus presumably with solute concentration in the film, the restriction of D being independent of C limiting the integration of eq. (2) becomes unrealistic at elevated pressures. A new equation must be obtained which accounts for the variation of D with C and C with pressure and solute composition for mixtures to permit integration of eq. (2). As data on solubility and solute interaction with polymers and other solutes become available in the future, it will be desirable to return to an integration of Fick's first law with the proper pressure and concentration dependence of D applied.

2. EVALUATION OF THE CONSTANTS P_o' AND A

To evaluate the characteristic constants P_o' and A in eq. (12), we substitute eqs. (6) and (12) into eq. (2) and carry out the integration:

$$J = (P_o'/AL)(e^{Ap_1} - e^{Ap_2}) \tag{18}$$

Substituting $p_1 = p_2 + \Delta p$ into eq. (18)

$$J = (P_o'/AL)\,[e^{A(p_2+\Delta p)} - e^{Ap_2}] = (P_o'/AL)\,(e^{\Delta p} - 1)e^{Ap_2} \tag{19}$$

Let $K = e^{\Delta p} - 1$, then

$$J = (P_o'/AL)\,Ke^{Ap_2} \tag{20}$$

Taking logarithm on both sides:

$$\text{In } J = \ln\,(P_o'K/AL) + A_{p_2} \tag{21}$$

If Δp is kept constant, a plot of ln J vs. p_2 should give a straight line. P_o' and A can then be evaluated from the slope and the intercept. Some of the P_o' and A values are given in Table I.

P_o' of nonpolar organic gases can be correlated with their boiling points (7). Recently, a similar correlation was made between the permeation rate and the number of carbon atoms in paraffin molecules (13). The Hildebrand solubility parameter has been used in correlating transport parameters (1,3,4,7,14,15). However, we usually have the problem of getting accurate values of the Hildebrand solubility parameter for permeates and polymer.

3. A MODIFIED FICK'S FIRST LAW

If the value of A is small, eq. (12) can be approximated by a linear equation:

$$P = P_o'(1 + Ap + \tfrac{1}{2}A^2p^2) \tag{22}$$

TABLE I

P_o' and A Values for Permeation through Polyethylene

	$A \times 10^2$	$P_o' \times 10^9$ $\left(\frac{\text{C.C. STP cm}}{\text{sec}\quad \text{cm}^2\quad \text{cmHg}}\right)$	Source
I. Unannealed Polyethylene (Visqueen); volume fraction of amorphous phase = 53%.			(7)
CH_4	3.55	0.845	
C_2H_4	3.28	1.12	
C_2H_6	3.43	1.30	
C_3H_8	2.65	2.30	
C_4H_8	2.31	3.60	
C_4H_{10}	2.21	3.32	
N_2O	2.31	3.40	
CCl_2F_2	2.21	0.955	
II. Annealed Polyethylene (Enjay)			(8)
CH_4	0.0321	0.190	
C_2H_4	0.0590	0.663	
N_2	0.0058	0.128	

Fick's law [eq. (2)] can then be written as

$$J = P_o'(\Delta p/L) + P_1(\Delta p/L) \tag{23}$$

It should be noted that P_o', the permeability at $p_2 = 0$, is independent of the pressure, and

$$P_1 = (P_o'/2)\, A(p_1 + p_2) \tag{24}$$

Therefore, at high pressures, gaseous permeation is considered to follow a modified Fick's first law in which the variation of permeability with pressure has been taken into account. This modified Fick's law can be written in the following forms:

$$J = (P_o'e^{Ap})\,(\Delta p/L) \tag{25}$$

or

$$J = (P_o' + P_1)\,(\Delta p/L) \tag{26}$$

In the second form P_1 may be called a second permeation constant. It is essentially a correction term for pressure. For organic gases which are very pressure dependent, that is, with high values of A, eq. (12) may be expanded to contain terms of order higher than 2 to give better accuracy in estimating permeability. For simple organic gases such as hydrogen, nitrogen, and oxygen, A will be zero and one will have the original Fick's first law.

4. THE EFFECT OF Δp AND THE CRITERIA FOR INTEGRATED P TO BE EQUAL TO DIFFERENTIAL P

As discussed previously, the permeability coefficient is found to be a function of pressure at high upstream and downstream pressures with constant Δp. It is expected that there should also be a variation in permeability when Δp is varied. However, for small Δp, such as 15 lb./sq. in., the permeability can, for all practical purposes, be considered independent of Δp (2,7). This finding has an important bearing on the data of permeability at various pressures because although integrated permeability is obtained at a certain Δp, it is actually equal to the more significant differential permeability (7). Mathematically this is illustrated as

$$\bar{P} = (1/\Delta p) \int_{p-1/2\Delta p}^{p+1/2\Delta p} P_o' e^{Ap}\, dp \tag{27}$$

Integrating eq. (27) we obtain

$$\bar{P} = P \frac{\text{Sinh } (A\Delta p/2)}{A\Delta p/2} \tag{28}$$

In order for the integrated permeability to be equal to the differential permeability, the condition requires that

$$A\Delta p < 0.5 \tag{29}$$

For a large Δp, it was found (2) that when Δp was increased, P decreased at constant p_1 and increased at constant p_2. In this case, keeping p_2 constant means increasing p_1 and the plasticizing effect on the upstream side of the membrane. This results in an increase of film swelling and permeability. On the other hand, keeping p_1 constant means decreasing p_2. This leads to a decrease of the plasticizing effect and film permeability (Fig. 2).

The effect of Δp on permeability was also investigated recently by Stern and Jobbins (16) with Δp varying over a wider range, up to 100 atm. The data were represented by an experimental equation similar to eq. (12).

$$\bar{P} = P_o'' \exp\,(m\Delta p) \tag{30}$$

It should be noted that using a high Δp may result in an appreciable film compression. This may in turn affect the permeability value calculated from eq. (7) because of the possible change in film morphology and film thickness. The former is still, more or less an unknown factor, although we might deduce some general ideas from the permeation data of films under stretching. The latter, however, can be readily calculated from film compressibility data.

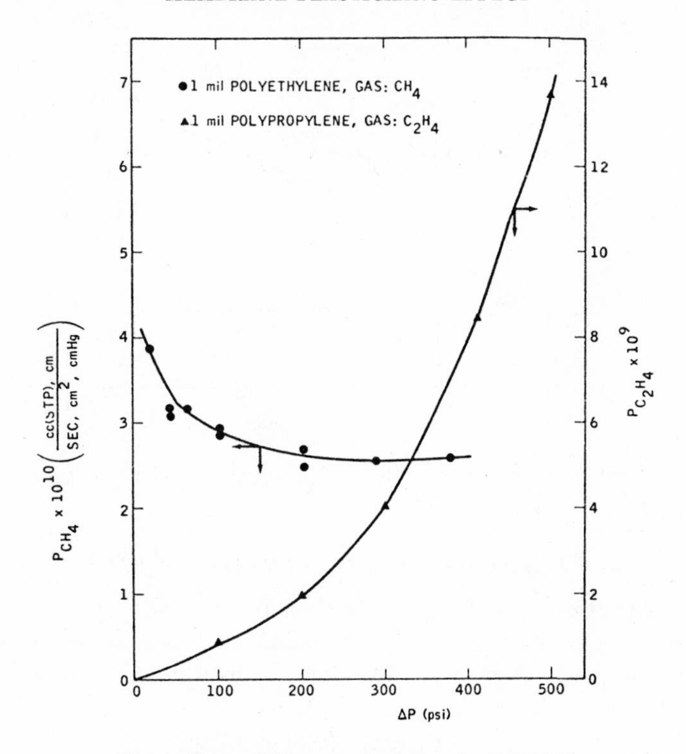

Fig. 2. Permeability as a function of ΔP [Li(2)].

When a large Δp is used, we may find unusual effects due to the structure of the supporting materials next to the film; Stern (16) reported that the permeability value decreased somewhat when the supporting material was changed from tissue paper to a porous metal plate. This might be mainly due to the film being pressed tightly against the holes in the metal plate, resulting in a reduction of film area available for mass transfer or in a change of diffusion path length.

5. SUBCRITICAL PRESSURE VERSUS SUPERCRITICAL PRESSURE

As mentioned before, the exponential relationship between permeability and pressure usually holds approximately up to the critical pressure of the permeating gas. Above the critical pressure, the permeability increases less rapidly with pressure as illustrated by the data of Figure 3 for methane and ethylene. At the present time, there are no general equations available that correlate data in both pressure regions.

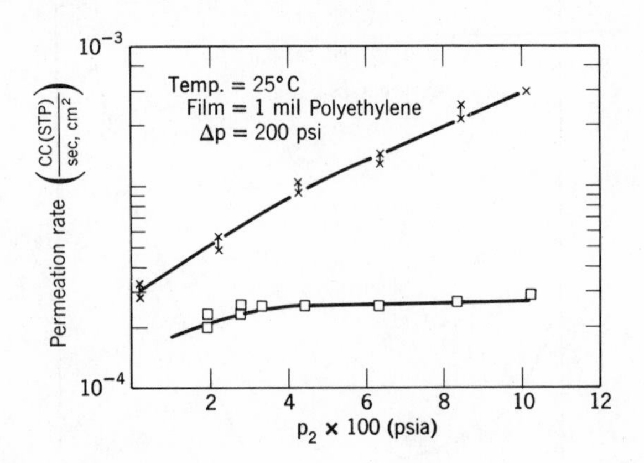

Fig. 3. Permeation of methane and ethylene through polyethylene [Li and Long(3)].

6. *PRESSURE PROFILE INSIDE A FILM*

The easiest way to calculate a hypothetical pressure profile inside a membrane is as follows: for a given set of data, namely J, p_1, and p_2, we can obtain P_o' and A as described before. Then substitute X for L and p_{2x} for p_2 in eq. (18):

$$J = (P_o'/AX)\,(e^{Ap_1} - e^{Ap_{2x}}) \tag{31}$$

Dividing eq. (31) by eq. (18) and rearranging terms, we obtain

$$\frac{X}{L} = \frac{e^{Ap_1} - e^{Ap_{2x}}}{e^{Ap_1} - e^{Ap_2}} \tag{32}$$

The hypothetical pressure, p_{2x}, was calculated from the above equation for each value of X/L. It should be noted that

$$p_{2x} = p_1 \qquad \text{when } X/L = 0$$

and

$$p_{2x} = p_2 \qquad \text{when } X/L = 1$$

The hypothetical pressure profiles calculated for methane and ethylene have small curvature. This is in contrast to the concentration profiles computed for liquid permeation as discussed later.

7. *EFFECT OF PRESSURE ON SOLUBILITY*

The effect of pressure on the solubilities of several hydrocarbon and inorganic gases in both subcritical and supercritical pressure regions was recently investigated (2,8). The solubility curves of all the gases studied can

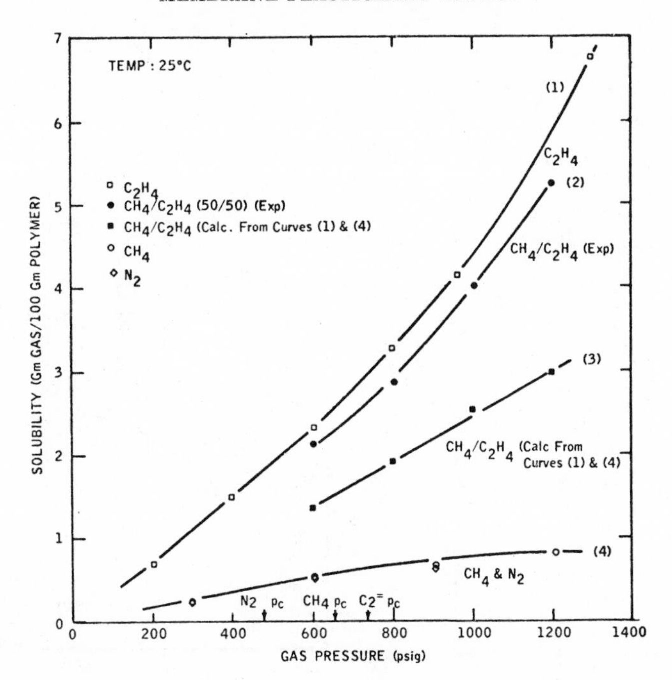

Fig. 4. Solubilities of methane, ethylene, nitrogen, and methane-ethylene mixture in polyethylene [Li and Long (3)].

be approximated by a straight line at least up to the critical pressure of the gases as illustrated by the results with polyethylene films (Fig. 4). This means that Henry's law holds roughly in the subcritical pressure region, whereas an appreciable deviation of it occurs at high pressures. Similar results were obtained with molten polyethylene (17,18).

From the various solubility curves summarized in Figure 5, we see that the solubility coefficient, which is the slope of the curves, may be correlated with concentration, depending on the shape of the curve, by the following equations:

$$H = H_{o_1}e^{\alpha p} \tag{33}$$

$$H = \beta H_{o_2}/(\beta + p) \tag{34}$$

$$H = H_{o_3}{}^{-\alpha p} \tag{35}$$

$$H = H_{o_4} \tag{36}$$

where α and β are correlation constants. It should be noted that it usually is easy to do curve fitting, but to obtain correlation equations with general

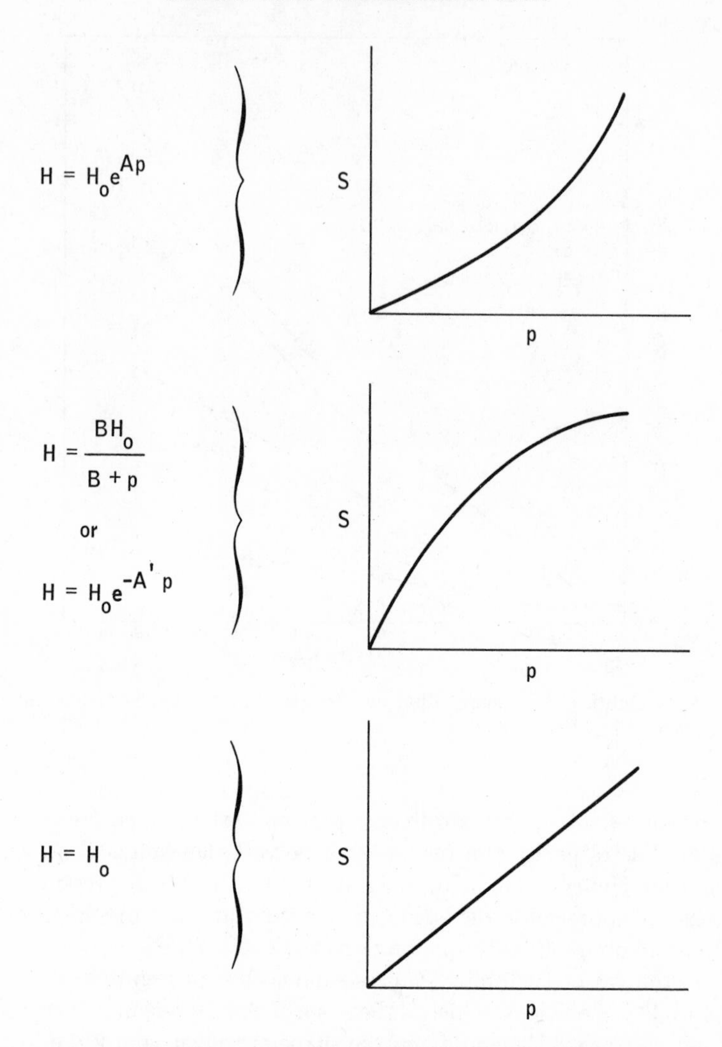

Fig. 5. Pressure effect on solubility coefficient.

correlation constants applicable to all cases is a different matter. For example, we can use eq. (34) to correlate many concave solubility curves, and thus obtain different values of β and H_o for different curves. It is then desirable to study the behavior of β and H_o to find ways to predict their values at different conditions.

The corresponding state principle was recently proposed by Stern, *et al.*

(9) to correlate H_o in eq. (33). Specifically, log H_o is correlated as a function of $(Tc/T)^2$:

$$\log H_o = -5.64 + 1.14\,(Tc/T)^2 \tag{37}$$

This was tested on various data with varying degrees of success. It should be noted that the critical temperature as well as critical pressure of a gas may change when the gas in a free state is adsorbed onto a certain surface or absorbed into a material. The correlation would be better if such a change of critical properties could be introduced into the equation when more understanding is generated about the condensed state of permeates inside a polymer.

A parameter p_s was defined (9) as the pressure at which a 5% deviation from Henry's law of solubility occurs. p_s is derived from the exponential relationship between solubility and concentration. It therefore cannot be used in the cases where other solubility-pressure relationships apply. Anyway, it seems that the 5% criterion is too stringent because it is quite common in polymer sorption studies for data scattering to be larger than 5%. Besides, we have found that sometimes due to the difficulty in applying uniform thermal treatment to a sheet of polymeric film, even the data from two pieces of film cut from the same polymeric sheet may differ much more than 5%.

8. BEHAVIOR OF THE DIFFUSION COEFFICIENT

The diffusion coefficient at high pressure can be obtained from the available data of permeability and solubility. The diffusion coefficient of many organic gases and vapors at low pressures, that is, 1 atm upstream pressure and near zero downstream pressure were measured by Stannett, *et al.* (1), Rogers (19), Richman and Long (20), Kokes and Long (21), Laurence and Slattery (22), and others. In all of these cases, the diffusion coefficients are strongly dependent on concentration. In addition, better solvents were found to plasticize the polymer more strongly, giving a more marked dependence of transport parameters upon concentration.

Kuppers and Reid suggested that a diffusion coefficient based on an activity gradient might not vary with concentration (23). Morrison (24) later found experimentally that the diffusion coefficient based on an activity gradient for benzene-polyvinyl acetate system does not vary as significantly with concentration as the Fick's diffusion coefficient. However, both kinds of diffusion coefficients are quite concentration dependent for the systems of acetone and propylamine diffusing through polyvinyl acetate.

9. EFFECT OF SUBJECTING POLYMER TO STRESS ON TRANSPORT PARAMETERS

If the absorption of vapor or gas causes the polymer to swell, then the configuration of the polymer molecules will change. These configurational changes, governed by the retardation time of the polymer molecular chains, are not instantaneous. Mears studied the diffusion of vapors in polymers to assess the separate contributions of concentration dependence and time dependence to the observed behavior of diffusion coefficient (25). He showed that this could be done by studying the time lag measured in unsteady-state permeation. Usually the stress results in an initial rapid permeation. The reason appears to be that when a vapor comes in contact with a membrane, it instantly swells the surface layer of the membrane. This creates a stress which renders the underlying membrane temporarily more permeable until the equilibrium density is restored by a redistribution of the polymer segments.

Stress can be imposed semipermanently on a membrane by stretching it to a high degree. Stretching of polyolefins to draw ratios equal to or greater than 500% was found to lead to substantial decrease of permeability to organic liquids by Bixler and Michaels (26) and also to organic gases by Michaels, *et al.* (27). Likewise, Davis and Taylor (28) found that the diffusion coefficient decreased considerably and the activation energies increased greatly with increasing draw ratio for the sorption of dyes into Nylon 66. Peterlin *et al.* (29) recently found that, for drawn polyethylene, the diffusion process was characterized by a drastically reduced D_o, high activation energies, and a large concentration dependence. The solubility process was also affected but to a smaller extent. For example, at 25°C, the diffusion coefficient is about 150 times larger in the undrawn membrane compared to the drawn, whereas the solubility is only six times greater. They also observed that the plasticizing effect of sorbed molecules enhances the relaxation of the polymeric molecular chains, which results in a gradual increase of solubility with an increasing number of sorption cycles (Fig. 6). However, regardless of how many sorption cycles that are used, we would expect to find a residual effect of configurational change, because the high-degree of stretching causes some permanent deformation of the polymer.

10. EFFECT OF TEMPERATURE

According to Mears (11), Fujita (30), Li and Henley (7), and Li and Long (8), the exponential relationship between concentration and diffusivity usually holds at low temperatures and the apparent activation energy for diffusion decreases with increasing temperature. Data on CO_2 and C_2H_4 shown in Table II indicate that, at 25°C, the Arrhenius relationship holds

TABLE II

Arrhenius Relationship for CO_2 and C_2H_4 Permeating through Polyethylene Film at Sub- and Super-Critical Pressures

$$P = P_o'' e^{-E_p/RT}$$

Gas	E_p (Kcal/g-mole)	$P_o'' \times 10^5$ (C.C. STP, cm/sec, sq. cm, cmHg)	Highest Pressure Investigated (psig)	Temperature Range Investigated (°C)	Source
CO_2	5.98	5.63	1300	15.1 to 40.0	(7)
C_2H_4	5.07	3.84	870	25.0 to 48.0	(8)

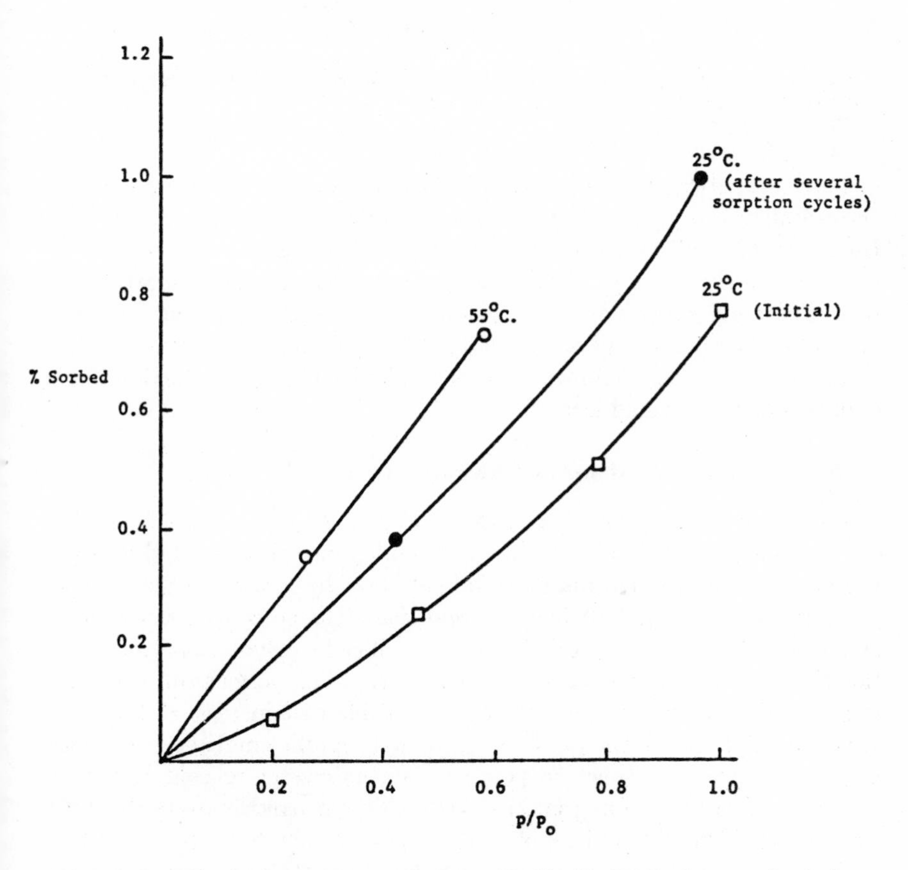

Fig. 6. Sorption isotherms for methylene chloride in 900% drawn polyethylene [Peterlin *et al.*(29)].

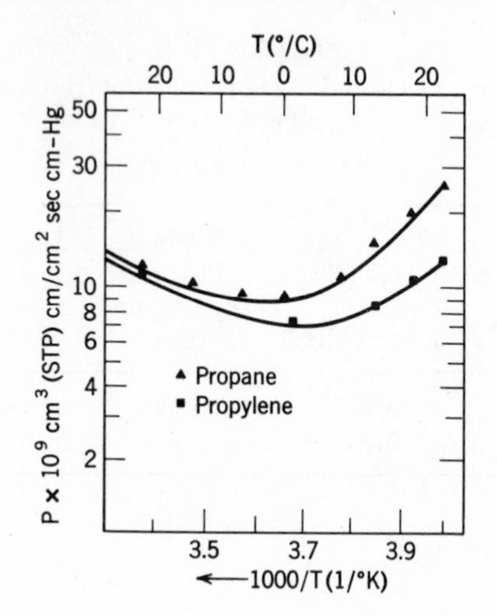

Fig. 7. Permeability of propane and propylene as a function of temperature [Henley and Dos Santos (31)].

in both subcritical and supercritical pressure regions. Similar results on permeability were reported by Stern and Jobbins (16). Rogers (19) and Henley and Dos Santos (31) have also noted rather abrupt changes in the temperature dependence of the transport process around the structural transition temperatures of the polymers (Fig. 7). Such abrupt variations in diffusivity, solubility, and permeability are mainly due to the changes in the microstructure of the polymer. Similar results have been obtained in liquid permeation as discussed later.

B. Permeation of Inorganic Gases and Vapors

Water vapor was observed to have a strong plasticizing effect. The equilibrium sorption isotherms of hydrophilic films such as ethyl cellulose were determined by Yasuda and Stannett (32) to be a function of vapor pressure. The diffusion coefficients were found to be concentration independent. Their work on the effect of emulsifier in polymer provides additional information about the diffusion process. The permeation constants were found to remain unchanged despite considerable increase of the water vapor sorption due to the presence of the hydrophilic emulsifiers. It would appear that the emulsifiers in polymers act as clustering centers for the water vapor sorption, but that the clusters do not participate in the diffusion process. The diffusivities of water vapor through nylon and cellulose, both hydrophilic films, were reported, by Stannett and co-workers (1), however, to increase with increasing concentration.

The work of Thornton, Stannett, and Szwarc (33) has furnished further evidence of clustering of water molecules. It was found that the permeation rate of liquid water through Mylar, a hydrophilic polymer, is greater than that of water vapor. This is interpreted as due to the clustering effect when saturation is approached.

The sorption of water vapor at high relative humidity is normally described by solution theory. Kwai discussed his work from this aspect (34) and pointed out the significance of the elastic contribution of polymers to the sorption isotherms.

Clustering of penetrant molecules in polymers is now generally regarded as the main cause for the nonideal sorption and diffusion behavior of water vapor. Clustering refers to the phenomenon of permeate molecules associating with their own kind within the polymer structure. This occurs when the first solvent molecule enters a polymer, loosening its structure and making it easier for subsequent molecules to come to the neighborhood of the first molecule rather than to go elsewhere.

A "clustering function" was derived by Zimm and Lundberg (35) to express the clustering tendency of the penetrant molecules.

$$G_{11} = (1/v)\iint[F_p(i \cdot j) - 1]\, di\, dj \tag{38}$$

The entire term in the integral is the probability of two permeating molecules, i and j, occupying the positions specified by the coordinates $(i \cdot j)$. It has also been shown

$$G_{11}/v_{pe} = -\psi_{po}(\partial \gamma_{pe}/\partial a_{pe}) - 1 \tag{39}$$

The quantity $\psi G_{11}/v$ is the mean number of permeate molecules in the neighborhood of a given permeate molecule; thus it measures the clustering tendency of the penetrant. In an ideal solution,

$$G_{11}/v_{pe} = 1 \tag{40}$$

By definition, poor solvents have high clustering functions while good solvents have low clustering functions.

Direct experimental evidence for the existence of penetrant clusters in polymers was provided by Barrer and Barrie (36) who measured the light scattering by ethyl cellulose films at high relative humidities, by Veith (37) from the studies of the dielectric property of polystyrene film, and by Mayne (38) from observing globules of water in polymeric films through microscopes.

It is interesting to compare the permeation constants of carbon dioxide and nitrous oxide from the viewpoint of plasticizing effect. Although both of these gases have the same molecular weight and similar molecular forms, the early work by Li and Henley and recent work by Stein and Jobbins all show that the nitrous oxide permeability varies with pressure, while carbon dioxide maintains a constant permeability at least in the low-pressure range such as from 17.7 to 162.5 psia. Moreover, nitrous oxide gives a much higher permeability than carbon dioxide at the same pressures. It seems that, because nitrous oxide is polar while carbon dioxide is nonpolar, it may be easier for nitrous oxide to be adsorbed on the polymer molecules and to exert the plasticizing effect. Also, adsorption depends on concentration or pressure; therefore, the permeability of nitrous oxide is pressure dependent and greater than that of CO_2.

III. PERMEATION OF SINGLE LIQUIDS

A. Concentration Dependence of Diffusivity

The liquid permeation refers to the process in which a liquid permeates through a membrane into a vacuum space on the other side (5,39). The steady-state permeation process may be described by Fick's first law. Unfortunately, for the permeation of organic compounds through hydrophobic films, the value of D usually depends very strongly on the solvent concentration in the plastic film. Many expressions have been proposed to relate D to the solubility of solvent in the film and to a diffusivity D_o obtained at zero concentration of solvents. The equations most widely accepted (1,3,40,41) are

$$D = D_o(1 + ac) \tag{41}$$

and

$$D = D_o e^{aC} \tag{42}$$

where D_o and a are constants at a given temperature. Eq. (42) holds for the case where D is more strongly concentration-dependent.

Substituting eq. (42) in eq. (2) gives the steady-state permeation rate for a single component

$$Q = (D_o/aL)\,[e^{aC_1} - e^{aC_2}] \tag{43}$$

in terms of the concentration of the permeating material in the plastic film at the upstream (C_1) and downstream (C_2) sides, the diffusivity D_o, the constant a, and the film thickness, L. The values of D_o and a for a number of permeate-polymer systems are summarized in Table III.

B. Effect of Crystallinity on the Plasticizing Constant

The value of a increases with crystallinity (42) as shown in Fig. 8. This explains the frequently encountered difficulty in reproducing permeation

TABLE III
Diffusion Constants for Hydrocarbons in Polymers

Hydrocarbon	Temperature (°C)	Diffusivity (Do cm²/sec)	Exponential Factor (a)	Crystallinity of Film (%)	Source
Polypropylene					(42)
n-Heptane	0	4.3×10^{-11}	87	56.2	
	25	2.3×10^{-10}	58	53.4	
	53	6.5×10^{-9}	33	51.9	
Methylcyclohexane	0.5	1.9×10^{-10}	49	50.3	
	26	1.6×10^{-10}	33	49.4	
	53	3.0×10^{-9}	20	46.7	
Toluene	−0.5	1.3×10^{-10}	71	55.2	
	24	1.8×10^{-9}	40	53.5	
	50	9.5×10^{-9}	22	51.4	
Polyethylene					(41, 67)
Benzene	23	3×10^{-8}	4.0	70 (DYNK)	
	23	4×10^{-9}	5.0	90 (Marlex)	
	25	11.8×10^{-9}	4.0	70	
n-Hexane	23	2×10^{-8}	5.0	70	
	23	3×10^{-9}	6.0	90	
	25	6.4×10^{-9}	5.0	70	
n-Octane	25	4.4×10^{-9}	4.0	70	
n-Decane	25	4.8×10^{-9}	4.0	70	
3-Methylpentane	25	4.8×10^{-9}	4.5	70	
Neohexane	25	3.4×10^{-9}	5.0	70	
Cyclohexane	25	4.8×10^{-9}	5.0	70	
Hexene-2	25	8.3×10^{-9}	4.0	70	
Carbon tetrachloride	25	3.7×10^{-9}	4.5	70	
p-Dioxane	25	5.1×10^{-9}	3.0	70	

data. For example, small changes in the crystallinity or, conversely, in the amorphous content of the polymer, will have large effects on the concentration of solvent in the polymer and, thus will have very large effects on the diffusivity. Therefore, it should be difficult to get two samples of polymer of the same general type to give the same permeation rates. Experience has shown that the permeation rates vary considerably for samples of polymer taken from different portions of the same roll of plastic film.

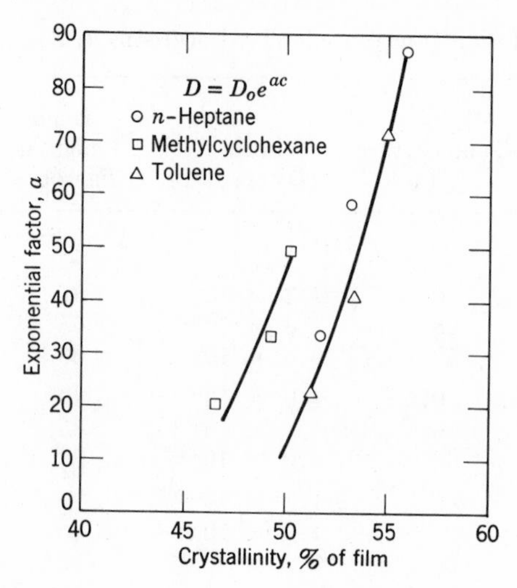

Fig. 8. Variation of exponential factor, a, with film crystallinity [Long (42)].

C. Effect of Temperature on the Plasticizing Constant

The values of a show that the effect of concentration on diffusivity decreases quite rapidly as temperature is increased; this is, however, expected since the D_o values, or diffusivities at zero concentration, increase very rapidly as temperature is increased. Therefore, we would expect the effect of further plasticizing by the presence of permeate to be less as temperature is raised.

D. Concentration Profiles through a Membrane

With a knowledge of the value of a and D_o, the effect of distance through the film on solvent concentration for steady-state permeation can be calculated (42,43). The results show that the solvent concentration on the upstream side of the film is nearly the overall bulk concentration of the solvent film. However, the concentration decreases very slowly wirh distance through the film to about 90% of its upstream value at a distance 75% of the way through the film. From there on to the downstream side of the film, the concentration drops very rapidly to a value depending on pressure. At higher temperatures the concentration profiles are similar but drop a little faster with distance through the film. Figure 9 shows such a distribution of permeate concentration as a function of film thickness.

In contrast to the concentration, the point diffusivity, D, is very large compared to D_o over the upstream 90% of the film thickness. From then on

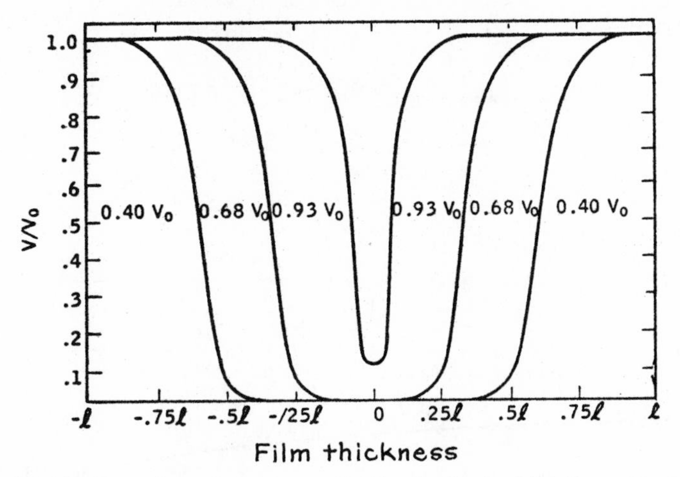

Fig. 9. Distribution of penetrant concentration in sheet of thickness 21 at various times [Buckley and Berger (43)].

to the downstream side of the film, this diffusivity drops very rapidly and almost linearly to the value of D_o or very near to it, depending on downstream concentration. Thus, essentially all the resistance to diffusion occurs near the downstream side of the film and the concentration profile resembles the two-zone process model proposed by Binning (39).

The slope of the concentration curve $\Delta C/\Delta L$ is the driving force for permeation at any point in the film and is thus inversely proportional to the diffusivity at a given temperature. However, the shape of the curve is independent of film thickness if plotted on a basis from $L = 0$ to $L = 1$ where $L = 1$ is the distance to the downstream side of the film.

The concentration profiles calculated by Long (42) are very similar to those which have been experimentally observed by Richman and Long (20) for the diffusion of methyl iodine into polyvinyl acetate. They are, however, steeper at the downstream side than those recently measured by Kim and Kammermeyer for the systems of Nylon 6-water, nylon-dioxane, cellulose acetate-water and polyethylene-dioxane (44). This may be mainly due to the difference in the values of a for different systems.

In gas permeation, the concentration profiles are even flatter, as shown before, indicating that the resistance is quite uniform throughout the entire film. The difference between liquid permeation and gas permeation is not in kind but rather in degree. The large difference between the concentration profiles calculated for liquid permeation and for gas permeation is merely due to the fact that the upstream and downstream concentrations in liquid permeation can be very different, resulting in concentration profiles of large curvature; whereas the two end concentrations in gas permeation

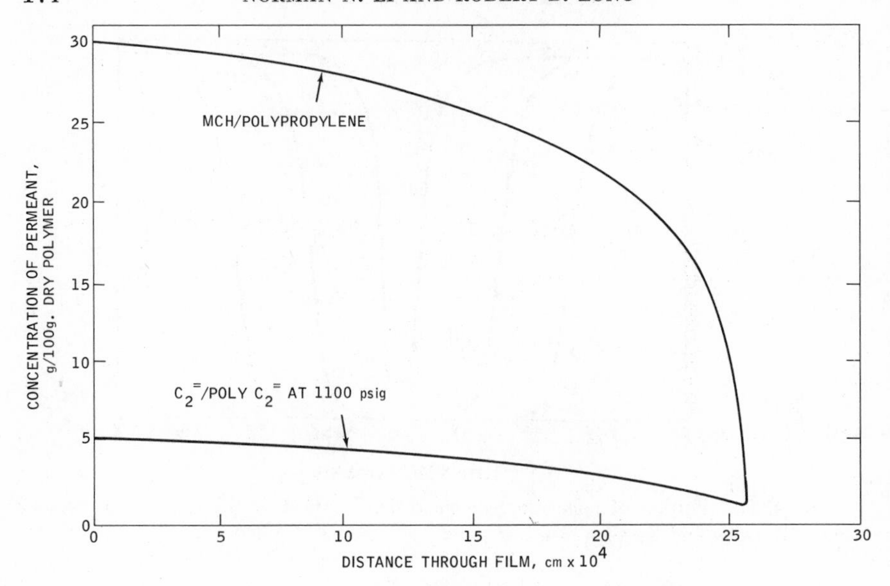

Fig. 10. Illustrative comparison of liquid and gas permeation.

are quite small, resulting in almost flat concentration profiles. These are compared graphically in Figure 10.

E. Effect of Downstream Concentration on Permeation Rate

From eq. (43), it is evident that as the permeate concentration at the upstream side of the membrane is increased, the permeation rate will increase far more rapidly than linearly with the concentration.

Therefore, a membrane exposed to a vapor at 95% of saturation will show a three- to fivefold lower permeability than the same membrane exposed to saturated vapor or liquid at the same temperature (42). Based on the same reasoning, for downstream concentrations up to 60% or more of the upstream concentrations, the permeation rate will be essentially unaffected; that is, reduced less than 1%. This is the fact that has led to the general speculation that permeation rate is independent of downstream pressure. As illustrated in Figure 11 when the downstream pressure is close to saturation, its effect on permeation rate is strongly felt.

F. Different Kinds of Diffusivity Reported in Literature

There are several kinds of diffusivity reported in the literature. This may sometimes cause confusion. A discussion of the most common ones is

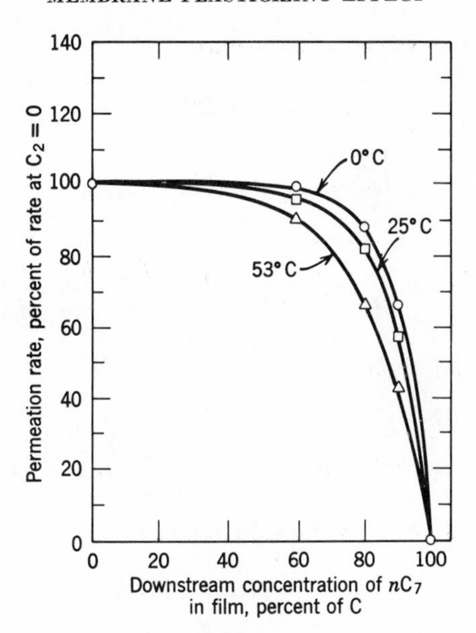

Fig. 11. Effect of downstream concentration on permeation of *n*-heptane through polypropylene film [Long (42)].

given below since no discussion of this kind has hitherto appeared in any review papers on membrane permeation.

At steady state, dynamic situation, the diffusion coefficient measured or calculated from permeation data of gas or liquid is the so-called intrinsic diffusion coefficeint, whereas the diffusion coefficient at the equilibrium state calculated from the rate of weight increase of the polymer sample includes mass flow. The mass flow occurs because the volume of the polymer sample, even if it can be swollen, is more or less restricted and the diffusion of one compound may produce nonhomogeneous concentration distribution, which results in mass flow. There is another kind of diffusion coefficient, called thermodynamic diffusion coefficient. It is related to the intrinsic diffusion coefficient by eq. (44):

$$(D_A)_T = D_A[(d \ln C_A)/(d \ln a_A)] \tag{44}$$

This relationship is derived as follows. The force F per mole of substance is

$$F = -(d\mu/dx) \tag{45}$$

The rate of transfer of compound A due to the force F is

$$FC_A/f = (C_A/f)(d\mu/dx) \tag{46}$$

where f is the resistance coefficient. Since

$$\mu_A = \mu_{oA} + RT \ln a_A \tag{47}$$

we have

$$d\mu_A/dX = RT[(d \ln a_A)/dX] \tag{48}$$

Therefore,

$$\frac{FC_A}{f} = \frac{C_A}{f} RT\left(\frac{d \ln a_A}{dX}\right) = \frac{C_A}{a} \frac{RT}{f} \frac{da_A}{dX} \tag{49}$$

If dC is used instead of da, then

$$\frac{FC_A}{f} = \frac{C_A}{a_A} \frac{RT}{f} \frac{da_A}{dC_A} \frac{dC_A}{dX} = \frac{RT}{f}\left(\frac{d \ln a_A}{d \ln C_A}\right)\frac{dC_A}{dX} \tag{50}$$

But the transfer rate can also be expressed as

$$J = -[D_A(dC_A/dX)] \tag{51}$$

Comparing eqs. (50) and (51), we obtain

$$D_A = \frac{RT}{f} \frac{d \ln a_A}{d \ln C_A} \tag{52}$$

$$\text{Define } (D_A)_T = RT/f \tag{53}$$

then

$$D_A = (D_A)_T \frac{d \ln C_A}{d \ln a_A} \tag{54}$$

The thermodynamic diffusion coefficient can therefore be calculated from eq. (54) when the intrinsic diffusion coefficient is known. It can also be evaluated from the following equation as discussed by Garrett and Park (45) and Wilkens and Long (46).

$$\ln (D_A)_T = \ln D_o + A'V - BV^2 \tag{55}$$

Similar equations were discussed by Chalykh and Vasenin (47).

Because of the complicated relationship between concentration and activity, the shape of the curve of $(D_A)_T$ can be quite different from that of D_A. If we plot $(D_A)_T$ versus V, a maximum value of $(D_A)_T$ will occur. This can be predicted from eq. (55) because the value of $(D_A)_T$ will first increase then decrease when the value of V becomes large enough to make the third term in the equation more dominating over the second term.

G. Effect of Solvent on Permeation Rate

As discussed before, the permeation rate of a compound through a film can be increased by using a solvent to swell the film. This is shown for liquid permeation by Coughlin and Pollack (48). The permeation of toluene through the films was enhanced by first swelling the films with various swelling agents as shown in Figure 12. There also appears to be a rough correlation between permeability of toluene and the boiling points and vapor pressures of the swelling agents.

H. Effect of Polymer Structural Change

The Arrhenius plots for saturated hydrocarbons permeating through polyolefin films show an abrupt change in the permeation rate curve near the glass transition temperature (8,42). Some typical data are shown in

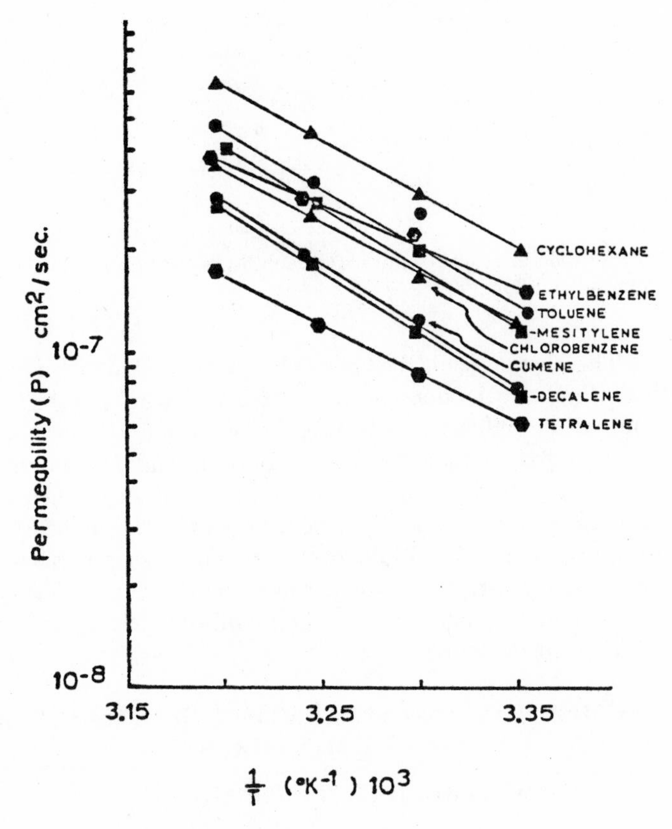

Fig. 12. Log_{10} (P) versus $1/t$ [Coughlin and Pollack (48)].

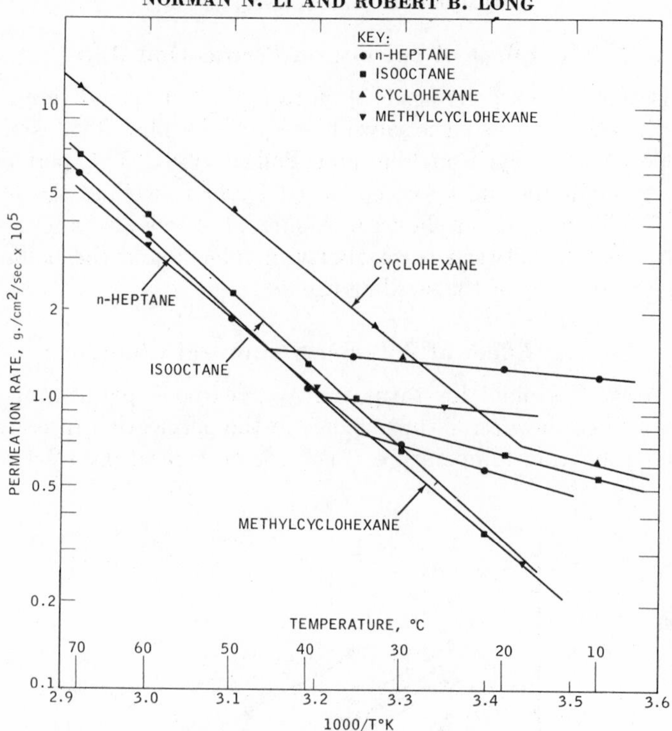

Fig. 13. Permeation rates for paraffins and naphthenes through 1 mil annealed polypropylene film [Li and Long (8)].

Figure 13. The change could be more abrupt than that observed in gas permeation. Similar behavior has been observed with oxygenated compounds and alkyl halides permeating through polystyrene, polyvinyl acetate, and methyl methacrylate (49). Both E and D_o undergo a significant change around the transition point of the polymer when the polymer changes from solid to the highly elastic state. The higher Arrhenius slopes at higher temperatures indicate that the increase in permeation rate with temperature must involve something more than diffusion. Some polymer-solvent interaction must also be involved, such as increasing the amorphous content of the polymer.

IV. PERMEATION AND SEPARATION OF GASEOUS AND LIQUID MIXTURES

A. Permeation Rate of Mixture

In general it is easier for a slightly soluble material to permeate through a polymeric film if a more soluble material with which it is completely miscible is also present in the film (2,8,13,42). Therefore, the permeation rate of a 50/50 weight mixture is equal to that of the faster permeating component of the mixture. This is illustrated by the permeation of mixtures of toluene-methyl cyclohexane, thiophene-benzene, and toluene-heptane through polypropylene as shown in Figure 14.

It is, however, not necessary that the plasticizing actions of permeates always supplement each other to result in an enhancement of permeation

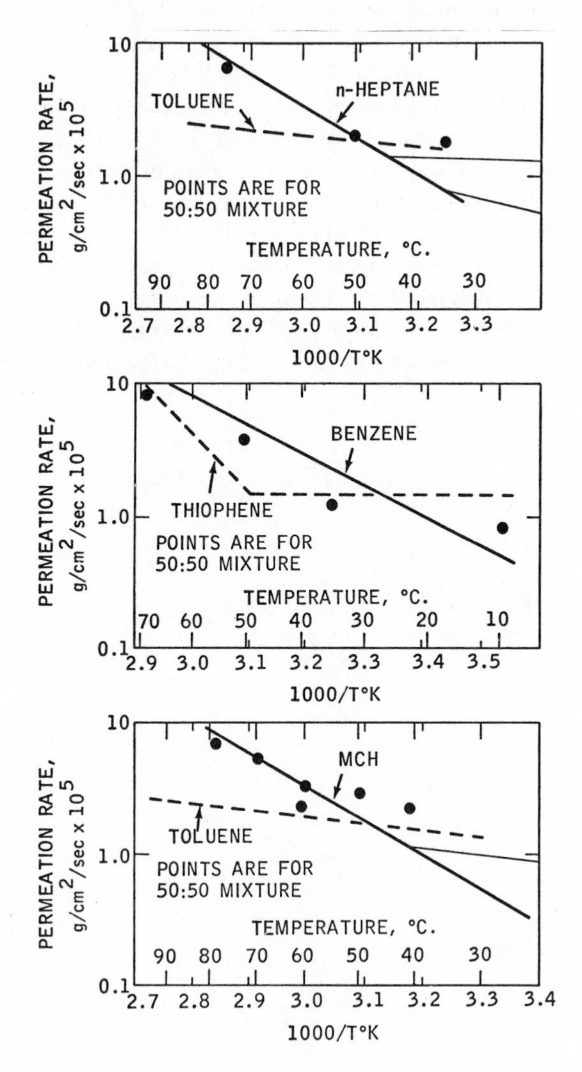

Fig. 14. Comparison of permeation rates through polypropylene (pure compounds versus mixture).

rate. For example, the presence of dimethyl butane actually decreases the permeation rate of cyclohexane. This may be due to the hindrance caused by the bulky molecular volume of dimethyl butane which predominates over any plasticizing action that it may have on the polymer membrane (13).

B. Separation of Mixtures

The separation factor in general cannot be calculated from the permeation data of pure compounds because of the plasticizing effect as is illustrated by some typical data in Figure 15. The interactions among components in a mixture and between permeates and polymer may be very strong. As previously discussed some components will solubilize others in the film giving them abnormally high permeation rates and a low separation factor as illustrated by the separation data of the benzene/isopropyl alcohol/polyethylene system in Figure 16. Similar results were obtained by Reilly (50) in separating an ethane/propane mixture and by Robeson (51) in separating an ethane/butane mixture. In addition, some permeates will change the structure of the polymer. Therefore, only in ideal systems will we be able to predict separation factors from the permeation rates of pure components.

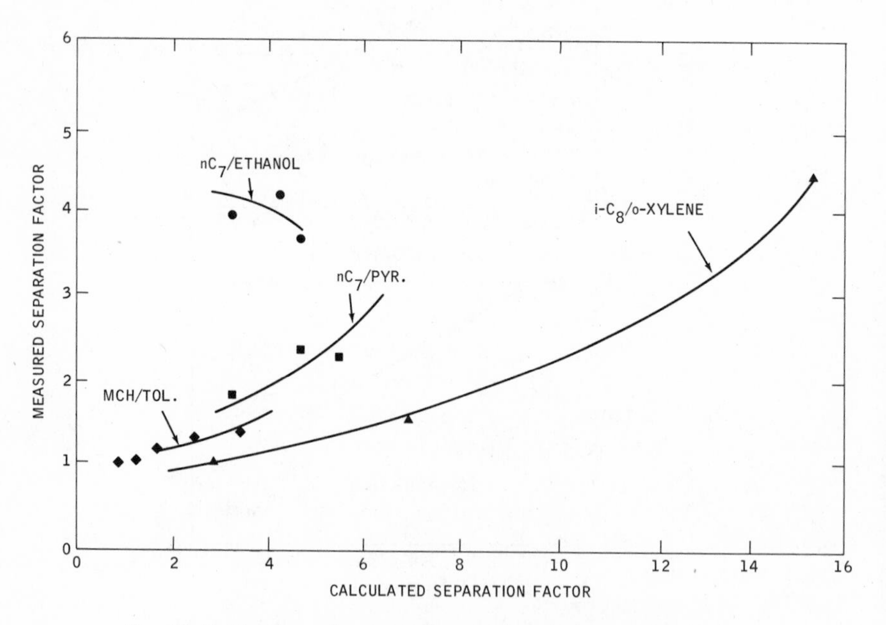

Fig. 15. Comparison of measured and calculated separation factors.

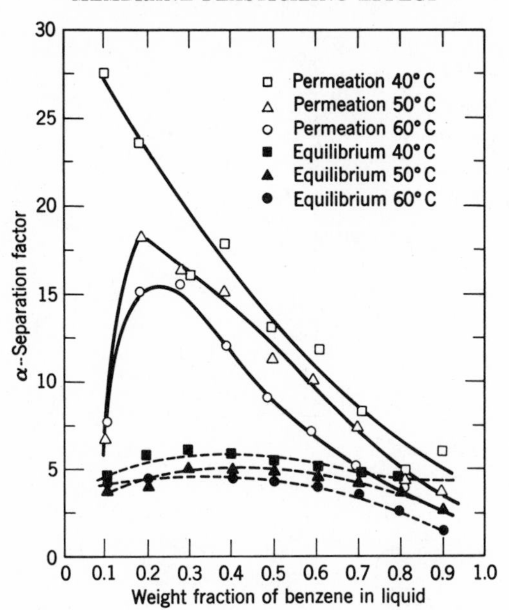

Fig. 16. Liquid composition against separation factors for benzene/isopropyl alcohol/polyethylene system [Carter and Jaganuadhaswamy (69)].

We have discussed previously that high crystallinity gives high value of the plasticizing constants defined by eqs. (18) and (43). This would mean a high plasticizing effect and low separation. But it is not axiomatic that in order to get high degree of separation, we should use membranes with a low degree of crystallinity, because then the amount of permeates dissolved in the film would be high. This condition by itself would mean low separation. Therefore, to optimize separation, the effect of crystallinity on the plasticizing constant should be balanced against the effect of permeate concentration inside the film.

Since the plasticizing effect is a function of film crystallinity, it is therefore important to pre-anneal the film to ensure its structure stability before using it for permeation and separation studies. The difference in separations between fresh and annealed films can be very large as shown by the case of the separation of heptane from toluene (Fig. 17). The usual annealing procedure is to submerge the film in a liquid held at high temperature for a specified time, for example at 85°C for 24 hr. The liquid used in annealing, unless it attacks the film, usually has no effect on the film crystallinity (2). In other words, it appears that the separation depends on the annealing process itself and not on the liquid used in the annealing process.

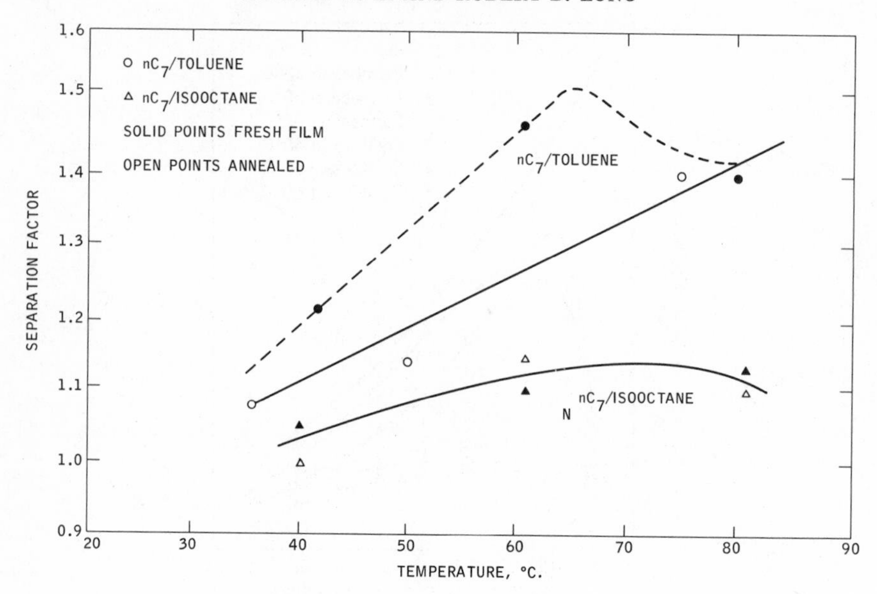

Fig. 17. Comparison of separation factors for annealed and fresh polypropylene.

C. Effect of Temperature on Separation

The degree of crystallinity can be changed simply by changing the temperature of the separation process. According to Eyring's hole theory of diffusion, the formation of "holes" in a polymer matrix requires enough energy to break down a number of secondary valence bonds. At high temperatures, large holes are produced. This results in increasing the permeation rate but decreasing the separation factor. For example, the separation factor for benzene relative to cyclohexane decreased from about 1.6 to 1.4 when the temperature was increased from 25° to 45°C for a 50/50 mixture (13).

D. The Solubility of a Mixture in the Polymer

Because the sorption by the polymer of the more soluble compound of the mixture increases the solubility of the less soluble compounds and therefore raises the total solubility of the mixture, it is important to study the sorption of the mixture so that a better insight can be obtained about the plasticizing effect.

The solubility of a mixture can be studied in one of two ways. One is to obtain the total solubility of the mixture; the other is to obtain the solubility of the individual component of the mixture. The total solubility curves of gases in polyolefin films were found to be much higher than those calculated from the solubilities of the individual gases at their partial pressures in the mixture (Fig. 4). Such a large difference means, of course, a strong plasticizing effect of the permeating compounds (2,8).

The compositions of the dissolved gaseous mixtures in polyolefin films were determined by the commonly used desorption experiment (2,4,51,52). Care must be taken to desorb all the gases inside the polymer because desorption usually takes much longer time for completion than sorption, due to the decreasing concentration difference, which is the driving force for desorption (see Fig. 18).

A separation factor, S, was calculated, which is defined as the concentration ratio of the more soluble component to the less soluble component in the product to the corresponding ratio for the two components in the raffinate. In the case of polyethylene, the separation factor predicted from the solubility data of pure methane and ethylene increases with increasing pressure, whereas the actual separation factor increases at a slower rate up to about 1000 psi, then begins to decrease, apparently

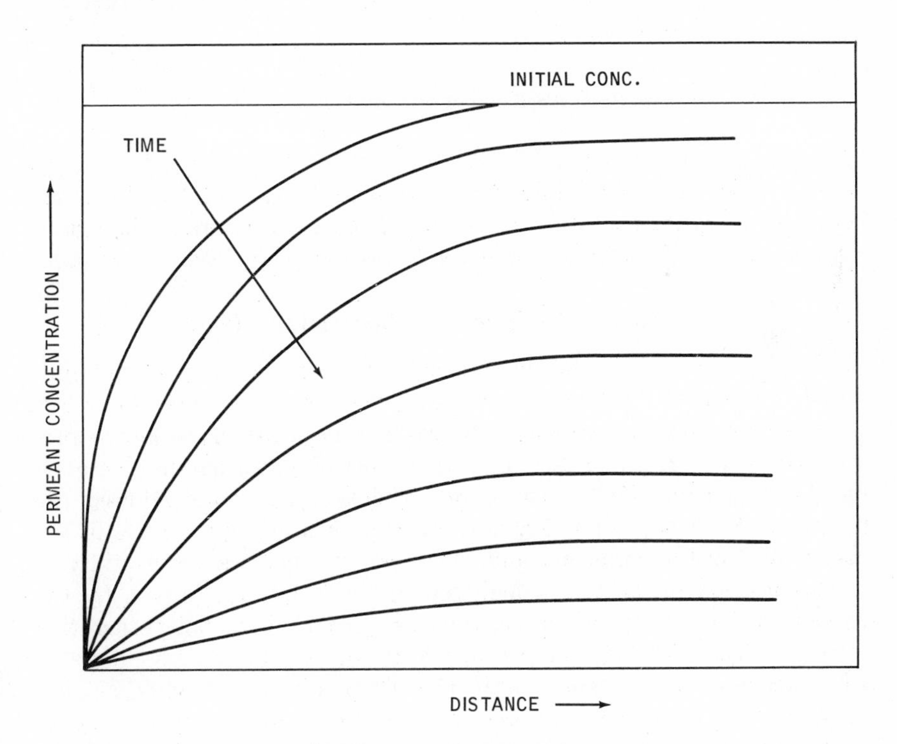

Fig. 18. Increase of plasticizing effect in polyethylene with pressure [Li(2)].

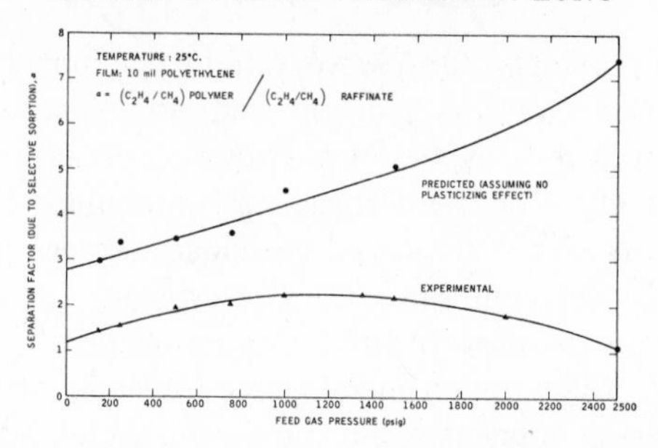

Fig. 19. Effect of plasticizing on separation by selective sorption in polyethylene [Li (2)].

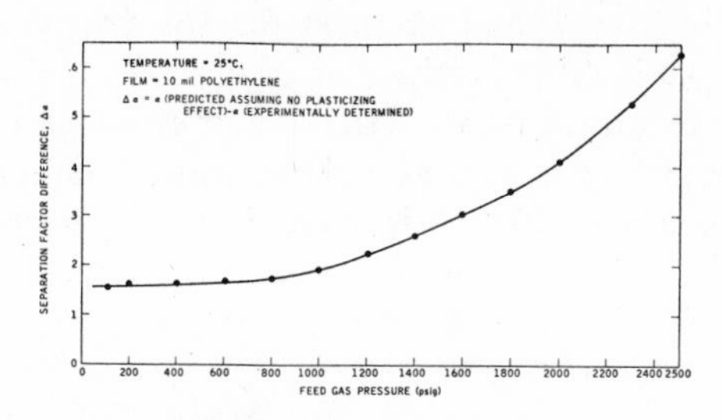

Fig. 20. Permeation out of a block of polymer.

because of severe film swelling at high pressures (Fig. 19). The difference between these two separation curves is a definitive measure of the plasticizing effect, which increases rapidly with pressure (Fig. 20).

E. Ways to Reduce the Plasticizing Effect

The plasticizing effect can be reduced if the film swelling can be minimized. A method for getting a high separation factor, therefore, is to select the upstream and downstream pressures or concentrations as low as permissible by a given Δp or Δc. The pressure or concentration drop across a membrane controls the permeation rate whereas the absolute values of the upstream and downstream pressures or concentrations govern the plasticizing effect of the permeates and, therefore, the membrane selectivity.

This method for getting a high separation factor was proven by the results from ethylene-methane separations which show that the separation factor decreases with increasing upstream and/or downstream pressure while keeping Δp constant (Table IV). Obviously an economic balance must be made among permeation rate, separation factor, number of stages, and compression costs to optimize the separation process conditions.

TABLE IV
Separation of Methane-Ethylene Mixture

Temperature: 25°C

Film Thickness: 10 mil Polyethylene
17 mil Polypropylene

Feed Composition: C_2H_4/CH_4 50/50 by weight

Upstream Pressure/ Downstream Pressure	Polyethylene Film[a]	Polypropylene Film[a]
800/300	63	58
700/200	63	61
600/100	65	63
500/0	71	64
100/0	75	66

[a] C_2H_4 conc. (wt %) in permeate.

The use of rigid films, such as Teflon, also reduces the plasticizing effect. Increasing film crystallinity is another way to reduce the effect. There are several methods for doing this, such as annealing the film (2,8,42), stretching the film (29,53–60), and grafting the film with another kind of film. Film annealing and stretching have been discussed before. As for film grafting, Meyers, Rodgers, Stannett, and Szwarc (59) found that the permeability of nitrogen through 60% vinylpyridine graft polyethylene is only one-tenth of its original value through polyethylene, whereas 30 to 40% styrene or acrylonitrile grafted to polyethylene resulted in a reduction of the permeability by one-third. Fels (60) found that, in using graft copolymer as the membrane material, the separations of benzene/hexane and toulene/heptane both increased. The permeability was also increased, which was a net result of decreasing diffusivity and increasing solubility. The decreasing diffusivity was interpreted as being due to the decreasing free volume inside the film with increasing percent of grafting, whereas the increase of solubility was due to the change of the Hildebrand's solubility parameter when two polymers were grafted together.

V. CLOSURE

The membrane separation technique has many potential applications. Its use is, in general, of special benefit for separating isomers, compounds having similar chemical or physical properties, and thermally unstable compounds. The technique is particularly useful whenever conventional separation methods cannot be used economically to get reasonable separations.

Process development is moving rapidly (2,3,15,61–66). Polymer chemists are now preparing membranes with suitable physical properties required by industrial applications and are beginning to tailor-make membranes for specific separations. Equipment design and manufacturing have also been developed to an advanced level. Many processes are now in pilot-plant stages.

For the final commercialization of the process, there are, however, many variables that must be optimized in order to optimize the process conditions. This certainly requires the knowledge of the interactions among permeates inside the membrane and between the permeates and the membrane.

This chapter serves as a review and summary of the currently available information on such interactions, or stated in a general term, on the plasticizing effect of permeates. Our knowledge in this respect is still very incomplete. Fundamental studies aimed at obtaining more information on the plasticizing effect of permeates on membrane permeation and separations are needed.

VI. SYMBOLS

a = Parameter characterizing the concentration dependence of diffusion coefficient

A_{pe} = Activity of permeate

A = Parameter characterizing the pressure dependence of permeability coefficient

A' = Constant used in eq. (55) to characterize thermodynamic diffusion coefficient

B = Constant used in eq. (55) to characterize thermodynamic diffusion coefficient

C = Concentration

D = Diffusion coefficient

D_A = Intrinsic diffusion coefficient of Compound A

E_p = Activation energy of permeation, cal/g mol

$(D_A)_T$ = Thermodynamic diffusion coefficient of Compound A

f = Frictional coefficient

F = Force per mole of substance for diffusion

F_{po} = The probability parameter for a certain polymer used in Zimm's clustering function

G_{11} = Clustering function

H = Henry's law constant [CC (STP)/CC polymer $\times$ cm$\cdot$Hg]

H_o = Constant defined by eq. (14)
H_{o1}, H_{o2}, H_{o3} and H_{o4} = Constants defined by eqs. (33), (34), (35), and (36), respectively
J = Mass transfer rate [CC (STP)/time, area]
L = Film thickness
M = Parameter characterizing the Δp dependence of permeability
p = Pressure
p_{2x} = Hypothetical downstream pressure at X
P = Permeability coefficient
$\bar{P}$ = Integrated permeation coefficient
P_1 = Second permeability coefficient defined by eq. (24)
P_o = Constant defined by eq. (14)
P_o' = Constant defined by eq. (12)
P_o'' = Parameter characterizing the temperature dependence of permeability
Δp = $p_1 - p_2$
Q = Total amount of permeate from $t = 0$ until $t = t$
R = Gas constant
S = Separation factor
t = time
T = Temperature (°K)
v = Partial molecular volume of permeate in polymer
V = Volume fraction of polymer (or solvent)
W = Weight of absorbed gas in polymer (mg)
X = Distance between film surface and any point inside the film
α = Parameter characterizing the pressure dependence of H as defined by eq. (33)
β = Parameter characterizing the pressure dependence of H as defined by eq. (34)
γ = Volume fraction activity coefficient
μ = Chemical potential
ψ = Volume fraction

Subscripts

1 = High pressure (or concentration) side
2 = Low pressure (or concentration) side
C = Critical
g = Gas
i,j = Two different permeates
o = At standard state
pe = Permeate
po (or p) = Polymer

References

1. V. Stannett, M. Szwarc, R. L. Bhargava, J. A. Meyer, A. W. Myers, and C. E. Rogers, "Permeability of Plastic Films and Coated Paper to Gases and Vapors," *TAPPI Monograph Series*, No. 23 (1962).
2. N. N. Li, *Ind. Eng. Chem., Prod. Res. Devel.*, **8,** 282 (1969).
3. N. N. Li, R. B. Long, and E. J. Henley, *Ind. Eng. Chem.*, **57,** 18 (1965).
4. H. Z. Friedlander and R. N. Rickles, *Anal. Chem.*, **37,** 27A (1965).
5. N. N. Li and R. B. Long, "Membrane Processes," *Perry's Chemical Engineering Handbook* (5th ed.), McGraw-Hill, New York, (in press).
6. S. B. Tuwiner, L. P. Miller, and W. E. Brown, *Diffusion and Membrane Technology*, Reinhold, New York, 1962.
7. N. N. Li and E. J. Henley, *A.I.Ch.E. Journal*, **10,** 666 (1964).
8. N. N. Li and R. B. Long, *A.I.Ch.E. Journal*, **15** (1969).
9. S. A. Stern, J. T. Mullhanpt, and P. J. Garies, *A.I.Ch.E. Journal*, **15,** 64 (1969).
10. A. Kishimoto and K. Matsumoto, *J. Phys. Chem.*, **63,** 1529 (1959).
11. P. Mears, *J. Polymer Sci.*, **27,** 391 (1958).
12. Y. Nishijima and G. Oster, *J. Chem. Phys.*, **27,** 269 (1957).
13. R. Y. M. Huang and V. J. C. Lin, *J. Appl. Polymer Sci.*, **12,** 2615 (1968).
14. R. N. Rickles, *Ind. Eng. Chem.*, **58,** 19 (1966).
15. A. S. Michaels and H. J. Bixler in *Progress in Separation and Purification*, E. S. Perry, Ed., Vol. 1, Interscience, New York, 1968.
16. S. A. Stern and R. M. Jobbins, *ACS Polymer Preprints*, **10** (2), 1078, September 1969. Also, discussions following Stern's presentation.
17. P. L. Durrill and R. G. Griskey, *A.I.Ch.E. Journal*, **12,** 1147 (1966).
18. J. L. Lundberg, M. B. Wilk, and M. J. Huyett, *J. Polymer Sci.*, **57,** 275 (1962).
19. C. E. Rogers, *ACS Polymer Chem. Preprints*, **6** (1), 412 (April 1965).
20. D. Richman and F. A. Long, *J. Amer. Chem. Soc.*, **82,** 509 (1960).
21. R. J. Kokes and F. A. Long, *J. Am. Chem. Soc.*, **75,** 6142 (1953).
22. R. L. Laurence and J. C. Slattery, *J. Polymer Sci.*, **5,** 1327 (1967).
23. J. R. Kuppers and C. E. Reid, *J. Appl. Polymer Sci.*, **4,** 124 (1960).
24. M. E. Morrison, *A.I.Ch.E. Journal*, **13,** 815 (1967).
25. P. Mears, *J. Appl. Polymer Sci.*, **9,** 917 (1965).
26. H. J. Bixler and A. S. Michaels, paper presented at 53rd Natl. Meeting A.I.Ch.E., Pittsburgh, Pa., May 1964.
27. A. S. Michaels, W. R. Vieth, and H. J. Bixler, *J. Appl. Polymer Sci.*, **8,** 2735 (1964).
28. G. T. Davies and H. S. Taylor, *Textile Res. J.*, **35,** 405 (1965).
29. A. Peterlin, L. Williams, and V. Stannett, *J. Polymer Sci.*, **5,** 957 (1967).
30. H. Fujita, A. Kishimoto, and K. Matsumoto, *Trans. Faraday Soc.*, **56,** 424 (1960).
31. E. J. Henley and M. L. Dos Santos, *A.I.Ch.E. Journal*, **13,** 1117 (1967).
32. H. Yasuda and V. Stannett, *J. Polymer Sci.*, **57,** 907 (1962).
33. E. R. Thornton, V. Stannett, and M. Szwarc, *J. Polymer Sci.*, **28,** 465 (1958).
34. T. K. Kwai, *J. Polymer Sci.*, **A1,** 2977 (1963).
35. B. H. Zimm and J. L. Lundberg, *J. Phys. Chem.*, **60,** 425 (1956).
36. R. M. Barrer and J. A. Barrie, *J. Polymer Sci.*, **28,** 377 (1958).
37. H. Veith, *Koll. Z.*, **152,** 36 (1957).
38. J. E. O. Mayne, *J. Oil Color Chemical Assoc.*, **40,** 183 (1957).
39. R. C. Binning, R. J. Lee, J. F. Jennings, and E. C. Martin, *Ind. Eng. Chem.*, **53,** 45 (1961).

40. H. W. Chandler and E. J. Henley, *A.I.Ch.E. Journal*, **7**, 295 (1961).
41. D. W. McCall, *J. Polymer Sci.*, **26**, 151 (1957).
42. R. B. Long, *I. & E. C. Funaamentals*, **4**, 445 (1965).
43. D. J. Buckley and M. Berger, *J. Polymer Sci.*, **56**, 175 (1962).
44. S. N. Kim and K. Kammermeyer, Preprint, A.I.Ch.E. National Meeting, Atlanta, Ga., February 15–18, 1970.
45. T. A. Garret and G. S. Park, *J. Polymer Sci.*, **C**, No. 16, 601 (1967).
46. J. B. Wilkens and F. A. Long, *Trans. Faraday Soc.*, **53**, 1146 (1957).
47. A. E. Chalykh and R. M. Vasenin, *Vysokomolekulyarnye Coedineniya*, **8**, 1908 (1966).
48. R. W. Coughlin and F. A. Pollak, *A.I.Ch.E. Journal*, **15**, 208 (1969).
49. S. N. Zhurkov and G. Ya. Ryskin, *Zhur. Tekn. Fiz.*, **24**, 797 (1954).
50. J. W. Reilly, Sc.D. Dissertation, Stevens Institute of Technology, Hoboken, N. J. (1965).
51. L. M. Robeson, Ph.D. Dissertation, Univ. of Maryland, College Park, Maryland (1967).
52. A. S. Michaels, W. R. Vieth, and J. A. Barrie, *J. Appl. Physics*, **34**, 13 (1963).
53. S. W. Lasoski, Jr. and W. H. Cobbs, Jr., *J. Polymer Sci.*, **57**, 275 (1962).
54. W. W. Brandt, *J. Phys. Chem.*, **63**, 1080 (1959).
55. M. H. Walters, *J. Polymer Sci.*, **A1**, 3091 (1963).
56. R. W. Roberts, Ph.D. Dissertation, State University of Iowa, 1962.
57. H. Yasuda, V. Stannet, A. Peterlin, and H. L. Frisch, paper presented to 53rd Meeting of A.I.Ch.E., 1964.
58. A. B. Krewinghans, Sc.D. Dissertation, Massachuestts Institute of Technology, Cambridge, Mass. (1966).
59. A. W. Myers, C. E. Rodgers, V. Stannett, and M. Szwarc, Modern Plastics, 157 (May, 1957).
60. Morton Fels, Ph.D. Dissertation, University of Waterloo, Waterloo, Canada, 1968.
61. *Chem. Eng. News*, p. 46, July 18, 1966.
62. K. Kammermeyer, in *Progress in Separation and Purification*, Vol. 1, E. S. Perry, Ed., Interscience, New York, 1968, p. 297.
63. A. S. Michaels, H. J. Bixler, and P. N. Rigopulos, "New Concepts and Techniques for Hydrocarbon Separation," *7th World Petroleum Congress Proceedings*, Vol. 4, p. 21 (1967).
64. D. A. Pattison, *Chem. Eng.*, **75** (12), 38 (1968).
65. B. H. Sanders and C. Y. Choo, *Petroleum Refined*, **39**, 133 (1960).
66. S. A. Stern, "Membrane Processes," *Ind. Proc. Symp.*, Birmingham, Ala., 196 (1966).
67. D. W. McCall and W. P. Slichter, *J. Am. Chem. Soc.*, **80**, 1861 (1958).
68. T. Graham, *Phil. Mag.*, No. 32, S. 4, p. 401 (1866).
69. J. W. Carter and B. Jagannadhaswamy, in the proceedings of "Symposium on the Less Common Means of Separation," *Inst. Chem. Engrs.* (U.K.), 1963.

Separation and Purification by Reverse Osmosis

H. K. LONSDALE

Chemistry Department,
Gulf General Atomic, Incorporated,
San Diego, California

I. INTRODUCTION

During the past decade an essentially new method of separation and purification has evolved. Most commonly referred to as reverse osmosis, the method is one in which a semipermeable membrane is used to separate solutions at different thermodynamic activities of solvent and solute. Transport of one component, the solvent, is achieved by increasing the thermodynamic activity of the solvent in the more concentrated solution with the application of pressure. The name of the process derives from the fact that the normal osmotic flow of solvent, well known in biological membrane systems, for example, is reversed.

The development of the process has been supported largely by the Office of Saline Water of the United States Department of the Interior and by the State of California. Both agencies are interested in the application of the process to water desalination, for which it holds considerable promise. Thus, most of the experimental and theoretical studies and almost all of the equipment development and testing have been concerned with water desalination, purification, and renovation. The thrust of the development has been in the area of water-permeable, salt-impermeable membranes and much of the present review is of necessity devoted to this aspect. However, there is an increasing awareness among those active in the field that, with suitable membranes, reverse osmosis or variations thereof could develop into a new and useful unit operation for separation as well as purification. Among the numerous applications that are currently under development in addition to those aimed principally at water recovery are the concentration, purification, or separation of food products, pharmaceuticals, and chemicals. In some applications the membrane permeate is the desired product, while in others the product of value is the concentrate.

Reverse osmosis differs from conventional dialysis in that in reverse osmosis the solvent is purified of low molecular weight dissolved species by forcing the solvent under pressure through a membrane that is highly impermeable to the solutes. In dialysis low molecular weight solutes permeate the membrane at ambient pressure while larger molecules are retained; solvent flow is incidental. Reverse osmosis differs from ultrafiltration in a mechanistic rather than an operational sense: by convention, reverse osmosis membranes are those that do not differentiate between solutes and solvent (or between different solutes) on the basis of molecular size but rather because of the higher solubility and mobility of the solvent in the membrane. Further, by convention, ultrafiltration membranes are those that are essentially completely nonselective with respect to low molecular weight solutes so that the osmotic pressure difference between feed solution and ultrafiltrate is usually trivial. An interface exists, however, between reverse osmosis and ultrafiltration that is not sharply defined. For this reason, apparently, some investigators prefer the name "hyperfiltration" to "reverse osmosis." A review of ultrafiltration appeared in Volume I of this series (1) and a number of potential applications were cited in a recent paper (2).

Osmotic phenomena have, of course, been known for centuries but it was not until the work of Reid and his students at the University of Florida that serious attention was devoted to a search for membrane materials that could effectively desalt water by the reverse osmosis process (3,4). Following the discovery of the Florida group that cellulose acetate membranes were both highly permeable to water and highly impermeable to salts, Loeb and Sourirajan, working at UCLA, discovered a way to prepare modified cellulose acetate membranes that were characterized by an asymmetric structure and an extremely large water permeability (5).

These membranes were subsequently shown to possess a salt-rejecting, very thin dense skin on the order of 10^{-5} cm thick which is integrally supported by a nonselective, open-cell porous structure (6,7). The Loeb-Sourirajan discovery is clearly a milestone in the technology of membrane separations. It had not previously been possible to prepare imperfection-free membranes thinner than about 10^{-3} cm and the low fluxes associated with such membranes precluded widespread use of membranes for separation and purification processes. The Loeb-Sourirajan membrane not only grossly altered this picture for one particular application but created a widespread interest that has resulted in significant additional developments in membrane preparation and in novel applications.

In 1966 a book (8) and an extensive review article (9) appeared which summarized reverse osmosis technology to that time. In this review we shall attempt to give a general introduction to the process and to describe recent developments in this rapidly growing field. Of particular interest are the development of several novel and promising approaches to reverse osmosis membranes and the results of extensive field tests of desalination units.

II. TRANSPORT IN REVERSE OSMOSIS MEMBRANES

A. Transport Models

There are two mathematical descriptions of reverse osmosis transport phenomena in general use. These are discussed in detail in several publications, and only a general description will be presented here.

The simpler of these two descriptions has been called the solution-diffusion model (8,10). In this picture, each permeating species is transported across the membrane under the influence of its chemical potential gradient. The transport rate of each component is determined by its diffusion coefficient and solubility in the membrane, the product of these being a conventional permeability coefficient. Two basic assumptions are made: first, that Fick's law is obeyed (and the diffusion coefficient is usually considered concentration-independent) and second, that the flow of each component is unaffected by the flows of other components, that is, that the flows are uncoupled. For most reverse osmosis membrane-solute-solvent systems with which detailed studies have been performed, these assumptions are supported by experiment. It is also conventional to assume that the chemical potential is continuous across the membrane-solution interface.

The chemical potential gradient for each component within the membrane is determined by the thermodynamic activity gradient and the pressure gradient. When the dilute-solution approximation is made that activity is proportional to concentration, the difference in chemical potential across the membrane for each component i is

$$\Delta\mu_i = RT\Delta \ln c_i + \bar{v}_i\Delta p \tag{1}$$

where R, T, and p are the gas constant, temperature, and pressure, respectively, c_i is the concentration of component i in the membrane, and $\bar{v}_i$ is the partial molar volume. For the solvent, it can be shown that the flux, J_1, is given by

$$J_1 = -\frac{D_1 c_1 \bar{v}_1(\Delta p - \Delta\pi)}{RT\Delta x} \tag{2}$$

in which D_1 is the diffusion coefficient of the solvent in the membrane, and Δp and $\Delta\pi$ are the pressure and osmotic pressure differences, respectively, across a membrane of thickness Δx.

While eq. (1) also applies to a solute, it is generally true that the concentration gradient term far outweighs the pressure gradient term and the solute flux, J_2, is given simply by

$$J_2 = -D_2(\Delta c_2/\Delta x) = -D_2 K(\Delta\rho_2/\Delta x) \tag{3}$$

In the right-hand equality the concentration of solute within the membrane c_2 has been replaced by the product of a distribution coefficient K and the external solution concentration ρ_2. The assumption has been made that K is concentration-independent and this is generally observed with uncharged membranes. In describing membrane semipermeability it is convenient to calculate either the solute "concentration reduction factor," defined as ρ_2'/ρ_2'' where $'$ and $''$ refer to feed and product solutions, respectively, or "solute rejection," defined as $(\rho_2' - \rho_2'')/\rho_2'$. In reverse osmosis the solute concentration in the product is determined only by the relative solute and solvent flows:

$$\rho_2'' = J_2\rho_1''/J_1$$

and the solution-diffusion transport eqs. (2) and (3) may be used to evaluate either semipermeability parameter in terms of the permeabilities and the net pressure difference, $\Delta p - \Delta\pi$. For example, solute rejection is given by

$$\text{Rejection} = \left[1 + \frac{D_2 K R T \rho_1''}{D_1 c_1 \bar{v}_1(\Delta p - \Delta\pi)}\right]^{-1} \tag{4}$$

Rejection improves with increasing net pressure difference $(\Delta p - \Delta\pi)$ because solvent flow is linear in the net pressure difference while solute flow is relatively independent of pressure.

When the solute is not highly rejected and when its molar volume is large, the $\bar{v}_2\Delta p$ term in eq. (1) becomes important. For example, taking the conditions $\Delta p = 50$ atm, $\bar{v}_2 = 50$ cc/mole, and $T = 25°C$, the ratio $\bar{v}_2 \Delta p / RT\Delta \ln c_2$ takes on values of 0.03, 0.1, and 0.3 as the solute rejection decreases from 96.7% to 64% to 29%. When the $\bar{v}_2 \Delta p$ term is included in the solute rejection equation, it takes the form (11)

$$\text{Rejection} = 1 - \frac{D_2 K(RT + \bar{v}_2 \Delta p)}{D_1 c_1 \bar{v}_1 (\Delta p - \Delta \pi) + D_2 KRT} \tag{5}$$

In practice, this refinement is usually not required because as the molar volume increases the solute rejection also increases, and usually more rapidly, so that the ratio $\bar{v}_2 \Delta p / RT\Delta \ln c_2$ is generally small.

The diffusion coefficient and solubility of water in most reverse osmosis membranes are not in general strongly dependent on either solute concentration or pressure (although these and other similar quantities are, of course, dependent on temperature) and it is convenient to combine the quantities multiplying $(\Delta p - \Delta \pi)$ in eq. (2) into a "membrane constant." This constant, referred to as A, is then

$$A \equiv J_1/(\Delta p - \Delta \pi) = -D_1 c_1 \bar{v}_1 / RT\Delta x \tag{6}$$

Similarly, it is convenient and usually safe to use a "solute permeation constant," B, given by

$$B \equiv J_2/\Delta \rho_2 = -D_2 K/\Delta x \tag{7}$$

With these definitions, the solute rejection is given by

$$\text{Rejection} = \frac{(A/B)(\Delta p - \Delta \pi)}{(A/B)(\Delta p - \Delta \pi) + \rho_1''} \tag{8}$$

Thus, two parameters are adequate to describe membrane performance when $(\Delta p - \Delta \pi)$ is fixed: the membrane constant is a measure of the water flow, and A/B is a measure of the solute rejection.

A second and more elegant description of transport in reverse osmosis membranes has as its basis the thermodynamics of irreversible processes (12). One of the basic principles of irreversible thermodynamics is that the flow of each component is determined not only by the thermodynamic force acting on that component but on the forces exerted on all other components as well. Thus all the flows are coupled. A simple manifestation of flow coupling exists in flow of salt water through a pipe: a pressure differential causes a flow of water and the salt is carried along without concentration change. In a two-component system the flux equations have the form

$$J_1 = L_{11} X_1 + L_{12} X_2 \tag{9}$$

$$J_2 = L_{21} X_1 + L_{22} X_2 \tag{10}$$

where the Ls are flow coefficients relating the fluxes and the thermodynamic driving forces represented by the X terms. The appropriate force in this case is the negative gradient of the chemical potential:

$$J_1 = -L_{11} \text{ grad } \mu_1 - L_{12} \text{ grad } \mu_2 \tag{11}$$

$$J_2 = -L_{21} \text{ grad } \mu_1 - L_{22} \text{ grad } \mu_2 \tag{12}$$

and useful expressions for the flows can be obtained by substituting for the μs with eq. (1). According to the Onsager reciprocal relation, the cross coefficients are equal, that is, $L_{21} = L_{12}$.

Instead of the two flow coefficients that appear in the uncoupled flow equations, that is, D_1c_1 and D_2K in eqs. (2) and (3) or A and B in eqs. (6) and (7), the three L coefficients appear in the coupled flow equations and reverse osmosis results may sometimes appear in these terms. Kedem and Katchalsky (13) have transformed the coupled flow equations into more useful forms which show the interaction of flows in terms of three related coefficients that are experimentally more accessible than the L coefficients. Instead of water and solute flows, the coupled flows they consider are the somewhat more convenient total volumetric flow, J_v, and solute flow, J_s. The relationships are

$$J_v = L_p(\Delta p - \sigma \Delta \pi) \tag{13}$$

$$J_s = \bar{c}_2(1 - \sigma) J_v + \omega \Delta \pi \tag{14}$$

where L_p is the hydrodynamic permeability, $L_p = (J_v/\Delta p)_{\Delta\pi=0}$; ω is the solute permeability in the absence of volumetric flow, $\omega = (J_s/\Delta\pi)_{J_v=0}$; $\bar{c}_2$ is an average solute concentration within the membrane; and σ is the coupling or Staverman reflection coefficient (14), $\sigma = (\Delta p/\Delta\pi)_{J_v=0}$. In most practical cases the volumetric flow is approximately equal to solvent flow because $J_v = \bar{v}_1 J_1 + \bar{v}_2 J_2$ and $J_2 \ll J_1$. Thus when $\sigma = 1$ the hydrodynamic permeability is identical to the membrane constant, A, of eq. (6). These relationships and the coefficients L_p, ω, and σ are now frequently found in the literature of biophysics and the three coefficients have been experimentally determined for a number of membrane-solute-solvent systems (15).

In this phenomenological treatment no attempt is made to define specific interactions between solvent and solute, for example, within the membrane. Spiegler (16) has proposed such a detailed model in which solvent-solute, solvent-membrane, and solute-membrane interactions are taken into account. In this model, which was recently extended by Merten (17) and by Spiegler and Kedem (18), flows of solvent and solute are related to three frictional coefficients which express the three interactions. Only in a few detailed studies of membrane transport have these models been used and the frictional coefficients determined (19–22). Two general conclusions to arise from these studies are that solvent-solute interaction within the membrane is greater than that in free solution, sometimes by an order of magnitude or more, and that solute-membrane interactions greatly exceed solvent-membrane interactions. The latter generalization is at least partly the result of the types of systems commonly examined: hydrophilic membranes where the solvent is water and the solute is an electrolyte. As sophisticated membrane transport studies become more widespread, it appears that the use of the irreversible thermodynamic treatment and of the frictional model will become more general.

As a practical matter it has been shown several times that with the highly selective cellulose acetate membranes most commonly used in reverse osmosis, solvent-solute flow coupling is generally not important (9,20,23). In this case $\sigma = 1$ and $L_{12} = L_{21} = 0$ and eqs. (11) and (12) reduce to eqs. (2) and (3).

B. Solute Exclusion

Regardless of the model used to describe membrane transport, it is clear that solute permeation is minimized when the solute is highly excluded from the membrane, that is, when K in eq. (3) or $\bar{c}_2$ in eq. (14) is small. In the case of salt-water solutions, solute exclusion will be favored with membranes of low dielectric constant. If the membrane is too hydrophobic, however, it is also relatively impermeable to water. Based on a number of studies of membrane materials for reverse osmosis desalination membranes, it appears that membranes that sorb 10–20 wt % water at saturation have the potential for combining high water permeability and high salt exclusion. The cellulose 2.5-acetate used to make water desalination membranes sorbs 0.16 g H_2O/cc membrane at saturation.

Fixed-charged membranes of the type used in electrodialysis exclude ions because of the electrostatic repulsion between the fixed charges and the ions of like sign (coions). The phenomenon is generally referred to as Donnan exclusion. Because of their ion-exclusion properties, these membranes have also been examined as reverse osmosis membranes. From considerations of charge neutrality and equilibrium between membrane and solution, it can be shown that the distribution coefficient of coions between a fixed-charge membrane and the solution is given by

$$K_{co} = \frac{-\chi + \sqrt{\chi^2 + 4\rho_2^2}}{2\rho_2} \tag{15}$$

where χ is the fixed-charge capacity. In the case where $\rho_2 \gg \chi$, K_{co} approaches unity and when ρ_2 approaches 0, K_{co} approaches ρ_2/χ. If water and solute are strongly coupled and if the membrane flux is high, the solution that permeates the membrane is just that contained within the membrane. Even though the distribution coefficient for counterions can be much greater than unity (generally equal to $1/K_{co}$), electroneutrality requires that the number of equivalents of counterions transported cannot exceed the number of equivalents of coions, and the solute rejection is given in this case by $1 - K_{co}$.

III. MEMBRANES

As we have noted, the invention by Loeb and Sourirajan of cellulose acetate membranes effectively 1000–2000 Å thick took place about ten years ago. Since that time, a large number of other membrane materials have been examined and several completely new approaches to reverse osmosis membranes have been conceived and developed. However, most of the commercially available reverse osmosis equipment utilizes the Loeb-Sourirajan type cellulose acetate membrane and only minor improvements in the method of preparation and in performance characteristics have been achieved. In spite of intensive research no membrane material with more favorable overall properties has become practical. We shall therefore briefly review this unique membrane before describing more recent developments.

A. Modified Membranes of the Loeb-Sourirajan Type

The Loeb-Sourirajan membrane is prepared (5,24) by casting a solution of cellulose acetate (degree of substitution 2.5) in a good solvent to which one or more poor solvents or nonsolvents has been added. In the original disclosure, the solvent system consisted of acetone, water, and magnesium perchlorate, and the membranes were cast at approximately 0°C. Shortly after casting, the membrane is immersed in a nonsolvent, usually water, where it gels. During this process a thin skin forms on the side of the membrane exposed to the atmosphere during casting, and the remainder of the membrane becomes an open-cell porous matrix. Various proposals have been made to explain the skin formation step. However, neither the equilibrium phase diagram of the multicomponent system nor the kinetics of nucleation and precipitation of the cellulose acetate phase have been worked out in detail, and it is sufficient for the moment to assume merely that gelation of the skin occurs as the first bit of solvent evaporates. The rest of the membrane precipitates as water is imbibed during the immersion step and here both water and cellulose acetate form continuous phases. The membrane as a whole sorbs 60–70% water by weight at saturation, most of it being held by capillarity in the pores.

In the as-cast state, these membranes reject only a small amount of sodium chloride, although they are highly impermeable to most species with molecular weight greater than 200. The semipermeability is markedly improved by annealing in water at temperatures between 60 and 90°C. During this annealing process the permeability to water decreases by about an order of magnitude while the permeability to sodium chloride decreases by two to three orders of magnitude so that the rejection of NaCl under typical reverse osmosis conditions increases from about 25% to 98–99%. The entire membrane shrinks somewhat during the annealing step (25) and a sufficient densification of the skin apparently takes place to grossly reduce its permeability and increase its permselectivity. It has recently been shown that aqueous annealing of normal, dense cellulose acetate membranes, that is, those cast from acetone and allowed to dry completely in air, produces similar changes in permselectivity (26,27). It has been shown by several groups, most completely by Sourirajan and co-workers (28,29), that the annealing step produces a family of membranes whose properties vary continuously with annealing temperature. It is possible, then, to tailor these membranes for particular applications.

Since the original work of Loeb and Sourirajan, some useful modifications to their procedure have been developed. Sourirajan and Govindan (30) found an improved formulation of the casting solution. At about the same time, Manjikian (31,32) found that these modified cellulose acetate membranes could be successfully cast from the mixed organic solvent acetone and formamide. Since that time, a number of other suitable solvent systems have been found (33) but none of these appears to offer further improvement in membrane performance. The upper limit to the achievable water permeability has been appreciably raised (with concomitant increase in the molecular weight cutoff) by casting membranes from a solvent high in formamide concentration (34). These very high flux cellulose acetate membranes exhibit virtually no selectivity for molecules as large as sucrose, for example, and some of them readily pass species with molecular weight of 10,000. They are therefore not within our definition of reverse osmosis membranes, but they have found application in ultrafiltration and dialysis. Furthermore, while modified membranes of the Loeb-Sourirajan type can be operated at pressures in excess of 100 atm without failure, the very high flux membranes tend to collapse at high pressures and also to plug and foul at high flow rates so that, in practice, their use is limited to pressures of only a few atmospheres and water flow rates typical of those observed with reverse osmosis membranes at high pressure (i.e., 10^{-3} g/cm^2-sec).

The structure of modified cellulose acetate membranes of the Loeb-Sourirajan type has been examined in some detail by Riley, Gardner, and Merten using the electron microscope (6,7). Typical photomicrographs of the air-dried surface, the nonair-dried or bottom surface, the cross section, and the cross section at the air-dried surface are shown in Figure 1. The membrane in the last photograph was prepared under conditions designed to produce a skin somewhat thicker than the norm. Typically, skin thicknesses are in the range 0.1–0.2 μ. In the figure, the skin is clearly visible and is on the order of 0.5 μ thick. There is no sharp line of demarcation between the porous part of the membrane and the skin but rather the pores become progressively smaller toward the air-dried surface until they vanish entirely and the membrane forms a continuum. There is a smooth graduation in pore size throughout the cross section. An electron photomicrograph of the cross section of the upper one-third of a modified membrane is shown in Figure 2.

The literature describing membranes of graded porosity goes back at least 30 years (35–37). The techniques used by the early researchers were similar to that used by Loeb and Sourirajan: the polymer was dissolved in a mixture of good and poor solvents and freshly cast membranes were usually immersed in a nonsolvent. Most of this work was done with cellulose nitrate and while there is no evidence that salt-rejecting membranes were prepared, it was evident that the membranes were asymmetric. This suggests that the "good solvent-poor solvent" technique may be fairly general and that this technique should be examined for the fabrication of other kinds of polymer membranes for a variety of potential separations and purifications. The general application of the technique will not be straightforward, but will probably depend in a complex way on undefined equilibrium and diffusional properties of the polymer-solvent system. The Loeb-Sourirajan technique has already been applied to cellulose ethers and other esters (38) but the membranes were inferior to cellulose acetate for water desalination.

The vast majority of the reverse osmosis applications reported in the literature to date have been concerned with water desalination. Most of our understanding of membrane behavior is based on experiments with aqueous solutions of sodium chloride. Because of the favorable and, thus far, unique properties of the modified cellulose acetate membranes of the Loeb-Sourirajan type, much of the reverse osmosis research has been carried out with this type of membrane and an extensive body of information has been published.

Typical results of water flux and sodium chloride rejection are presented as a function of applied pressure in Figure 3. The dilute salt solution used in that study had a low osmotic pressure (0.1% NaCl, $\pi \simeq 0.8$ atm) and the intercept of the flux versus applied pressure curve is virtually at the origin. Because these membranes have a thin dense skin of unknown thickness, it is not possible to obtain the water permeability, D_1c_1, from such results.

Fig. 1. Electron photomicrographs of modified cellulose acetate membranes showing the (a) top surface, (b) bottom surface, (c) cross section, and (d) cross section including the dense skin. Membrane samples (a)–(c) were freeze-dried and photographs were made of palladium-shadowed carbon replicas. Sample (d) was epoxy impregnated, ultramicrotomed, and examined in transmission. These photomicrographs were prepared by R. L. Riley and J. O. Gardner of Gulf General Atomic.

Fig. 2. An electron photomicrographic montage showing the gradation in porosity. The membrane was freeze-dried and photographs were made of a palladium-shadowed carbon replica. The photomicrograph was prepared by R. L. Riley and J. O. Gardner of Gulf General Atomic.

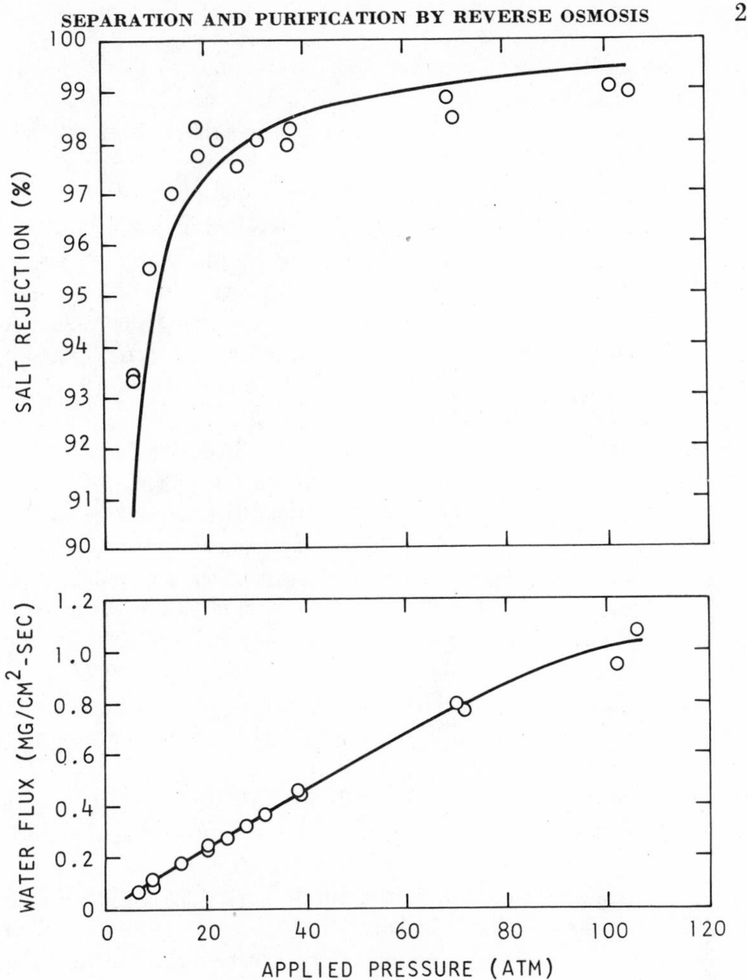

Fig. 3. Salt rejection and water flux versus applied pressure for a modified cellulose acetate membrane.

Rather, the water permeability observed in direct osmosis or reverse osmosis measurements with dense membranes has been used to calculate the skin thickness. The results are in satisfactory agreement with electron photomicrographs of the type shown in Figure 1; a thickness of 0.1–0.2 μ is now generally accepted as representative. The total membrane thickness is typically 100 μ. The significance of the contribution of Loeb and Sourirajan to membrane technology can be appreciated by comparing the water fluxes in Figure 3 with those reported by Reid and Breton (4) in the original study of cellulose acetate membranes for reverse osmosis: the thinnest membranes cast that exhibited high NaCl rejection (11 μ) yielded a water flux of only 1.5×10^{-5} g/cm²-sec* at 41 atm pressure or about 1–2% of the flux achieved with a modified membrane of the same selectivity. The flux ratio is inversely proportional to the thickness ratio, as expected, only if the effective thickness of the modified membrane is used.

The salt rejection in Figure 3 depends on applied pressure in accordance with eq. (4). The rejection increases with increasing pressure, asymptotically approaching 100%. As noted in Section II, the salt flow is essentially independent of pressure while the water flow is linear in the pressure so that with increasing pressure, water flow continues to increase while salt flow remains nearly invariant. The line drawn through the points was calculated from eq. (4), using a water permeability of 2.6×10^{-7} g/cm-sec and a salt permeability, D_2K, of 1.2×10^{-10} cm²/sec. These values are typical of those observed in direct osmosis or sorption-rate experiments (27).

As we have noted, the properties of the modified membrane can be controlled by regulation of the temperature at which it is annealed in water. The as-cast membrane rejects a small amount of sodium chloride, for example, when prepared by the Loeb-Sourirajan technique. With increasing temperature of aqueous annealing, the water permeability falls and the semipermeability increases smoothly, beginning at about 60°C. Details of the mechanism of this process are lacking although it is known that the membrane shrinks on annealing, with no evidence of second-order transitions in the dilatometric data (25) and, based on x-ray diffraction results, there is apparently no increase in crystallinity in these generally amorphous membranes. A certain minimum time of annealing is essential but only minor changes occur with increased annealing time.

The effects of annealing temperature on properties are shown in Figure 4. Plotted here is the salt rejection parameter A/B of eq. (8) versus the membrane constant A with the annealing temperatures indicated in the figure. The data were obtained at 800 psi with 1% NaCl. It happens that the curve in the figure can be fit rather well by an expression of the type $y = a/x$ so that the quantity A^2/B is essentially a constant of the membrane, nearly independent of heat-treatment temperature. This quantity is a useful figure of merit for membrane performance: as A^2/B increases the degree of semipermeability or solute rejection associated with a given solvent flux increases. Some improvements in membrane quality in terms of the A^2/B parameter have been achieved (30–32) since the original Loeb-Sourirajan formulation was disclosed.

In the uncoupled flow expressions, the solute rejection is determined by the permeability of the membrane to solvent and solute. Presented in Table I are some recently determined solute permeabilities along with cal-

* Multiply water flux in g/cm²-sec by 2.12×10^4 to convert to gal/ft²-day.

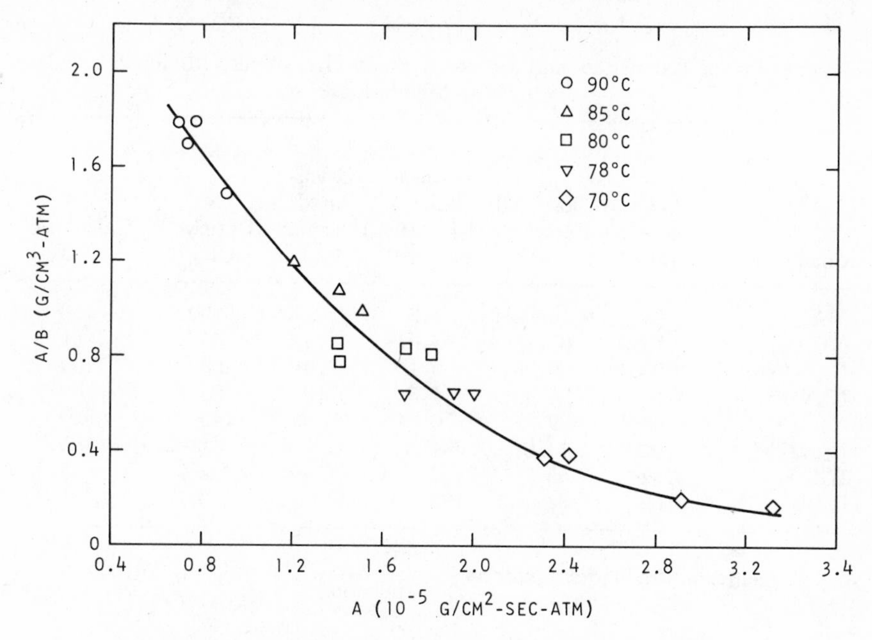

Fig. 4. Salt rejection parameter A/B versus membrane constant A.

culated and observed solute rejections (27). The membrane constant in the reverse osmosis tests was $1.0 \pm 0.1 \times 10^{-5}$ g/cm^2-sec-atm. For convenience in making the comparison, the concentration reduction factors are also tabulated. The solute permeabilities were determined on dense membranes, cast from an acetone solution and allowed to dry in air, by measuring the rate of sorption or desorption of solute. In this way, meaningful results can be obtained even if the membranes are not free of imperfections. The reverse osmosis data of Table I were obtained with modified membranes of the Sourirajan-Govindan formulation (30), annealed at 85°C and tested at 102 atm applied pressure and 25°C. The calculated concentration reduction factors exceed the observed values by a factor or two or so. However, both Saltonstall and co-workers (39) and other studies (27) have shown that the permeability of the dense membranes to sodium chloride, and presumably to other solutes as well, is sensitive to sample history and, in particular, is grossly reduced by aqueous annealing. For example, annealing at 85°C in water reduces the permeability to NaCl by a factor of seven while the permeability to water is reduced by only about a factor or two. With modified membranes the effects of annealing are even more dramatic. Other variables in membrane casting have also been shown to be important so that comparisons of the ratio of solvent to solute permeability in dense

TABLE I
Comparison of Calculated and Observed Solute Rejections, Modified Cellulose Acetate Membranes

Compound	Net Pressure $\Delta p - \Delta\pi$ (atm)	Observed Rejection (%)	Solute Permeability, $D_2K \times 10^{10}$ (cm²/sec)	Calculated Rejection (%)	Observed CRF[a]	Calculated CRF
NaCl	94.3	98.0–98.9	1.2	99.34	50–90	150
NH_4Cl	93.9	97.1	3.0	98.3	35	61
NaH_2PO_4	98.4	99.93	0.18	99.91	1400	1100
$NaNO_3$	96.9	96.4	4.5	97.6	28	42
$NaHCO_3$	97.1	99.27	0.34	99.82	140	560
Urea	102	44.6	34	82	1.8	5.7
H_3BO_3	102	55	37	81	2.2	5.3
$C_6H_3OHCl_2$	102	−34	190	16	0.75	1.2

[a] Concentration reduction factor $\equiv \dfrac{1}{1 - \text{rejection}}$.

membranes with that in modified membranes should be considered qualitative only. With modified membranes annealed at temperatures below about 80°C (or not annealed at all), such a comparison is almost meaningless because flow coupling becomes increasingly important with decreasing annealing temperature (40), and eqs. (2)–(4) no longer hold. The Staverman reflection coefficient, σ, for NaCl is 0.98 or greater for highly selective modified cellulose acetate membranes, but values of 0.7 and less have been observed with membranes annealed below 70°C (41).

A summary of additional solute rejection and water flux results obtained with highly annealed membranes is presented in Table II. The tests were performed at 102 atm. Some trends are obvious in Tables I and II: divalent ions are in general more highly rejected than monovalent ions and low molecular weight nonelectrolytes are not highly rejected. The negative rejection (i.e., enrichment) of phenol is of particular interest because it is one of the few cases studied in sufficient detail to permit evaluation of water-solute and solute-membrane interactions. While cellulose acetate is highly permeable to phenol, the negative rejections can only be explained on the basis of significant solvent-solute flow coupling (11,21). The same conclusion applies to 2,4-dichlorophenol in Table I and probably to a number of other nonelectrolytes.

Some of the limitations of the modified cellulose acetate membranes have been studied in detail. These include membrane compaction, hydrolysis of

TABLE II

Selectivity of Modified Cellulose Acetate Membranes to Several Solutes

Solute	Concentration (wt %)	Rejection (%)	Membrane Constant (10^{-5} g/cm^2-sec-atm)	Reference
NaCl	5.25	98.1	1.21	42
NaBr	5.25	98.0	0.88	42
KCl	5.25	95.8	1.02	42
$NaClO_4$	5.25	86.3	1.07	42
NH_4ClO_4	5.25	77.4	0.98	42
NH_4NO_3	0.25	80.3	1.30	42
$CaCl_2$	5.25	99.1	0.98	42
Na_2SO_4	5.25	99.3	0.74	42
HNO_3	0.25	51.3	1.26	42
Sodium lauryl sulfate	0.172	98.2	1.77	42
Sucrose	5.25	99.7	1.58	42
Tetramethyl ammonium chloride	5.25	99.6	1.16	42
Tetraethyl ammonium chloride	5.25	99.4	1.26	42
Phenol	0.0080	−22	0.91	11
Dextrose	0.094	99.74	1.15	27
Na_2HPO_4	1.5	99.98	0.95	27
NH_4OH	0.0106	25	1.07	27
2,4-dichlorophenoxy acetic acid (2,4-D)	0.0035	92.8	0.87	27

the cellulose ester, and membrane imperfections. Compaction is a term applied to the compression of the porous membrane at high pressure and the associated decrease in porosity and permeability. After exposure to pressure, the membrane porosity decreases irreversibly from about 60% to 5–10% as deduced from water sorption and gross thickness measurements. Two apparently related phenomena accompany this thickness reduction. First, the water flux becomes nonlinear in the net pressure at pressures on the order of 70–100 atm. Second, the membrane permeability decreases with time at a rate that increases with both applied pressure and temperature. The flux decline appears to be the result of membrane creep in which the thickness of the thin skin increases as the very fine pores just beneath the skin close. The main evidence for this mechanism lies in the fact that the permeability to both water and salts decreases at about the same rate, that is, semipermeability is invariant. The time dependence of the compaction phenomenon is consistent with a creep mechanism in that the flow decays with time according to a power law. Typical results (43) are presented in terms of log (membrane constant) versus log (time) in

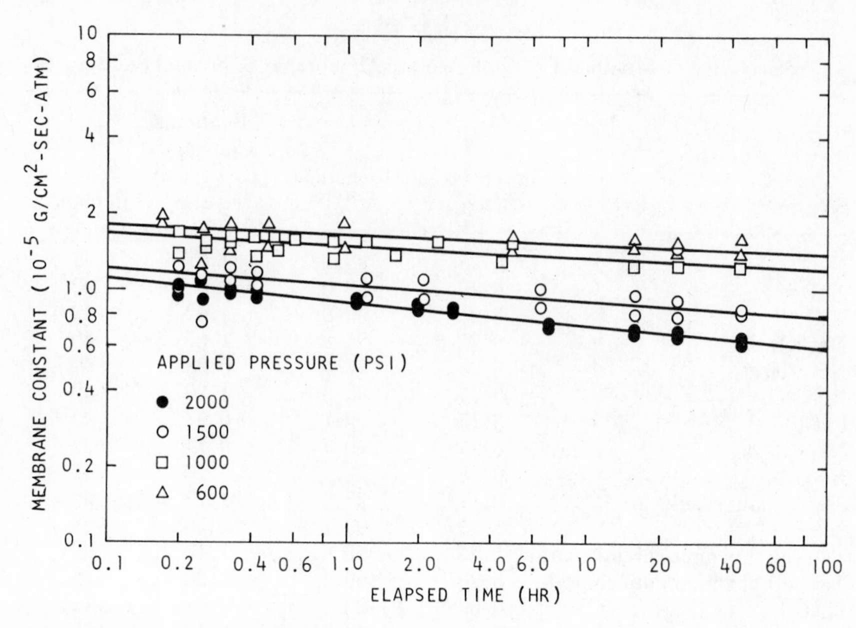

Fig. 5. Change in membrane constant with time for modified membranes annealed at 80°C.

Figure 5. Plots of this type have been found to be linear over four orders of magnitude in time. The compaction "slope," that is, the slope of the log-log plot, is dependent on pressure, as shown, but is also dependent on the temperature at which the membrane is annealed: higher compaction slopes are associated with lower annealing temperatures and higher initial membrane constants. The activation energy for this compaction phenomenon is also consistent with a creep mechanism, being on the order of 50 kcal/mole (44). Most reverse osmosis units are operated at pressures of 50 atm or less where the compaction phenomenon is not severe. Typical values of the compaction slopes at 25°C and the effect on integrated water production over varying periods of time τ are presented in Table III (27). It is assumed that the instantaneous flux has the form $J = J_i t^{-m}$, where J_i is the initial flux and m is the compaction slope. For example, if J_i is measured at an initial time $t_i = 1$ hr and the slope is 0.10, the total water produced after 10^4 hr (1.1 yr) will be only 44% of what would be expected without compaction.

The lifetime of modified cellulose acetate membranes is determined by hydrolysis. A detailed study (45) of the pH and temperature dependence of the hydrolysis kinetics has been combined with the trends of water and NaCl

TABLE III

Relative Integrated Water Production versus Compaction Slope

Compaction Slope (m)	τ/t_i		
	10^2	10^3	10^4
0.10	0.70	0.56	0.44
0.07	0.78	0.66	0.56
0.04	0.86	0.79	0.72
0.01	0.96	0.94	0.92

permeability with acetyl content to yield membrane lifetime predictions (46). A summary of some of these is given in Table IV. The membrane life is defined here in terms of the NaCl permeation constant, B; the times required for B to increase by 20, 50, 100, and 400% of its original value are presented. Because the membrane constant is much less dependent on acetyl content, the change in B is approximately equal to the change in (1 − rejection). For the most part, these lifetime predictions have been borne out by experiment (44). Membrane hydrolysis is minimized in the pH range 4–6 and most reverse osmosis units with cellulose acetate membranes are operated within this range, usually by acid addition to lower

TABLE IV

"Lifetime" of Cellulose Acetate Membranes: Time for NaCl Permeation Constant to Increase by 20, 50, 100, and 400% at 23°C

pH	Hydrolysis Rate Constant, k_1 (sec^{-1})	20% Δt	50% Δt	100% Δt	400% Δt
2	4.8×10^{-9}	15d [a]	33d	42d	130d
3	6×10^{-10}	120d	270d	1.2y	2.8y
4	2×10^{-10}	1.0y [b]	2.2y	3.7y	8.2y
5	1.8×10^{-10}	1.1y	2.4y	4.1y	9.5y
6	3×10^{-10}	180d	1.4y	2.5y	5.7y
7	1.3×10^{-9}	55d	120d	210d	1.3y
8	1.3×10^{-8}	5.5d	12d	21d	49d
9	1.3×10^{-7}	13h [c]	1.2d	2.1d	4.9d
10	1.3×10^{-6}	1.3h	2.9h	5.0y	12h

[a] d = days.
[b] y = years.
[c] h = hours.

the pH of natural waters. It has recently been shown that biological degradation of cellulose acetate membranes can be significant under certain circumstances. Cantor and Mechalas (47) cultured species of bacteria which attacked the membranes, resulting in a relatively rapid loss in semipermeability, apparently as the result of enzymatic hydrolysis.

As the degree of acetylation of cellulose acetate is increased, it becomes less hydrophilic and the permeability to most species decreases. A summary of the permeability to water, D_1c_1, and sodium chloride, D_2K, is shown in Figure 6 (8). The permselectivity increases with acetyl content, and this would appear to offer an additional degree of freedom in membrane tailoring. However, because of the limited number of solvents for cellulose acetates with degree of substitution other than 2.5, it has proven difficult to prepare high flux membranes of the Loeb-Sourirajan type. Recently, Saltonstall and co-workers have reported on the preparation of two such membranes of improved salt-rejection capability (48). One is prepared by blending two cellulose acetates with approximate degrees of substitution of 2.5 and 2.8 to produce a membrane with an average degree of substitution of about 2.7. The other membrane is prepared by increasing the overall degree of substitution by esterification with methacrylate or other groups. Both the polymer blend and the cellulose acetate methacrylate membrane typically exhibit 99.5% rejection of NaCl under conditions where 98% rejection is observed with cellulose 2.5-acetate membranes. These new membranes thus appear to be promising for seawater desalination, an application where at least 99% salt rejection is required to produce potable water in a single-pass operation.

B. Thin Films

A number of other types of reverse osmosis membranes have been developed in the past few years. In one of these, cellulose acetate is still the membrane material but a composite membrane is prepared consisting of a very thin film supported by a finely porous matrix (49,50). The structure is thus similar to that of the Loeb-Sourirajan membrane but in the composite membrane the porous supporting membrane is prepared first and the thin film is later applied to one surface either by transfer from another surface or by direct application. Such composite membranes offer at least one inherent advantage: the thin film and the porous support can be prepared from different materials, each optimized for a specific purpose. Some of these composite membranes prepared on a laboratory scale have exhibited excellent performance as illustrated in the reverse osmosis results with 0.9 wt % NaCl in Table V. The data were obtained with composite membranes consisting of a thin film of cellulose 2.5-acetate

TABLE V

Reverse Osmosis Results with Thin-Film Composite Membranes. Thin Films of Eastman 398-10 Cellulose Acetate Supported on Millipore VFWP Membranes

Thin Film Thickness (Å)	Applied Pressure (psi)	Water Flux (gal/ft²-day)	Salt Rejection (%)	Membrane Constant (10^{-5} g/cm²-sec-atm)
Acetone Solvent				
2800	1560	9.0	99.81	0.44
2800	1560	9.7	99.81	0.46
Methyl Acetate Solvent				
3000	840	4.5	99.47	0.39
3000	840	4.8	99.10	0.42
3000	840	4.5	99.35	0.39
3000	840	4.3	99.10	0.37

supported on a Millipore filter of 250 Å pore diameter (51). The rejections are superior to those in Tables I and II; this is apparently not because of imperfections in the modified membranes (which reject >99.9% of NaH_2PO_4, for example), but reflects the higher intrinsic selectivity of the thin films. The techniques developed for preparing these composite membranes should be of use in the development of other kinds of supported ultrathin polymeric films for other types of membrane separation processes.

C. Hollow Fibers

The use of fine hollow fibers of suitable polymers represents a distinct approach to reverse osmosis separation and purification. These fibers are quite fine but the wall thickness is nevertheless significantly larger than the thin skin on the modified cellulose acetate membrane. Thus, although the water transport rate per unit membrane area is substantially lower than that readily achieved with the modified membrane, high-water production rates in terms of volume of product water per unit volume of pressure vessel are possible because of the high fiber-packing density. The approach was apparently first conceived by Skeins and Mahon (52). Among the polymers they studied was cellulose triacetate, which, as noted in Figure 6, has a lower water permeability but higher selectivity than the 2.5-acetate used in the modified Loeb-Sourirajan membrane. Hollow fibers of nylon have been developed by Mattson, Tomsic (53), and co-workers. A photomicrograph of these fibers in cross section is shown in Figure 7. A summary of water (D_1c_1) and NaCl (D_2K) permeability results for a series of nylon

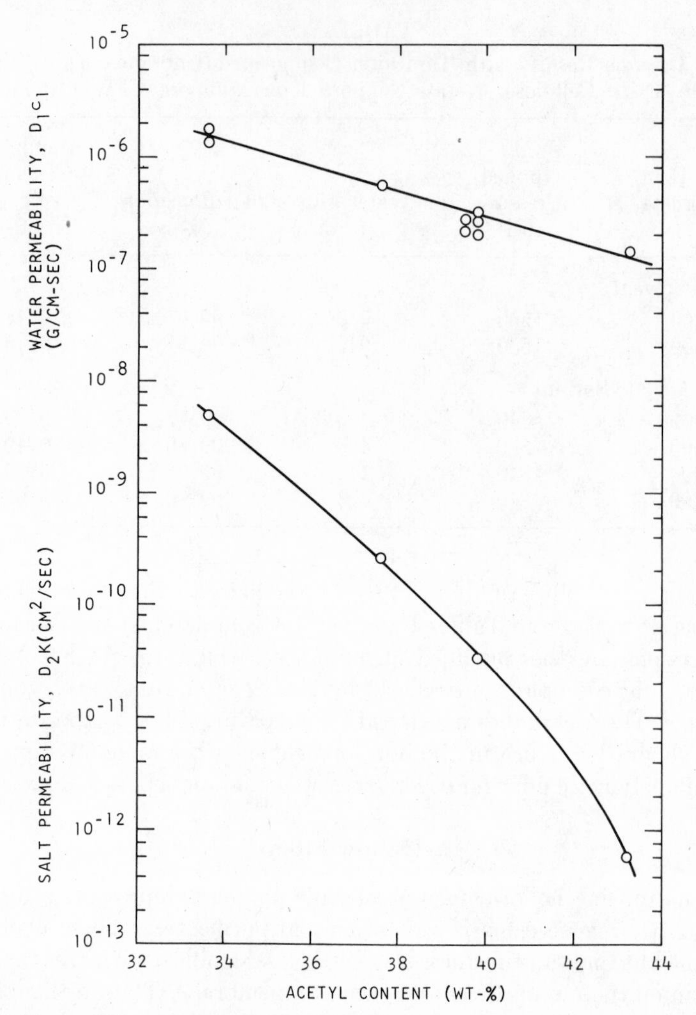

Fig. 6. Permeability to water (D_1c_1) and NaCl (D_2K) of cellulose acetate membranes versus acetyl content.

membranes is presented in Table VI (54). It may be noted by comparing these results with those of Figure 6 that nylons in general are less permeable to water than cellulose 2.5-acetate, which is consistent with their lower hydrophilicity. Furthermore, the ratio of water permeability to salt permeability is not as favorable for the nylons and this is consistent with the observations that these hollow fibers do not reject NaCl and other low molecular solutes as well as the cellulose acetate membranes.

Fig. 7. Cross section of nylon hollow fibers. Photograph supplied by Mr. W. E. Hansen, Du Pont Company.

The fibers are typically on the order of 50–100 μ outside diameter and are made with sufficient wall thickness to be self-supporting at the high external pressures required for reverse osmosis processing. In the two applications of this approach thus far described (52,53), cellulose triacetate and nylon fibers have been used with typical wall thicknesses of 10 μ. The fiber walls are apparently not "skinned" so the effective film thickness is also on the order of 10 μ. However, because of the extremely high membrane area per unit volume of membrane, reasonably high water production rates per unit volume of pressure vessel are achievable. Typical data cited for nylon hollow fibers are: packing density of 10,000 ft^2 of membrane/ft^3 of pressure vessel, water flux of 0.1 gal/ft^2-day at 600 psi, and specific water production rate of 1,000 gal/ft^3-day. From the flux, thickness, and pressure data, we can calculate the water permeability of the nylon fiber from the solution-diffusion model. The result, about 2.5×10^{-7} g/cm-sec, is consistent with some of the substituted 6/6 nylon permeability data of Table VI. The fibers are apparently not subject to the compaction phenomenon commonly observed with modified membranes. In the case of the nylon fibers at least, membrane hydrolysis is apparently insignificant over a wide pH range and membrane life is predicted to exceed that of

TABLE VI
Water and Salt Permeability of Selected Nylon Films as Measured in Direct Osmosis at 25°C

Material	Film	Thickness (μ)	D_1c_1 (g/cm-sec)	D_2K (cm²/sec)
Du Pont Zytel-31 nylon (polyamide 6/10)	Extruded	29	5.6×10^{-9}	$<1.3 \times 10^{-12}$
Du Pont Fe-2355 nylon copolymer	Extruded	22	2.2×10^{-8}	9.2×10^{-11}
Du Pont Zytel-42	Extruded	24	7.2×10^{-9}	7.9×10^{-11}
Du Pont Zytel-2353	Extruded	50	6.9×10^{-9}	1.3×10^{-10}
Du Pont Zytel-2380	Extruded	45	3.7×10^{-9}	1.1×10^{-10}
Du Pont Zytel-61 nylon copolymer	Ethanol cast	52	1.2×10^{-8}	1.6×10^{-10}
Du Pont Zytel-61-P nylon (Zytel-61 with plasticizer)	Ethanol cast	25	3.2×10^{-8}	3.5×10^{-10}
Belding Corticelli nylon 809 (alkoxy-alkyl substituted nylon 6/6)	Ethanol cast	85	5.0×10^{-7}	1.26×10^{-8}
Belding Corticelli nylon 819 (alkoxy-alkyl substituted nylon 6/6)	Ethanol cast	105	5.2×10^{-7}	1.1×10^{-8}
		26	4.4×10^{-7}	1.0×10^{-8}
Belding Corticelli nylon 829 (alkoxy-alkyl substituted nylon 6/6)	Ethanol cast	85	6.8×10^{-7}	1.6×10^{-8}

cellulose acetate. A limitation to the nylon hollow fibers is their semi-permeability. Rejection of sodium chloride appears to be on the order of 50% under conditions where 98% is observed with modified cellulose acetate membranes. Using the uncoupled solution-diffusion model, a rejection of 50% at 40 atm net pressure difference corresponds to a ratio of permeabilities, D_1c_1/D_2K, of about 30. The ratio observed in direct osmosis for the highly water permeable nylons of Table VI is about 40.

Tailoring of the nylon hollow fibers (53) to vary flow and selectivity is possible as shown in Figure 8. Using species with molecular weights 1540, 6000, and 20,000, relationships were established between membrane flux and solute rejection for a series of membranes.

D. Dynamically Formed Membranes

Another entirely new approach to the preparation of reverse osmosis membranes is that of dynamically formed membranes currently being pursued by Johnson, Kraus, and co-workers at Oak Ridge National Labor-

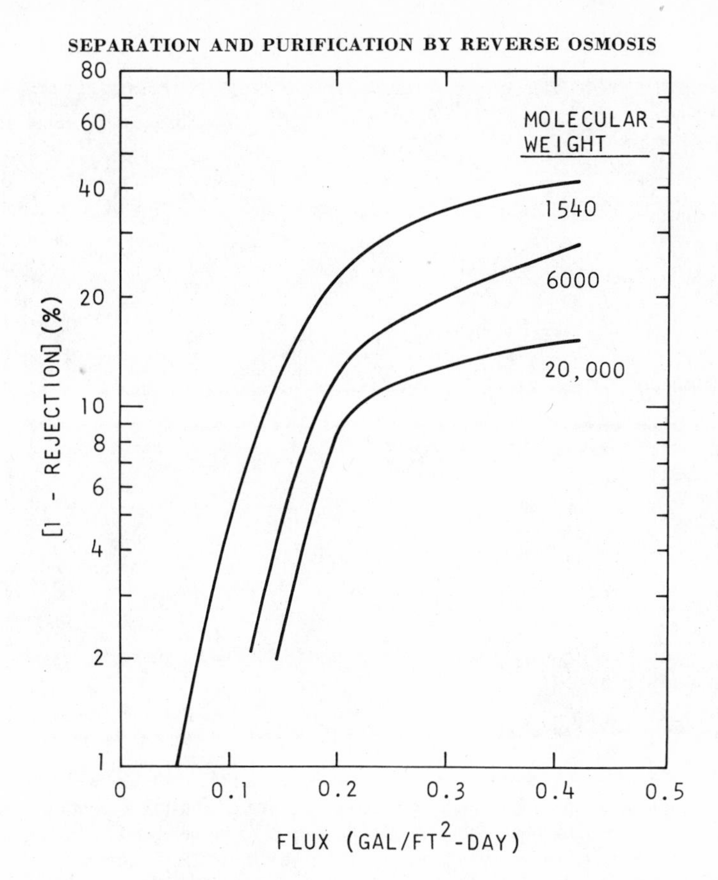

Fig. 8. Relationship between solute rejection and water flux for nylon hollow fibers.

atory (55–58). These membranes are prepared *in situ* on the surface of a finely porous support; they are formed dynamically by passing a pressurized solution containing any one of several dissolved or colloidal species over the porous support. Some of the materials from which membranes have been successfully prepared are hydrous oxides of metals such as zirconium and thorium, polyelectrolytes and neutral polymers, and the colloidal and suspended matter present in a variety of waste waters. Some of the porous supports found effective include ceramics, porous carbon tubes, finely porous silver frits, and Millipore filters. A cross section of a hydrous zirconium (IV) oxide membrane dynamically formed on one surface of a ceramic tube is shown in Figure 9 (58). The membrane-forming material has penetrated a few microns into the surface and the process is probably similar to the formation of filter mats. Water flow through membranes of this type usually decreases with time and solute rejection

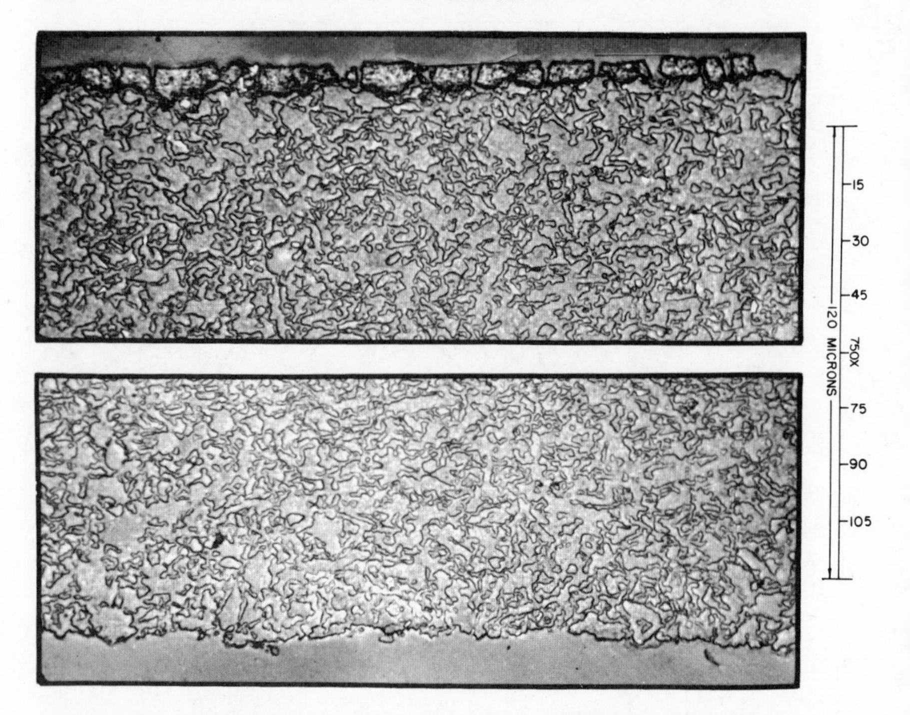

Fig. 9. Photomicrograph of the cross section of a ceramic tube on which a hydrous Zr (IV) oxide membrane had been dynamically formed. The upper section is the exterior (membrane had been dynamically formed. The upper section is the exterior (membrane) surface; the lower section is the interior (effluent) surface. Photograph courtesy of Dr. J. S. Johnson, Jr., ORNL.

improves with time, analogous to filter cake phenomena. However, many such dynamically formed membranes are sufficiently finely porous to reject sodium chloride. It is not possible, of course, to detach these membranes in order to determine their intrinsic properties. Effective membrane thickness is poorly defined and the membranes are not highly reproducible. A good deal of useful information has been inferred, however, by reverse osmosis measurements. Some of the membranes clearly reject salt by a Donnan exclusion mechanism by virtue of a considerable fixed-charge capacity. Divalent coions are more highly excluded and divalent counterions more highly sorbed than monovalent ions of the same sign, and this is reflected in the salt rejections. Some results of an 800 psi test of a hydrous zirconium (IV) oxide membrane prepared in acid medium on a porous carbon tube are presented in Figure 10. As we have noted, if solute and solvent flows are strongly coupled, solute rejection, R, is given by $1 - K_{co}$

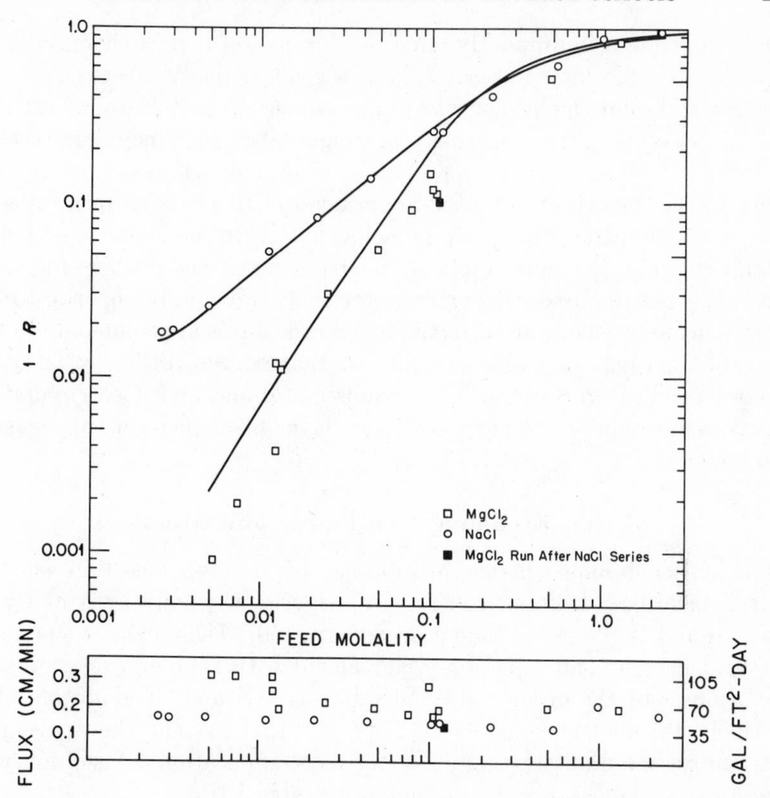

Fig. 10. Effect of feed concentration on the salt rejection and water flux exhibited by a hydrous Zr (IV) oxide membrane. Photograph supplied by Dr. J. S. Johnson, Jr., ORNL.

in the case of high solvent flux, or $1 - R = K_{co}$. Because K_{co} is proportional to the external solution concentration at low concentrations, a log-log plot of $1 - R$ versus concentration should be linear. For a monovalent coion such as Na^+, the slope should be unity and for a divalent coion (Mg^{2+}) the slope should be two. The curves in Figure 10 were drawn in this way using a fixed-charge capacity of 0.45 meq/cc water for both salts and applying a suitable correction for the concentration dependence of the activity coefficients of NaCl and $MgCl_2$ in solution (58). In this case, at least, solute rejection appears to be almost completely determined by solute exclusion.

In some cases, however, electrolytes are rejected by uncharged membranes such as polyvinylpyrrolidone or polyvinylpyridine, so that Donnan exclusion is apparently not the only important mechanism. Hydrous zirconium oxide membranes dynamically formed on porous silver, carbon, or poly(vinyl chloride) supports exhibit a good correlation between solute rejection and solute molecular weight for a series of alcohols from methanol to a poly(ethylene glycol) of molecular weight 6000. Such membranes could be useful for separation or fractionation of organic solutes.

One of the important potential advantages of this type of membrane is that very high water fluxes may be achievable with moderate rejection of sodium chloride. For some applications, particularly those where high salt rejection is not required, these membranes may prove to be superior to the Loeb-Sourirajan membranes. Initial membrane fluxes are commonly in the range of hundreds of gallons/ft^2-day so that concentration polarization effects (discussed in Section IV) are substantial and high feed circulation velocities are required to keep boundary layer phenomena within reasonable bounds.

E. Ion-Exchange Membranes and Others

Conventional ion-exchange membranes of the type used in electrodialysis have been examined in reverse osmosis, originally by McKelvey, Spiegler, and Wyllie more than a decade ago (60). These exhibit a concentration-dependent salt rejection in agreement with Donnan exclusion considerations, but the permeability to water is low and, in addition, only relatively thick membranes have been prepared so that the achievable water fluxes are uninterestingly low for most applications. A summary of some reverse osmosis tests is presented in Table VII.

Berg and co-workers (61) have developed a graphitic oxide membrane capable of rejecting sodium chloride in reverse osmosis. The membrane is prepared by depositing the graphitic oxide on one surface of a suitable porous support such as a Millipore filter. Some typical reverse osmosis results are presented in Figure 11. The test was performed at 41 atm applied pressure with 0.5% NaCl. While the membranes exhibit lower selectivity and lower water flux than modified cellulose acetate membranes, they are considerably more hydrolytically stable and could find application in highly acidic or basic media.

A series of unfired glasses that reject salt have been developed by Guter, Littman, and others (62). As with graphitic oxide, the selectivity and water permeability are both well below the marks commonly achieved with cellulose acetate but the "membranes" are stable under a variety of conditions deleterious to cellulose acetate.

A number of other new kinds of reverse osmosis membranes are currently being developed under sponsorship of the Office of Saline Water. These include a crosslinked polyvinylpyrrolidone system, hydroxyethyl meth-

TABLE VII

Membrane Constant and Salt Rejection of Ion-Exchange Membranes

Membrane	Thickness (μ)	Δp (psi)	Feed Solution (wt %)	Salt Rejection (%)	Membrane Constant (10^{-7}g/cm^2-sec-atm)	Water Flux (gal/ft^2-day)	Reference
Nalfilm-1 (cationic)	101	1400	0.585 NaCl	61	0.55	0.1	60
Nalfilm-2 (anionic)	101	1000	0.0585 NaCl	91	0.080	0.01	60
Nalfilm-2 (anionic)	101	1000	0.585 NaCl	90	0.11	0.02	60
Nalfilm-2 (anionic)	101	1000	5.85 NaCl	46, 42	0.12, 0.15	0.001	60
Nalfilm-2 (anionic)	101	1000	5.55 $CaCl_2$	72	0.13	0.01	60
AMF (anionic)	151	1000	5.85 NaCl	73	0.28	0.02	60
AMF (anionic)	151	1000	5.55 $CaCl_2$	94	0.35	0.02	60
AMF (anionic)	151	1000	7.1 Na_2SO_4	75	0.46	0.03	60
Permaplex C-10	750	1200	0.585 NaCl	19	4.6	0.8	60
AMF C-310	288	810	0.60 NaCl	77, 78	2.2, 2.4	0.3	25
AMF C-103 Fe	148	810	0.60 NaCl	81, 86	0.48, 0.51	0.06	25
Cl 2.5T	170	1050	0.60 NaCl	59, 80	0.66, 0.85	0.1	25
PRC Ion X-Negative	50	600	0.21 NaCl	46	31	2.5	43
PRC	50	600	0.21 NaCl	25	70	6.1	43

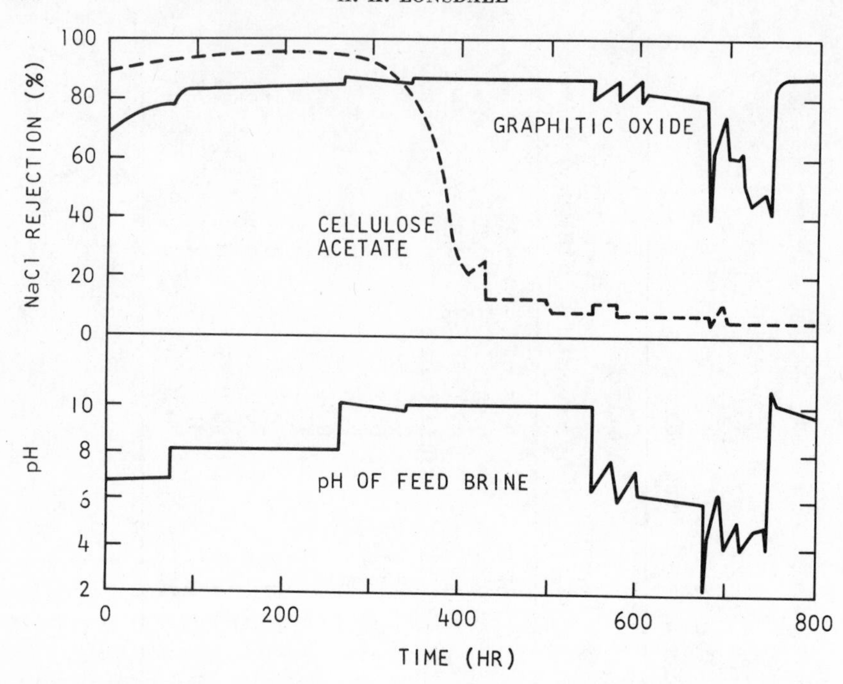

Fig. 11. Sodium chloride rejection by graphitic oxide and cellulose acetate membranes with feed brines of varying pH.

acrylate and other methacrylate-based polymers, certain esterified polysaccharides, surface-modified celluloses, a variety of block and graft copolymers, and others. These studies are described in the OSW Saline Water Conversion Report for 1967 (63). To aid in the evaluation of such materials and to expedite the search for additional materials, some simple permeability measurements have been proposed (10) in which the transport properties can be determined even when it is not possible to prepare imperfection-free membranes. Such procedures have proven to be of value in surveying potential membranes before attempting to prepare high flux, effectively very thin membranes, a task which has proven to be quite difficult.

In closing this brief description of membrane materials it is worth noting that while modified cellulose acetate membranes combine high water permeability with high salt rejection, some of the other membranes may find important applications in areas other than water desalination. For example, those membranes that do not retard the flow of low molecular weight salts could be useful in desalting and dehydrating food products, proteins, pharmaceuticals, and other products. Salts would be removed along with water while the solutes of value were retained. A number of such applications are already under study although they have not been extensively reported in the literature as yet.

IV. BOUNDARY LAYERS

As water is transported through the reverse osmosis membrane the solute that is retained tends to accumulate at the brine-membrane interface, creating a concentration polarization effect. This effect is well known in membrane separation processes. It has several consequences, all of which are usually considered negative: the high local solute concentration leads to increased solute transport and the high local osmotic pressure reduces the driving force for solvent flow, and both of these effects lower solute rejection. In addition, the solubility of slightly soluble compounds can be exceeded under certain conditions resulting in a scaly deposit on the membrane surface. The theory and implications of this polarization effect have been worked out in detail for a variety of flow conditions and flow channel configurations (64–68). Because even in a well-stirred system the liquid velocity approaches zero at the walls, the only mechanism for transporting the solute away from the interface is molecular diffusion. From a mass balance consideration, the convective flow of solute to the membrane by solvent flow must be balanced by diffusive flow back toward the bulk of the solution and through the membrane:

$$J_2 = B\Delta\rho_2 = J_1\rho_2 - D_2^0(d\rho_2/dy) \tag{16}$$

where y is a length parameter in the solution perpendicular to the membrane surface and D_2^0 is the molecular diffusivity of the solute in the solution. If the membrane is considered solute impermeable for simplicity, eq. (16) can be integrated to yield

$$\rho_2 = \rho_2^0 \exp\,(J_1\Delta y/D_2^0) \tag{17}$$

where ρ_2 is the interfacial solution concentration and ρ_2^0 is the bulk solution concentration. In turbulent flow, Δy is the boundary layer thickness. In laminar flow, Δy is a characteristic channel dimension such as the hydraulic radius. The magnitude of the effect under reverse osmosis conditions can be appreciated by substituting typical values. A typical water flux is 10^{-3} g/cm^2-sec (21 gal/ft^2-day), Δy is on the order of 10^{-2} cm for turbulent flow, and D_2^0 for NaCl is 1.6×10^{-5} cm^2/sec at 25°C. For these conditions the polarization modulus, ρ_2/ρ_2^0, is 2 and the polarization effect is already substantial. Increasing the water flux by a factor of two leads to a modulus of about 4, which is frequently intolerable in the desalination of natural waters. The polarization moduli for solutes with diffusion coefficients less than that of NaCl are, of course, higher and the modulus in a laminar flow system with a hydraulic radius substantially greater than 10^{-2} cm can be extremely large. Many natural waters and most waste streams contain a small amount of either colloidal matter or macromolecules, the diffusion coefficient of which in water is on the order of 10^{-7}–10^{-8} cm^2/sec (69). The interfacial concentrations of such species will obviously be very high relative to bulk values and it is commonly observed that membranes easily become fouled with either inorganic or organic films when operated on only slightly turbid feed waters. Some progress has been made in preventing membrane fouling by feed pretreatment and in removing fouling layers after deposition by various cleaning procedures, but this is probably the most serious present limitation in reverse osmosis as well as other membrane separation processes.

Problems associated with boundary layers should be reduced in systems in which membrane flux, J_1, is very small such as in hollow fibers. It is clear from eq. (17) that ρ_2/ρ_2^0 will be small if J_1 is on the order of 10^{-5} g/cm^2-sec and if turbulent flow were maintained throughout the system. Little information is available on fouling in hollow fiber reverse osmosis systems but presumably to some degree fouling will be common to all membrane separation processes.

V. FIELD TESTS

As a water purification process, reverse osmosis has reached the pilot plant stage. The results of field tests with pilot plants not only add to our information on membrane performance but also indicate some of the difficulties in working with natural waters and demonstrate the relative merits of the different engineering configurations. The most intensively investigated of these configurations are depicted in Figure 12. The plate-and-frame design is well known in unit operations. A number of water desalination pilot plants were assembled with this design, including the first by Loeb and Milstein (70), but work in this area has apparently declined recently. In the tubular designs, modified membranes are either inserted into or cast directly within long support tubes which serve as pressure vessles. The spiral-wound module has a kind of plate-and-frame design wherein flexible materials are used and the membrane sheets are rolled up into right circular cylinders to conserve pressure vessel volume. In the hollow fiber designs, the very fine hollow-spun fibers previously described are either wound continuously around a mandrel or are utilized in a device similar in configuration to a U-tube single-end heat exchanger. Pressurized feed solution is applied to the outside of the fiber and the product is withdrawn through the fiber bore.

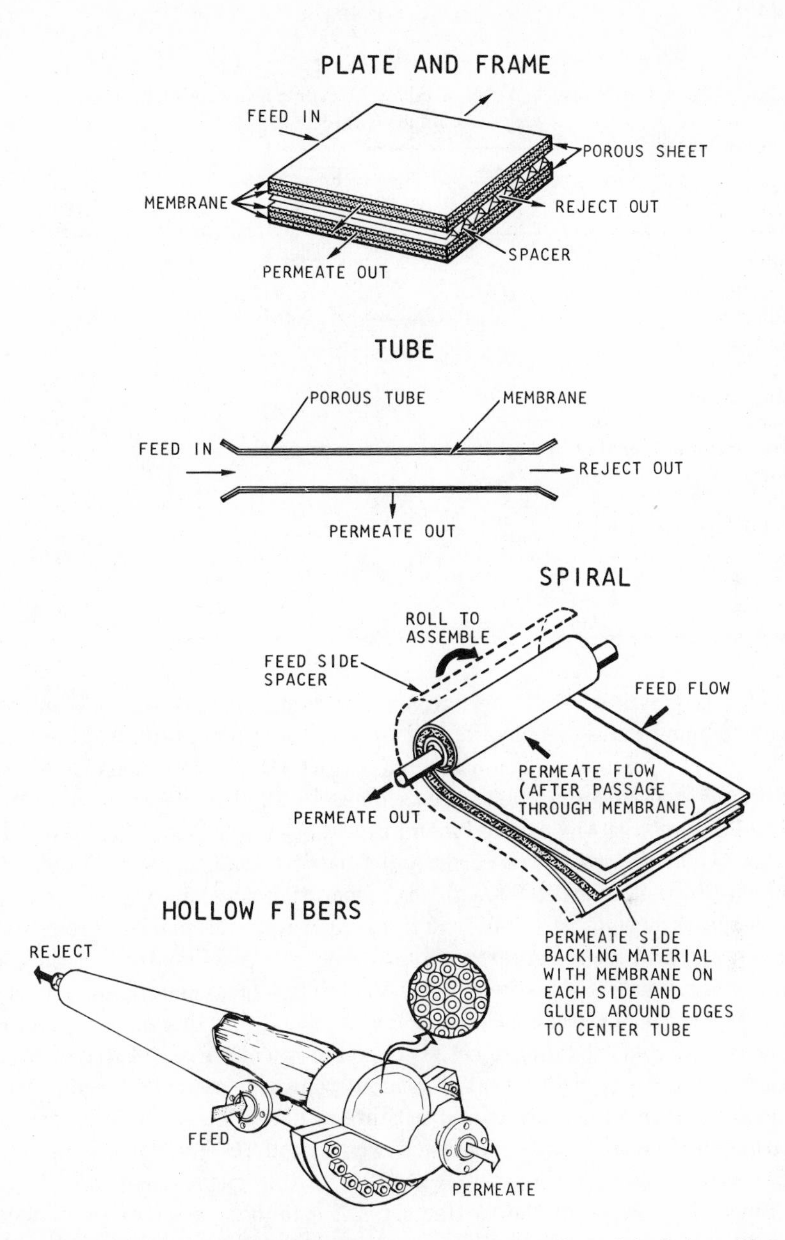

Fig. 12. Schematic representations of several reverse osmosis module designs. Photograph of hollow fiber unit supplied by Mr. W. E. Hansen, Du Pont Company.

TABLE VIII
Feed and Product Water Analysis. Reverse Osmosis Field Test at Coalinga, California

	Feed Water (ppm)	Product Water (ppm)
Sulfate	1227.2	48.1
Sodium	543.0	62.0
Chloride	266.0	78.0
Bicarbonate	158.6	36.6
Calcium	116.6	11.6
Magnesium	75.1	6.7
Silica	35.0	12.0
Iron and alumina (R_2O_3)	12.2	2.4
Boron	2.5	1.5
Iron	0.04	0.04
Total dissolved solids	2417.6	265.6
Total hardness (as $CaCO_3$)	604.3	56.9
pH	7.6	7.5

The best indication of the long-term performance of modified cellulose acetate membranes can be obtained from the Coalinga, California, tubular pilot plant designed and constructed by UCLA (71,72). This plant was first operated in June 1965 and as of early 1969 it was still operating, although not with the original membranes; plant availability has exceeded 98%. Capacity was increased several times from an initial value of 5000 gal/day to a nominal 10,000 gal/day. Operating pressure was 41 atm, and typically somewhat over 60% of the feed water was passed through the membrane and recovered as potable water. A typical analysis of the feed and product waters is given in Table VIII. The rejection of dissolved solids based on the feed water concentration was about 89%; based on the average concentration of feed and reject streams the rejection was 93–94%. Membranes with a variety of heat-treatment temperatures were used so that the flux and salt rejection varied within the plant. The average water flux during the first 21 months operation was 21 gal/ft^2-day, corresponding to an average membrane constant, A, of 2.5×10^{-5} g/cm^2-sec-atm.

Since the startup of the Coalinga plant a number of other pilot plants have been constructed. The largest plant reported on to date has been a 50,000 gal/day unit of the spiral module design, constructed and operated in San Diego, California, for the Office of Saline Water (73). This unit was put into operation initially at 10,000 gal/day in December 1967 and in July 1968 it was expanded to 50,000 gal/day; it was operating at that

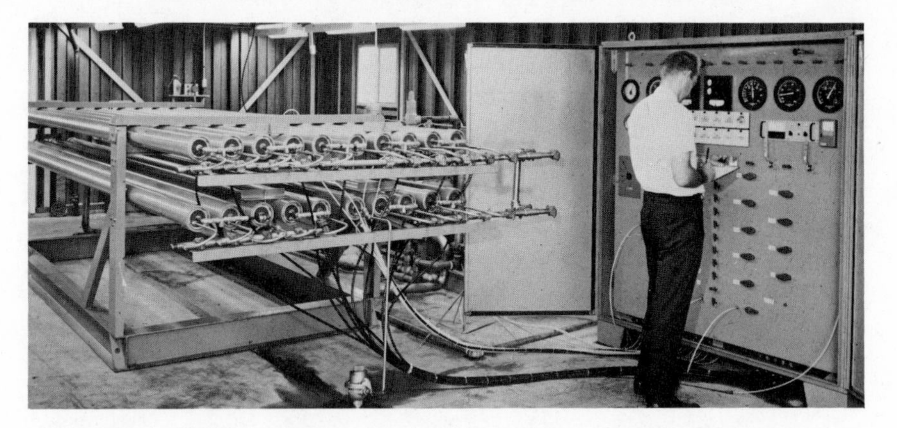

Fig. 13. Photograph of a 50,000 gallon per day reverse osmosis plant.

level as of early 1969. A photograph of the unit is presented in Figure 13. The plant operated at an applied pressure of 41 atm and 75% of the 4500 ppm brine was recovered as 350 ppm product. Except during the periods of planned experimentation, plant availability was virtually 100%. A typical analysis of the feed, brine, and product waters is shown in Table IX. The rejection of total dissolved solids based on the feed water concentration was slightly greater than 93%; based on the feed-brine average concentration, rejection was greater than 97%. Product water flux was about 10 gal/ft^2-day, corresponding to an average membrane constant of 1.2×10^{-5} g/cm^2-sec-atm. The water production rate per unit volume of pressure vessel (an important consideration in unit costs) was about 1500 gal/ft^3-day.

Some indication of the utility of membrane tailoring can be obtained by comparing the Coalinga and San Diego pilot plants. The water at Coalinga was of lower salinity and it was high in sulfate ion, to which the membrane is highly impermeable. In the San Diego test, the feed water was largely NaCl so that less permeable, higher rejection membranes were required to achieve potable product water. The rejection of Cl^- at Coalinga (based on the feed-brine average concentration) was only 81% while in the San Diego test it was 97%. Potable water was produced in both plants, but the water flux in the latter case was only about one half because the membranes were annealed at a higher temperature to increase salt rejection.

Other pilot plants using spiral-wound modules have been operated on a secondary sewage effluent at the Pomona, California, site of the County Sanitation Districts of Los Angeles County (74). The purpose of such tests has been to study the possibility of upgrading the quality of municipal waste waters for reuse. One of the pilot plants has operated at 27 atm with

TABLE IX
Typical Water Analysis of 50,000-GPD Pilot Plant

	Concentration (mg/1)		
	Feed	Brine	Product
Cations			
Sodium	900	3,700	106
Calcium	357	1,500	5.6
Magnesium	221	1,020	3.4
Potassium	26	103	3.8
Iron	7	27	0.1
Manganese	4.1	17	0.0
Anions			
Chloride	2,260	8,900	180
Sulfate	612	2,580	2.9
Bicarbonate	61	18	1.8
Silica	35	58	6.4
Fluoride	1.1	3.6	0.1
Nitrate	0.9	3	0.0
Boron	0.3	0.5	0.2
Total alkalinity ($CaCO_3$)	50	15	1.5
Total hardness ($CaCO_3$)	1,800	7,940	28
Dissolved solids [a]	4,490	17,620	300
Conductivity (μmho/cm at 20°C)	7,150	23,000	685
pH	6.0	4.6	6.0

[a] Residue upon drying at 180°C.

80 to 85% recovery of the feed water. Typical analyses of feed, brine, and product waters are presented in Table X. The feed water was the effluent from an activated sludge treatment process that had been further treated by passing it through an activated carbon column to remove most of the oxygen-demanding organic species.

An additional important application of reverse osmosis for water reclamation has been examined in field tests with several of the engineering designs by the Pulp Manufacturers Research League of Appleton, Wisconsin. Wiley, Amerlaan, and Dubey have reported favorably on the concentration of waste liquors of the pulp and paper industry and the renovation of waste waters for reuse (75). Highly annealed modified cellulose acetate membranes are virtually impermeable to most of the color-producing lignins and other oxygen-demanding substances. Up to 80% recovery of reusable water is possible. The waste can be concentrated to 8–10 wt % at which point disposal by evaporation and burning is feasible.

TABLE X

Water Quality Data from Reverse Osmosis Unit Operating on Secondary Effluent

	PO_4 (mg/1)		COD[a] (mg/1)		NH_3-N (mg/1)		NO_3-N (mg/1)		TDS[b] (mg/1)	
	Average	Range	Average	Range	Average	Range	Average	Range	Average	Range
Feed	30.9	21.6 –37.8	10.8	6.7–14.4	9.2	2.5–26.0	2.4	0.8– 5.0	623	530– 748
Product water	0.57	0.13– 1.2	1.7	0.0– 3.4	1.7	0.9–3.4	0.8	0.4– 1.2	73	30– 113
Brine	177	73 –240	43.8	26.6–60.4	94	19 –240	7.5	2.8–13.7	3402	2861–3889
Percent reduction based on feed	98.2		84		82		67		88	

[a] Chemical oxygen demand.
[b] Total dissolved solids.

Several studies have been made of the application of reverse osmosis to various facets of food processing including concentration, separation, and water recovery (76). In the existing method of fruit juice concentration, for example, many of the important aroma-imparting molecular species are evaporated along with the water during low-temperature, low-pressure distillation. Merson and Morgan (77) have reported that highly selective modified cellulose acetate membranes have a low permeability to most of the important alcohols, esters, and aldehydes in apple juice and to the largely aromatic oil-soluble aromas of orange juice. These and other juices have been successfully concentrated and the flavor of the reconstituted juices was found to be excellent. As normally prepared, the membranes are virtually impermeable to mono- and disaccharides. Sugar retention is therefore complete. The osmotic pressures of these concentrated fruit juices are high: four-fold concentrated orange juice has an osmotic pressure of 93 atm so that pressures well in excess of 100 atm are required to achieve significant water removal rates. Certain operational difficulties are encountered including membrane compaction, fouling, and concentration polarization effects. Lowe and co-workers (78) have described a plate-and-frame device for use in reverse osmosis concentrating of foods. Other applications in this area include concentration of maple sap in the production of maple syrup (79), the fractionation and concentration of whey (80), and the concentration of egg white. Reverse osmosis for food processing is the subject of a recent review by Merson, Ginnette, and Morgan (81).

A number of desalination field tests of nylon hollow fiber membrane have recently been reported (52). Results of one such test on an agricultural runoff water are shown in Table XI. The test was carried out at 42 atm applied pressure and at 24°C; the water recovery was 24%. With the raw water, individual ion rejections were not as high as with modified cellulose acetate membranes (cf. Table IX), but pH adjustment improved the rejection of most species.

Dynamically formed membranes have been applied to the recovery of water from municipal wastes and from pulp mill wastes (82). Membranes were formed on ceramic or graphite tubes from the suspended matter present in the water and no feed additives were necessary. With 1% sulfite liquor, rejection based on optical density measurements on feed and product waters was 90–96% after several days of continuous operation and there was substantial rejection of oxygen-demanding species and of total dissolved solids. Membrane constants were in the order of $2\text{–}4 \times 10^{-5}$ g/cm^2-sec-atm at 27 atm applied pressure. It would appear that other interesting and useful applications of these novel membranes to water recovery and renovation, concentration of valuable products, and other separation and purification processes will follow in the near future.

TABLE XI

Field Test Results of Hollow Fiber Desalination Unit at Firebaugh, California

	Run 1		Run 2 (with acidification)	
Ion	Feed (ppm)	Product (ppm)	Feed (ppm)	Product (ppm)
SO_4^{2-}	3,265	36	3,515	101
Na^+	1,410	329	1,665	200
Cl^-	449	372	450	212
Ca^{2+}	328	18.9	336	7.7
HCO_3^-	295	165	49.0	25.6
Mg^{2+}	167	7.3	173	3.2
SiO_2	47.1	32.8	47.3	32.5
NO_3^- (as N)	17.1	11.4	9.7	7.9
F^-	0.2	<0.1	0.2	<0.1
K^+	3.4	0.8	3.6	0.5
B	11.0	11.6	12.0	13.0
Fe^{2+}	<0.1	0.0	0.3	0.1
Total alkalinity	242	135	40.2	21.0
Total hardness	3,150	101	1,692	44.0
Total dissolved solids	5,926	958	6,000	654
pH	7.4	7.3	5.8	5.6

Acknowledgments

The author is indebted to the Office of Saline Water for support of some of the studies reported herein and to Dr. U. Merten for reviewing the manuscript.

References

1. A. S. Michaels, "Ultrafiltration," in *Progesss in Separation and Purification*, Vol. I, E. S. Perry, Ed., Interscience Publishers, New York, 1968.
2. A. S. Michaels, *Chem. Eng. Progr.*, **64**, 31 (1968).
3. E. J. Breton, Jr., "Water and Ion Flow Through Imperfect Osmotic Membranes," Office of Saline Water Research and Development Progress Report
4. C. E. Reid and E. J. Breton, *J. Appl. Polymer Sci.*, **1**, 133 (1959). No. 16, 1957.
5. S. Loeb and S. Sourirajan, UCLA Department of Engineering Report 60-60, 1960; *Advan. Chem. Ser.*, **38**, 117 (1962).
6. R. L. Riley, J. O. Gardner, and U. Merten, *Science*, **143**, 801 (1964).
7. R. L. Riley, U. Merten, and J. O. Gardner, *Desalination*, **1**, 30 (1966).
8. U. Merten, Ed., *Desalination by Reverse Osmosis*, The M.I.T. Press, Cambridge, 1966.
9. J. S. Johnson, Jr., L. Dresner, and K. A. Kraus, "Hyperfiltration (Reverse Osmosis)," in *Principles of Desalination*, K. S. Spiegler, Ed., Academic Press, New York, 1966, Chapter 8.

10. H. K. Lonsdale, U. Merten, and R. L. Riley, *J. Appl. Polymer Sci.*, **9**, 1341 (1965).
11. H. K. Lonsdale, U. Merten, and M. Tagami, *J. Appl. Polymer Sci.*, **11**, 1807 (1967).
12. A. Katchalsky and P. F. Curran, *Nonequilibrium Thermodynamics in Biophysics*, Harvard University Press, Cambridge, 1967.
13. O. Kedem and A. Katchalsky, *Biochim. Biophys. Acta*, **27**, 229 (1958).
14. A. J. Staverman, *Trans. Faraday Soc.*, **48**, 176 (1952).
15. W. D. Stein, *The Movement of Molecules across Cell Membranes*, Academic Press, New York, 1967.
16. K. S. Spiegler, *Trans. Faraday Soc.*, **54**, 1408 (1958).
17. U. Merten, "Transport Properties of Osmotic Membranes," in *Desalination by Reverse Osmosis*, U. Merten, Ed., The M.I.T. Press, Cambridge, 1966.
18. K. S. Spiegler and O. Kedem, *Desalination*, **1**, 311 (1966).
19. B. Z. Ginzburg and A. Katchalsky, *J. Gen. Physiol.*, **47**, 403 (1963).
20. G. Thau, R. Bloch, and O. Kedem, *Desalination*, **1**, 129 (1966).
21. U. Merten, H. K. Lonsdale, R. L. Riley, and M. Tagami, "Transport Processes In Cellulose Acetate Membranes," Proceedings of the NATO Advanced Study Institute on Synthetic Polymer Membranes, Ravello, Italy, September 11–19, 1966, to be published.
22. R. L. Riley, C. R. Lyons, and U. Merten, "Transport Properties of Crosslinked Polyvinylpyrrolidone Membranes," Proceedings of the National Meeting of the American Institute of Chemical Engineers, Technical Session on Tailored Polymeric Materials for Advanced Separations Processes, Washington, D.C., November 1969, *Desalination*, in press.
23. E. Klein, J. K. Smith, and R. P. Wendt, *J. Polymer Sci.*, to be published.
24. S. Loeb and S. Sourirajan, U.S. Patent 3,133,132 (May 1964); S. Loeb, S. Sourirajan, and D. E. Weaver, U.S. Patent 3,133,137 (May 1964).
25. U. Merten, H. K. Lonsdale, R. L. Riley, and K. D. Vos, "Reverse Osmosis Membrane Research," Office of Saline Water Research and Development Progress Report No. 265, March 1967.
26. P. A. Cantor, *et al.*, "Research and Development of New and Improved Cellulose Ester Membranes," Final Report on Office of Saline Water Contract No. 14-01-0001-1732, January 1969, to be published.
27. H. K. Lonsdale, C. E. Milstead, B. P. Cross, and F. M. Graber, "Study of Rejection of Various Solutes by Reverse Osmosis Membranes," Office of Saline Water Research and Development Progress Report No. 447, March 1969.
28. S. Kimura and S. Sourirajan, *A.I.Ch.E. J.*, **13**, 497 (1967).
29. S. Kimura and S. Sourirajan, *Ind. Eng. Chem., Process Design Develop.*, **7**, 197, 548 (1968).
30. S. Sourirajan and T. S. Govindan, *Proceedings First International Symposium on Water Desalination*, Washington, D.C., October 3–9, 1965, Vol. 1, p. 251.
31. S. Manjikian, *Ind. Eng. Chem. Prod. Res. Develop.*, **6**, 23 (1967).
32. S. Manjikian, S. Loeb, and J. W. McCutchan, *Proceedings First International, Symposium on Water Desalination*, Washington, D.C., October 3–9, 1965 Vol. 2, p. 159.
33. S. Manjikian and C. Allen, "Development of Reverse Osmosis Desalination Membranes," Office of Saline Water Research and Development Progress Report No. 378, December 1968.

34. C. J. van Oss, "Ultrafiltration Membranes," this volume; C. J. van Oss, C. R. McConnell, R. K. Tompkins, and P. M. Bronson, *Clin. Chem.*, in press (1969).
35. W. J. Elford, *Trans. Faraday Soc.*, **33,** 1094 (1937).
36. P. Grabar, *Compt. Rend.* **198,** 1640 (1934).
37. M. F. Vaughan, *Nature,* **183,** 43 (1959).
38. S. Loeb, U.S. Patent 3,364,288 (January 16, 1968).
39. C. W. Saltonstall, Jr., W. M. King, and D. L. Hoernschemeyer, *Desalination,* **4,** 309 (1968).
40. L. L. Markley, R. A. Cross, and H. J. Bixler, "Membranes for Desalination by Reverse Osmosis," Office of Saline Water Research and Development Progress Report No. 281, December 1967.
41. J. Jagur-Grodzinski and O. Kedem, *Desalination,* **1,** 327 (1966).
42. R. Blunk, UCLA Department of Engineering Report 64-28, 1964.
43. U. Merten, H. K. Lonsdale, R. L. Riley, and K. D. Vos, "Reverse Osmosis for Water Desalination," Office of Saline Water Research and Development Progress Report No. 208, April 1966.
44. U. Merten, H. K. Lonsdale, R. L. Riley, and K. D. Vos, "Reverse Osmosis Membrane Research," Office of Saline Water Research and Development Progress Report No. 369, March 1968.
45. K. D. Vos, F. O. Burris, Jr., and R. L. Riley, *J. Appl. Polymer Sci.*, **10,** 825 (1966).
46. K. D. Vos, A. P. Hatcher, and U. Merten, *Ind. Eng. Chem. Prod. Res. Develop.*, **5,** 211 (1966).
47. P. A. Cantor, B. J. Mechalas, O. S. Schaeffler, and P. H. Allen, III, "Biological Degradation of Cellulose Acetate Reverse Osmosis Membranes," Office of Saline Water Research and Development Progress Report No. 340, 1968.
48. C. W. Saltonstall, Jr., "Development and Testing of High-Retention Reverse Osmosis Membranes," paper presented at International Conference PURAQUA, Rome, Italy, February 17–22, 1969, to be published.
49. R. L. Riley, H. K. Lonsdale, L. D. LaGrange, and C. R. Lyons, "Development of Ultrathin Membranes," Office of Saline Water Research and Development Progress Report No. 386, May 1968.
50. L. T. Rozelle, J. E. Cadotte, R. D. Corneliussen, and E. E. Erickson, "Development of New Reverse Osmosis Membranes for Desalination," Office of Saline Water Research and Development Progress Report No. 359, October 1968.
51. R. L. Riley, H. K. Lonsdale, C. R. Lyons, and U. Merten, *J. Appl. Polymer Sci.*, **11,** 2143 (1967).
52. W. E. Skeins and H. I. Mahon, *J. Appl. Polymer Sci.*, **7,** 1549 (1963); H. I. Mahon, U.S. Patent 3,228,876 (January 11, 1966).
53. R. J. Mattson and V. J. Tomsic, *Chem. Eng. Prog.*, **65,** 62 (1969).
54. H. K. Lonsdale, U. Merten, R. L. Riley, and K. D. Vos. "Reverse Osmosis for Water Desalination," Office of Saline Water Research and Development Progress Report No. 150, June 1965.
55. A. E. Marcinkowsky, K. A. Kraus, H. O. Phillips, J. S. Johnson, Jr., and A. J. Shor, *J. Am. Chem. Soc.*, **88,** 5744 (1966).
56. K. A. Kraus, H. O. Phillips, A. E. Marcinkowsky, and J. S. Johnson, *Desalination,* **1,** 225 (1966).
57. K. A. Kraus, A. J. Shor, and J. S. Johnson, Jr., *Desalination,* **2,** 243 (1967).
58. A. J. Shor, K. A. Kraus, W. T. Smith, Jr., and J. S. Johnson, Jr., *J. Phys. Chem.*, **72,** 2200 (1968).

59. J. R. Kuppers, A. E. Marcinkowsky, K. A. Kraus, and J. S. Johnson, *Separation Science*, **2(S)**, 617 (1967).
60. J. G. McKelvey, Jr., K. S. Spiegler, and M. R. J. Wyllie, *Chem. Eng. Prog. Symposium Series, A.I.Ch.E.*, **55**, 199 (1959).
61. L. C. Flowers, D. E. Sestrich, and D. A. Berg, "Reverse Osmosis Membranes Containing Graphitic Oxide," Office of Saline Water Research and Development Progress Report No. 224, 1966.
62. F. E. Littman and G. A. Guter, "Research on Porous Glass Membranes for Reverse Osmosis," Office of Saline Water Research and Development Progress Report No. 379, December 1968.
63. Saline Water Conversion Report for 1967, U.S. Department of the Interior, U.S. Government Printing Office, Washington, D.C.
64. U. Merten, H. K. Lonsdale, and R. L. Riley, *Ind. Eng. Chem. Fundamentals*, **3**, 210 (1964).
65. T. K. Sherwood, P. L. T. Brian, R. E. Fisher, and L. Dresner, *Ind. Eng. Chem. Fundamentals*, **4**, 113 (1965).
66. P. L. T. Brian, *Ind. Eng. Chem. Fundamentals*, **4**, 439(1965). Also, M.I.T. Desalination Research Laboratory Report 295-7, May 1965.
67. P. L. T. Brian, "Mass Transport In Reverse Osmosis," in *Desalination by Reverse Osmosis*, U. Merten, Ed., The M.I.T. Press, Cambridge, 1966, Chapter 5.
68. W. N. Gill, C. Tien, and D. W. Zeh, *Ind. Eng. Chem. Fundamentals*, **4**, 433 (1965).
69. C. Tanford, *Physical Chemistry of Macromolecules*, Wiley, New York, 1961.
70. S. Loeb and F. Milstein, *Dechema Monograph.*, **47**, 707 (1962).
71. S. Loeb, *Desalination*, **1**, 35 (1966).
72. S. Loeb and J. S. Johnson, *Chem. Eng. Prog.*, **63**, 90 (1967).
73. R. G. Sudak, "River Valley Golf Course Test of a 10,000 GPD Reverse Osmosis Pilot Plant, Vol. I," Final Report on Office of Saline Water Contract No. 14-01-0001-1264, September 1968, to be published.
74. T. J. Larson, *Desalination*, **7**, 187 (1970).
75. A. J. Wiley, A. C. F. Amerlaan, and G. A. Dubey, *Tappi*, **50**, 455 (1967).
76. A. I. Morgan, Jr., E. Lowe, R. L. Merson, and E. L. Durkee, *Food Technol.*, **19**, 52 (1965).
77. R. L. Merson and A. I. Morgan, Jr., *Food Technol.*, **22**, 97 (1968).
78. E. Lowe, E. L. Durkee, and A. I. Morgan, Jr., *Food Technol.*, **22**, 915 (1968).
79. C. O. Willitts, J. C. Underwood, and U. Merten, *Food Technol.*, **21**, 24 (1967).
80. P. G. Marshall, W. L. Dunkley, and E. Lowe, *Food Technol.*, **22**, 969 (1968).
81. R. L. Merson, L. F. Ginnette, and A. I. Morgan, Jr., *Dechema Monograph.*, **63**, 179 (1969).
82. J. J. Perona, *et al.*, *Environ. Sci. and Technol.*, **1**, 991 (1967).

Progress in Inorganic Thin-Layer Chromatography

F. W. H. M. Merkus

Head, Pharmaceutical Department,
Roman Catholic Hospital,
Sittard, The Netherlands

I. TLC OF CATIONS ON CELLULOSE

A. Introduction

Many papers have been published on paper chromatography of cations. Despite the growth of TLC in organic analysis, the application to inorganic analysis has been somewhat limited. Reviews of the literature on inorganic TLC have been written by Seiler (1), Randerath (2), Pollard *et al.* (3,4), Garel (5), Lesigang-Buchtela (6), Takeuchi (7), Takitana and Kawanabe (8), Merkus (9), and Lederer (10).

TLC has, as is known, advantages over paper chromatography because of its greater sharpness of separation, its rapidity, and its sensitivity (see Fig. 1). In the past years a number of investigators have carried out thin-layer chromatographic separations of cations. As sorbent they have been using silica gel. The separation of cations on silica gel, however, cannot be performed with good results. Many cations are grouped together, because their R_f values are in the same region.

It is important to note that rather little use has been made of cellulose as a sorbent for the separation of cations by TLC. This fact is particularly interesting because all data for inorganic paper chromatography are applicable to inorganic TLC on cellulose and because by means of paper chromatography many problems in inorganic analysis could be solved. It will be demonstrated that the results of the separation of cations on cellulose are excellent and that TLC of cations on cellulose is capable of considerable utility.

B. Experimental Data for the Preparation of the Cellulose Layer

Thin layers of cellulose can be made by the usual applicators available for the general preparation of TLC coatings. It is very important that the glass plates to which the thin layer is to be applied be thoroughly clean. This is best accomplished by washing the plates in a concentrated solution of sodium carbonate followed by rinsing with distilled water. The use of some detergents and alcohol must be rejected because they contain fatty substances. In order to obtain a completely smooth layer, it is advisable to disperse the aqueous suspension of the cellulose powder thoroughly for about 1 min in a mechanical mixer. Freshly prepared plates should be always dried in air. Cellulose layers require no activation at higher temperatures. They may be left to dry out on the laboratory bench (see below).

The cellulose powder-water suspension is prepared by mixing 15 g powder in 90 ml distilled water. The cellulose powders involved in most studies on inorganic TLC are cellulose powder MN 300 or MN 300 HR (Macherey, Nagel & Co., Düren, Germany). These cellulose powders, suspended in

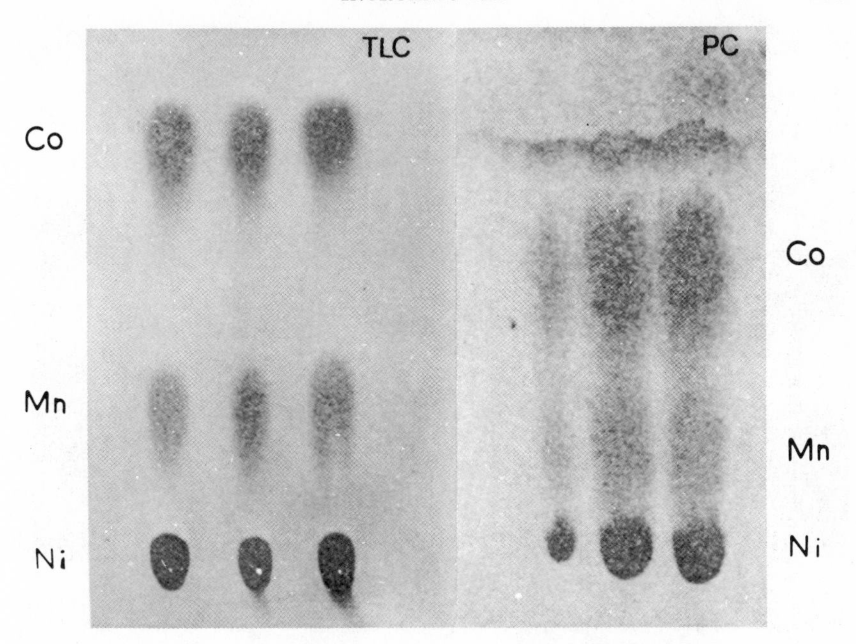

Fig. 1. TLC on cellulose MN 300 and PC on Whatman Nr. I in the same chamber and in the same time. The sharpness of separation on the cellulose layer is far better than the result on paper. Spray reagent: pyridylazonaphthol (0,2% in methanol), followed by exposure to NH_3 vapor. Solvent mixture: acetone-HCl (d = 1.19)-H_2O (90:3:7).

water, possess an exceptional adhesiveness for glass and, therefore, the coated layers are firmly bound to the glass plate support.

Merkus (9) has successfully prepared thin layers on small glass plates (20 × 5 cm) without a standard applicator by simply pouring the suspension over the glass plates (see Figs. 2 and 3). The quality of these thin layers is excellent. Most of his experiments, therefore, were carried out on cellulose layers made by this procedure. The photographs of separations made on this type of plate, shown in this chapter, demonstrate the quality of these layers. Precoated cellulose plates capable of rendering excellent separations are also available commercially, for example, Merck No. 5716. These plates are not expensive because they can be cut into small pieces for use.

The practice of activating cellulose layers as is often found in the literature is not recommended. The exposure of cellulose to high temperatures adversely affects its water-retaining capability. If the TLC process on cellulose layers is considered as partition chromatography, the water imbibed in the cellulose serves as the stationary phase. The cation, there-

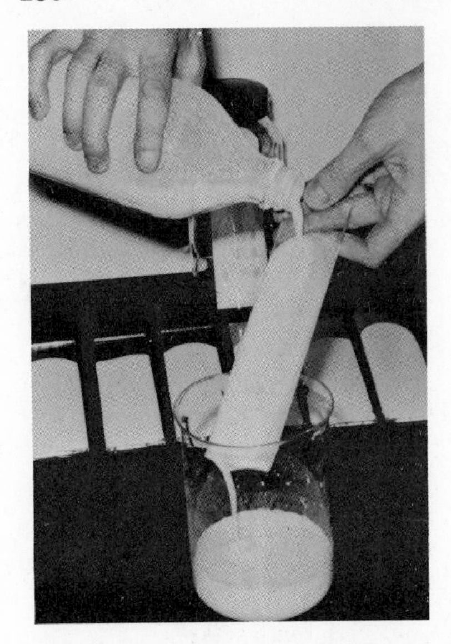
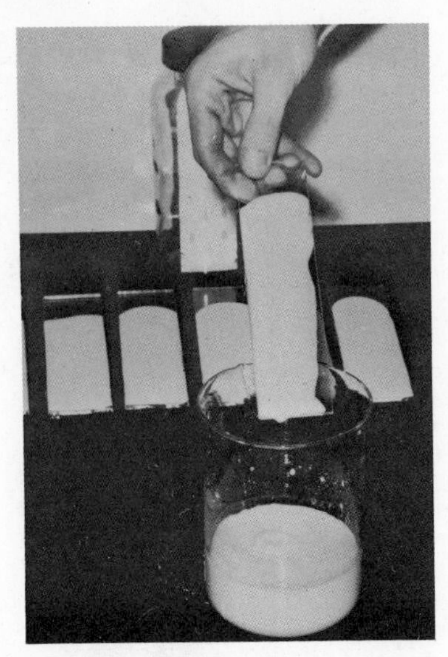

Figs. 2 and 3. Application of the cellulose layer by hand. These hand-made chromatoplates are of very good quality. The layer is remarkably stable and can be written upon with a pen [Merkus (9)].

fore, distributes itself between this stationary water phase and the mobile phase of choice. In this light, then, the use of the word "adsorbent" in referring to the cellulose layer is not applicable; "sorbent" would be more logical.

C. The Cellulose Layer as Stationary Phase

The starting material for cellulose MN 300 thin layers is a highly purified cellulose powder of which more than 95% of the fibers are 2 to 20 μ in length. The fibers in chromatographic paper are several times longer. Consequently, the surface area of the stationary phase in the cellulose powder is many times greater. These two properties of the cellulose powder cause an important limitation for TLC in the rapid diffusion of the solute along the fibers and an enhancement of the amount of solute which can be supported per unit area of the thin layer (11).

Photomicrographs of cellulose fibers of chromatographic paper and cellulose powder MN 300 for TLC are shown in Figures 4 and 5. The cellulose fiber consists of bundles of polysaccharide chains, some regions of which are highly organized and are crystalline by x-ray diffraction patterns. The x-ray data also show strong interchain hydrogen bonding in the crystalline regions. The remaining material (about 40%) is amorphous.

Fig. 4. Photomicrograph of cellulose fibers in chromatography paper (magnification about 100 ×) [Merkus (9)].

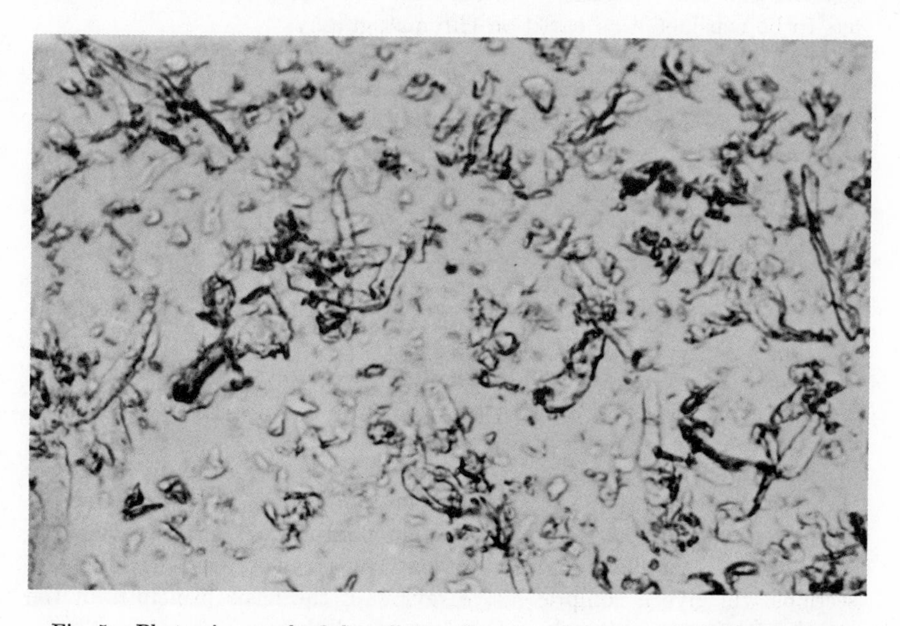

Fig. 5. Photomicrograph of the cellulose fibers in cellulose powder MN 300 (magnification about 500 ×) [Merkus (9)].

The amorphous regions are the active sites of the cellulose in the chromatographic process (12).

Paper chromatography and TLC on cellulose can be defined as chromatography in which the stationary phase is a volume phase. The water that is bound by the cellulose may penetrate into the amorphous regions by imbibition and the cellulose-water complex formed may be considered a "solution" of cellulose in water. The stationary phase is consequently a volume phase consisting of water. In order to understand a separation process, it is necessary to know which components of the mobile phase can be bound to the cellulose and influence the composition of the stationary phase.

A very important investigation on the mechanism of paper chromatography has been carried out by Schute (13,14,15). Schute's findings concerning the influence of the mobile phase on the stationary phase have been confirmed by other investigators [Ackerman and Michal (16,17,18)]. These studies show that liquids other than water can also be imbibed into cellulose. The degree of imbibition decreases in the sequence water, methanol, ethanol. The imbibition of acetone and *n*-butyl alcohol is negligible. If the water content of the mobile phase is low the mobile phase can withdraw water from the cellulose and, conversely, if the water content is high cellulose can take up water. The separation of cations on cellulose layers has to be considered as partition chromatography.

Some authors make use of an ion exchange process to explain certain separations. It is well known that purified cellulose contains a few carboxylic groups due to oxidation of the cellulose molecules. However, the ion exchange process can be neglected in these and most other studies because for the separation of cations on cellulose (paper or thin layer), a mobile phase containing a high concentration of hydrochloric acid is usually chosen.

II. MECHANISM OF THE SEPARATION OF CATIONS BY TLC ON CELLULOSE

A. Introduction

To separate cations by TLC on cellulose, it is important to know the theoretical aspects of the correlation between the R_f (or R_M) value of a cation and the composition of the solvent mixture. Recently, Merkus (9) summarized a number of important data on this subject. The following sections will give a comprehensive review of the basic principles of the mechanism for the separation of cations on cellulose.

B. The Relationship between the R_f Value and the Composition of the Solvent Mixture

Most investigators have tried to separate mixtures of cations by paper or thin-layer chromatography with solvent mixtures chosen empirically. An important and original study on the relationship between the R_f value of cations and the composition of the solvent mixtures has been carried out by Hartkamp and Specker (20,21). The findings of these experimenters with paper chromatography were confirmed and extended by Merkus (9) for TLC on cellulose layers. A review of the most important data on this subject is presented here.

It appears from the extensive literature on inorganic paper chromatography that most investigators have chosen organic solvent-HCl—H_2O as the solvent mixture for the separation of cations. Hartkamp and Specker (20,21) have studied tetrahydrofuran-HCl—H_2O. The solvent mixtures involved in Merkus' experiments were acetone-HCl—H_2O, methanol-HCl—H_2O, and *n*-butyl alcohol HCl—H_2O (9).

It is known that all cations are very well transported by water as the mobile phase and, accordingly, have R_f values close to unity. This is understandable from a consideration of the following relation:

$$R_f = \frac{1}{1 + \alpha(A_s/A_m)} \tag{1}$$

The partition coefficient, α, is equal to 1 in this case because the stationary phase and the mobile phase both consist of water. Since the cross-sectional area of the mobile phase (A_m) is much larger than the cross-sectional area of the stationary phase (A_s) the ratio of A_s/A_m is very small. The R_f value therefore is nearly 1.0 for all cations. However, when a solvent mixture consists of HCl, an organic solvent, and water, the influence of each individual component on the transport of each cation must be known in order to effectively separate mixtures.

In order to elucidate the relationship between the composition of the solvent mixture and the transport of a cation in each series of experiments, the organic solvent concentration was fixed at a constant level while the HCl- and H_2O concentrations were varied inversely (9,20). Hartkamp and Specker (20,21) have classified the cations in four transport groups. Each group is characterized by a typical sequential order of increase or decrease of the R_f value, caused by changing the composition of the solvent mixture. Hartkamp and Specker assume that the cation in the stationary phase (the cellulose-water-complex) is present as a hydrated cation, called the aquo-ion. In the mobile phase the cation can be present as an aquo-ion, a solvo-ion, or a chloro-complex. The ionic species is dependent on the properties of the cation and the composition of the solvent mixture. Four types of transports of the cation are to be distinguished here.

Normal Transport. The transport of the cation is solely dependent on the water concentration in the mobile phase. The cation is moving with the solvent mixture as a hydrated cation (aquo-ion) (e.g., Cr, Al, Ni). The equilibrium of eq. (2) is displaced to the left:

$$\begin{array}{ll} \text{mobile phase:} & M(H_2O)_n + n\,\text{Solv} \rightleftharpoons M(\text{Solv})_n + nH_2O \\ & \quad\upharpoonleft\downharpoonright \\ \text{stationary phase:} & M(H_2O)_n \end{array} \tag{2}$$

Inhibited Normal Transport. The hydrated cation is dehydrated by the protons and chloride ions of the solvent mixture. So the transport of the cation is inhibited by hydrochloric acid. The dehydration (decrease of the R_f value) increases with increasing ionic radius (Li < Na < K and Mg < Ca < Sr < Ba). (This applies for instance to Ba, Sr, K, Na.) The equilibrium of eq. (3) is displaced to the right:

$$\begin{array}{ll} \text{mobile phase:} & M(H_2O)_n + Cl(H_2O) \rightleftharpoons M(H_2O)_{n-1} + Cl(H_2O)_2 \\ & \quad\upharpoonleft\downharpoonright \\ \text{stationary phase:} & M(H_2O)_n \end{array} \tag{3}$$

Quasi Normal Transport. The cation is moving with the solvent mixture as a solvo-ion. Solvation of the cation by the organic solvent molecules (e.g., methanol) is responsible for the good transport of a cation, when a solvent mixture has been used which does not contain enough water to explain the relatively high R_f value (e.g., Mg, Li). The equilibrium of eq. (4) is displaced to the right:

$$\begin{array}{ll} \text{mobile phase:} & M(H_2O)_n + n\,\text{Solv} \rightleftharpoons M(\text{Solv})_n + nH_2O \\ & \quad\upharpoonleft\downharpoonright \\ \text{stationary phase:} & M(H_2O)_n \end{array} \tag{4}$$

Anomalous Transport. The cation is moving with the solvent mixture as a chloro-complex, soluble in the organic component of the solvent mixture. Increasing contents of hydrochloric acid and of organic solvent improve the transport of the cation (this applies to Cu, UO_2, Mn, Co, Fe, Cd, Zn, for instance). The equilibrium of eq. (5) is displaced to the right:

$$\begin{array}{ll} \text{mobile phase:} & M(H_2O)_n + p\,Cl + q\,\text{Solv} \rightleftharpoons MCl_p\text{Solv}_q + nH_2O \\ & \quad\upharpoonleft\downharpoonright \\ \text{stationary phase:} & M(H_2O)_n \end{array} \tag{5}$$

The four theoretically derived R_f curves for these transport groups of Hartkamp and Specker are in excellent agreement with the R_f curves of their experiments and with the results of the experiments of Merkus in inorganic TLC on cellulose layers.

A simplified summary of the mathematical derivations of Hartkamp and Specker is presented in this section, including a correction by Boumans (22) and some simplifications by Merkus (9). The solvent mixture acetone-HCl-H_2O will be used for this discussion. The correlation between the R_f value of a cation and the partition coefficient can be expressed by

$$R_f = \frac{1}{1 + \alpha(A_s/A_m)} \tag{1}$$

For dilute solutions the partition coefficient can be expressed by the ratio of the concentration of the cation in the stationary phase to that in the mobile phase:

$$\alpha = \frac{[M(H_2O)n]s}{[M(H_2O)n]m} \tag{6}$$

The partition coefficient is dependent on temperature and on the composition of the stationary and the mobile phases. Each series of experiments, however, were performed at the same time with the same TLC-plates (cellulose powder MN 300), and under the same circumstances. Therefore, the influence of temperature and other variables can be neglected. The partition coefficient may be considered to be dependent only on the properties of the stationary water and the components of the mobile phase, composed of an organic solvent (Solv), water (H_2O), and hydrochloric acid (HCl). Thus

$$\alpha = F([H_2O]m, [Solv]m, [HCl]m, [H_2O]s) \tag{7}$$

1. NORMAL TRANSPORT

The composition of the solvent mixture can be expressed in mole fractions. This will give a quantitative meaning to the influence of the individual components on the R_f value of the cation.

When in eq. (6) the concentration in moles per liter is changed to mole fractions (μ), a correction must be introduced. This was omitted by Hartkamp and Specker, but was corrected by Boumans (22) as follows:

$$\frac{\text{mol}}{\text{liter}} = \frac{\text{mole fraction}}{\text{mole volume of solvent}} = \frac{\mu}{V_o} \tag{8}$$

where V_o is equal to the molecular weight divided by the density.

$$[M(H_2O)n] = \frac{\mu M(H_2O)n}{V_o} \tag{9}$$

Substitution of eq. (9) into eq. (6) gives:

$$\alpha = \frac{[M(H_2O)n]s}{[M(H_2O)n]m} = \frac{\{\mu M(H_2O)n\}s}{\{\mu M(H_2O)n\}m} \cdot \frac{(V_o)m}{(V_o)s} \quad (10)$$

For pure water as stationary phase and pure acetone as mobile phase the ratio $(V_o)m/(V_o)s$ would be

$$\frac{58}{18} \cdot \frac{1.0}{0.8} = 4 \text{ (acetone, mol. wt. } = 58, \text{d} = 0.8\text{; water, mol. wt. } = 18, \text{d} = 1.0\text{)}$$

However, the mobile phase consists partly of acetone. For a mixture of acetone-HCl-H_2O the ratio is always smaller than 4. For most solvent mixtures a value of about 2 can be calculated. Thus, the correction of Boumans amounts to 2 to 4.

An exact measurement of the ratio A_s/A_m is impossible. From start to front the A_m is decreasing. Furthermore the A_s is dependent on the properties of the mobile phase because the mobile phase influences the content of water and other solvents in the stationary phase. Hartkamp and Specker have taken for this ratio the value of 1:10. With this value the experimental and theoretically calculated R_f curves were in good agreement.

Horner, Emrich, and Kirschner (23) have shown that paper can absorb water vapor tightly to 5–7% of its dry weight. With pure liquid water paper can take up an amount of water of 150–200% of its weight (interstitial water). This would make a value between 1:20 and 1:40 for the ratio A_s/A_m acceptable. In Merkus' experiments on cellulose layers the value for the correction of Boumans was about 2 and the ratio A_s/A_m was estimated at 1:20, From these data it can be concluded that it is quite possible that

$$\text{(correction factor of Boumans)} \cdot \frac{A_s}{A_m} = \frac{1}{10} \quad (11)$$

Eq. (11) will be substituted into eqs. (17), (19), (21), and (23). The formation of the aquo-ion takes place in n steps. For each step the probability (W) of the hydration is proportional to the water concentration

$$W_i = k_i \cdot \mu H_2O \quad (12)$$

The proportionality constant k_i takes into account the hydration energy of the step i. The total probability can be presented as the multiplication of the probabilities for n steps. The mobile phase can be written

$$W_m = k_m(\mu H_2O)_m{}^n \quad (13)$$

and the stationary phase:

$$W_s = k_s(\mu H_2O)_s{}^n \quad (14)$$

Since the mole fractions of the aquo-ions are proportional to the probability of formation of the aquo-ions when equilibrium is attained, eqs. (13) and (14) can be substituted in eq. (10).

$$\alpha = \frac{k_s}{k_m} \cdot \frac{(\mu H_2O)_s{}^n}{(\mu H_2O)_m{}^n} \cdot \text{(correction factor)} \tag{15}$$

If we accept that the stationary phase consists only of water, then $(\mu H_2O)_s$ is equal to 1 and supposing $k_s/k_m = 1$, eq. (15) can be rewritten as follows

$$\alpha = \frac{1}{(\mu H_2O)_m{}^n} \cdot \text{correction factor} \tag{16}$$

Substituting eqs. (16) and (11) into eq. (1), the R_f function for Normal Transport can be expressed by

$$R_f = \frac{1}{1 + [1/(\mu H_2O)_m{}^n] \cdot (1/10)} \tag{17}$$

2. INHIBITED NORMAL TRANSPORT

The normal transport of cations is favored by water as component of the mobile phase. The aquo-ion is very well transported by a mobile phase containing a great deal of water. However, there are cations (e.g., some alkali and alkaline earth cations) that are easily dehydrated by the hydrogen ions or chloride ions in the solvent mixture, competing with the cations for the water molecules in the solvent mixture. The hydration and dehydration is dependent on the ionic charge and the ionic radius. Cations such as Ba, Sr, K, and Na are easily dehydrated by a decrease of the water content and an increase in the HCl content of the solvent mixture. This results in a strong drop in the R_f value. The greater the ionic radius, the stronger the decrease in R_f value. The radius of the alkali and alkaline earth cations is: Li 0.60 Å; Na 0.95 Å; K 1.33 Å; Be 0.31 Å; Mg 0.65 Å; Ca 0.97Å; Sr 1.13 Å; Ba 1.35 Å.

The inhibition by HCl of the formation of aquo-ions of alkaline and alkaline earth cations in the mobile phase is expressed by Hartkamp and Specker (20,21) in the equation

$$\alpha = \frac{1}{\{(\mu H_2O)_m - n' \cdot (\mu HCl)_m\}^n} \cdot \text{(correction factor)} \tag{18}$$

Substituting this partition coefficient of the cations, classified into the inhibited normal transport-group, in eq. (1) gives the R_f function for the inhibited normal transport

$$R_f = \frac{1}{1 + [1/\{(\mu H_2O)_m - n' \cdot (\mu HCl)_m\}^n] \cdot 1/10} \tag{19}$$

(It must be noted here that in eqs. (18) and (19) the number n' of water molecules, which are withdrawn from the aquo-ions by HCl, is dependent on the composition of the solvent, that is, the total content of water and HCl and the nature of the organic component.)

Furthermore, it must be emphasized that these theoretical considerations are based on the assumption that the stationary phase consists only of water. There is, however, no doubt that HCl and some organic solvents, like methanol, can penetrate in the stationary phase.

3. QUASI-NORMAL TRANSPORT

When the transport of a cation in the mobile phase is favored by the molecules of the organic solvent (e.g., methanol), the equilibrium in eq. (4) is placed to the right. The R_f value of the cation is high, although the mobile phase does not contain enough water to explain the good transport. The solvation of the cation by the solvent molecules is responsible for the relatively high R_f value. It is understandable that solvation plays an important role especially in the case of methanol. The solvation of cations by higher alcohols, such as *n*-butyl alcohol, or ketones, such as acetone, is negligible.

The probability of the formation of a solvo-ion can be expressed by

$$W = k\{(\mu H_2O)_m + f(\mu \text{ Solv})_m\}^n \tag{20}$$

Factor f represents the properties of the organic solvent and is inversely proportional to the chain length of the organic solvent molecule. When f is small, the R_f curve for quasi-normal transport resembles the R_f curve for normal transport, because only aquo-ions are present in the mobile phase. Hartkamp and Specker have derived for the partition coefficient

$$\alpha = \frac{1}{\{(\mu H_2O)_m + f(\mu \text{ Solv})_m\}^n} \cdot \text{(correction factor)} \tag{21}$$

Substitution of this in eq. (1) gives the R_f function for quasi-normal transport:

$$R_f = \frac{1}{1 + [1/\{(\mu H_2O)_m + f(\mu \text{ Solv})_m\}^n] \cdot 1/10} \tag{22}$$

4. ANOMALOUS TRANSPORT

The R_f curves of the cations, classified under this transport group deviate strongly from R_f curves of the other three groups. In spite of a decrease in the H_2O content of the mobile phase, a remarkable increase in the R_f value of the cation can be noticed. The cation cannot be present as an aquo-ion in the mobile phase because the HCl and organic solvent contents dominate. The only explanation for this anomaly is the assumption that formation of a chloride-complex takes place. The formed metal-chloride complexes are easily soluble in organic solvents, such as acetone and tetrahydrofuran. Therefore, the transport of Cu, Co, Mn, and UO_2 is improved by an increase of the HCl and organic component in the solvent.

For the partition coefficient of these cations Hartkamp and Specker have derived:

$$\alpha = \frac{1}{\{(\mu H_2O)_m{}^n + k_{III}(\mu \text{ Solv})^q(\mu Cl)^p\}} \cdot \text{(correction factor)} \qquad (23)$$

Substitution of this equation in (1) gives the R_f function for this anomalous transport group:

$$R_f = \frac{1}{1 + [1/\{(\mu H_2O)_m{}^n + k_{III}(\mu \text{ Solv})^q(\mu Cl)^p\}]\cdot 1/10} \qquad (24)$$

The equilibrium constant k_{III} relates to eq. (5). This constant is large in the case of Fe, Zn, Ga, and Cd. Thus, these cations have high R_f values (0.9–1.0) as can be derived from eq. (24).

5. EXPERIMENTAL R_f CURVES

The experimental R_f curves of Hartkamp and Specker (21) in paper chromatography using tetrahydrofuran-HCl-H_2O as solvent mixture and the R_f curves of Merkus in TLC on cellulose MN 300 using acetone-HCl-H_2O show a striking agreement with the four theoretically derived R_f curves, drawn in Figures 6, 7, 8, and 9.

Some examples of the experimental R_f curves of Merkus are given in Figures 10, 11, and 12. No further R_f curves are illustrated here because of the resemblance of the R_f curves of the cations classified in the same transport group.

From these R_f curves it can be concluded that it is simple to separate a mixture of cations, classified in different transport groups. The desired separation can be obtained by a solvent mixture, composed by studying the relationship between the R_f value and the composition of the solvent mixture (see Figs. 6–9). The R_f values of cations of the same transport group are generally different enough to get a sharp separation. Even when the transport curves of two cations are congruent (Al-Cr(III) or Fe-Zn), the cations can be detected and identified by using a suitable selective reagent.

C. Relationship between the R_M Value and the Content of HCl in the Solvent

The theoretical basis of the relationship between the R_M values of cations in paper chromatography and the composition of a solvent mixture was first

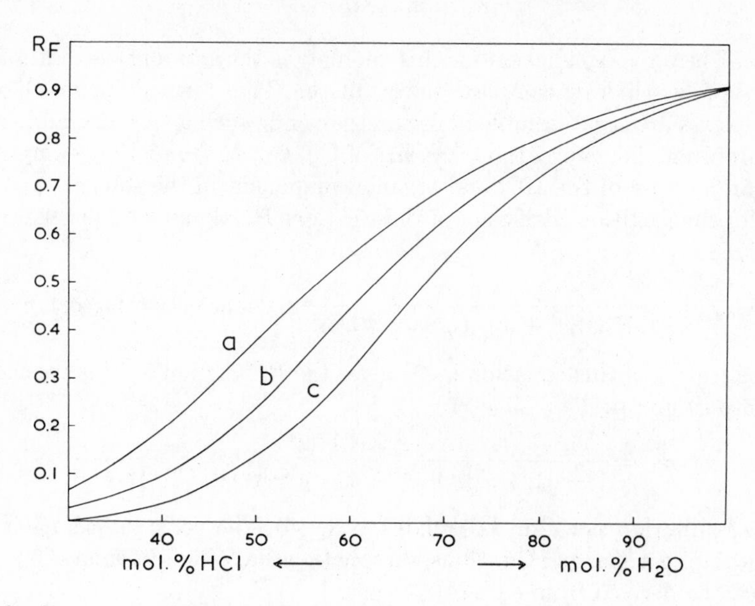

Fig. 6. R_f curves for normal transport; (*a*) $n = 4$; (*b*)$n = 5$; (*c*)$n = 6$ in eq. (15) (20).

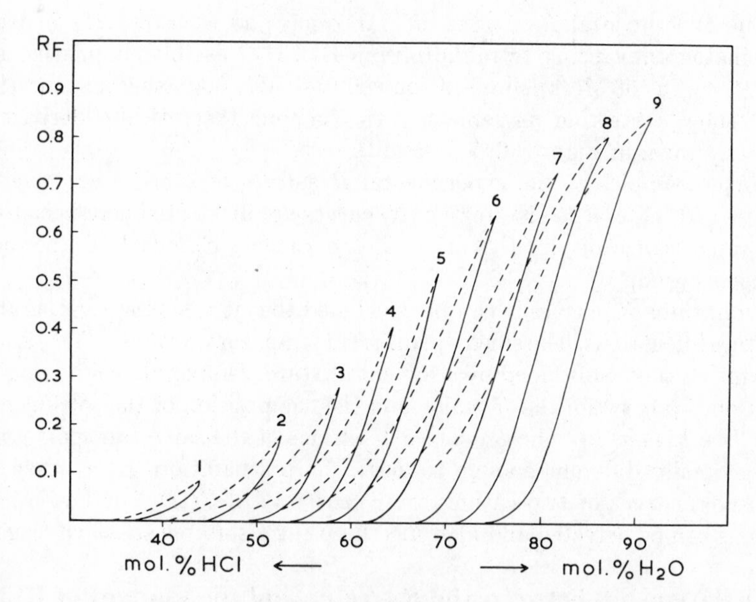

Fig. 7. R_f curves for inhibited normal transport. In eq. (17) $n = 6$ and $n' = 1$ for curve - - - and $n' = 3$ for curve ——; μ Solv = (1) 0.56; (2) 0.48; (3) 0.42; (4) 0.36; (5) 0.31; (6) 0.26; (7) 0.19; (8) 0.14; (9) 0.09 (20).

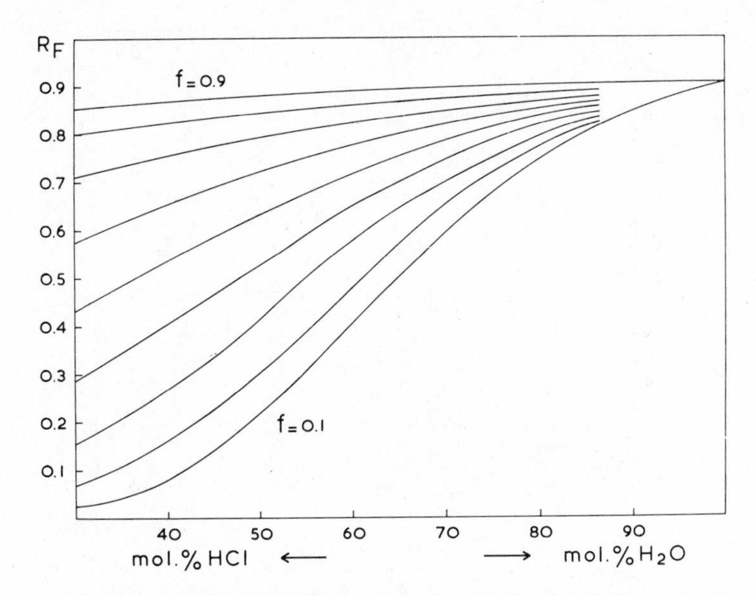

Fig. 8. R_f curves for quasi-normal transport (20).

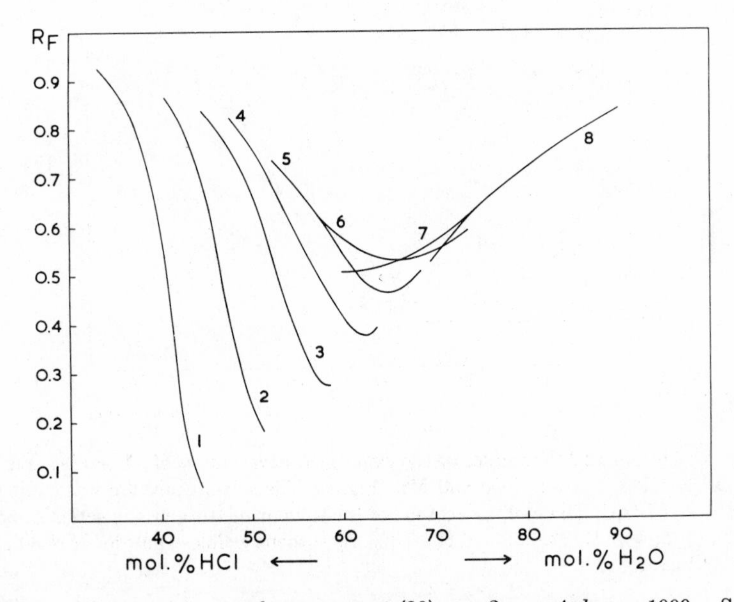

Fig. 9. R_f curves for anomalous transport (20) $p = 2$; $q = 4$; $k_{III} = 1000$; μ Solv = (1) 0.56; (2) 0.48; (3) 0.42; (4) 0.36; (5) 0.31; (6) 0.26; (7) 0.19; (8) 0.09.

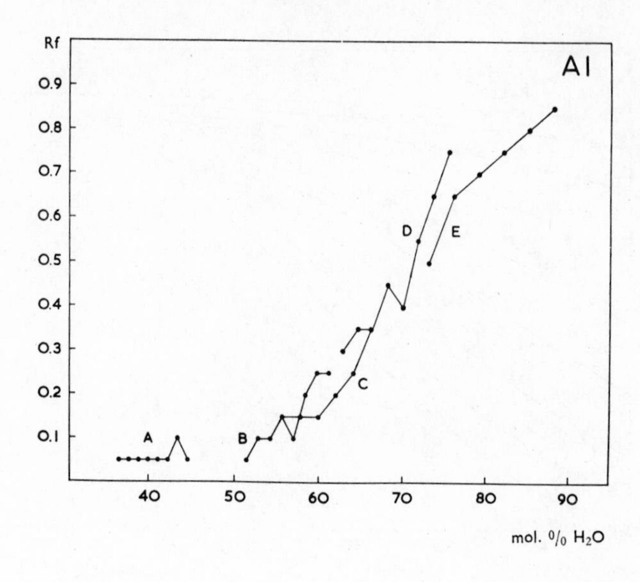

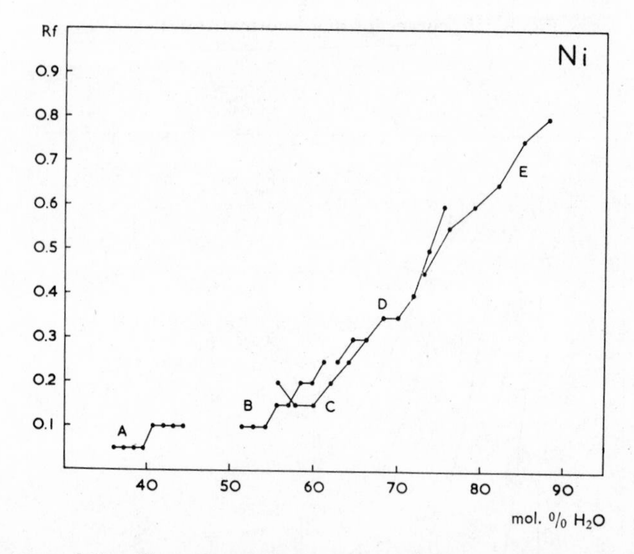

Figs. 10, 11, and 12. Examples of the experimental R_f curves of Al and Ni (Fig. 1), of Sr and Ba (Fig. 2), and of Co and Mn (Fig. 4). The solvent mixture was composed of acetone-HCl-H_2O. The mol. % acetone of the solvent mixture was in series A, 55%; B, 38%; C, 31.6%; D, 23.6%; and E, 10.3%. As sorbent cellulose powder MN 300 was used [Merkus (9)].

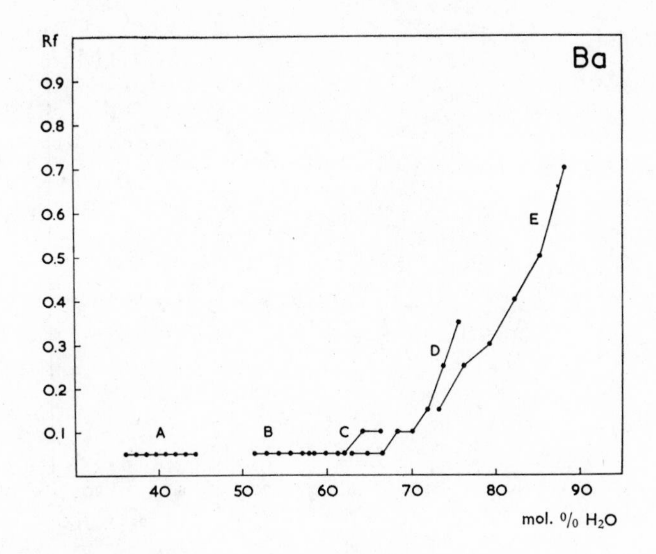

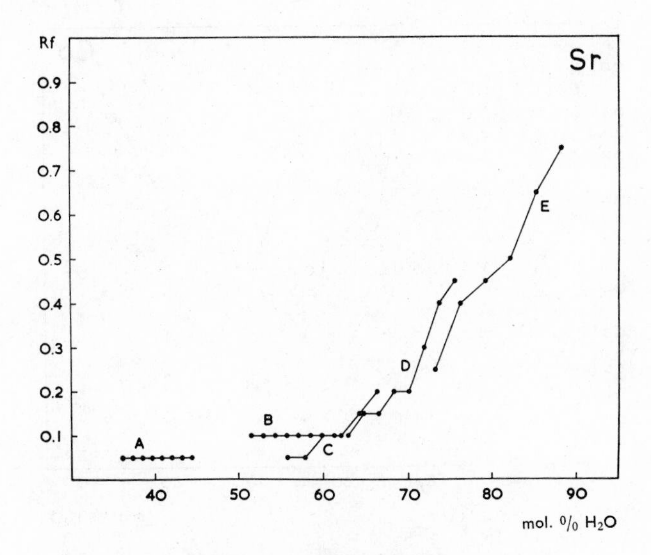

Fig. 11. (See page 248).

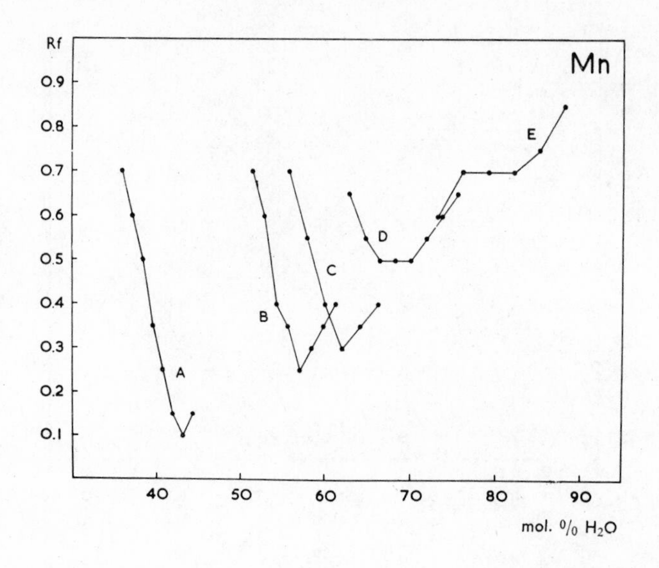

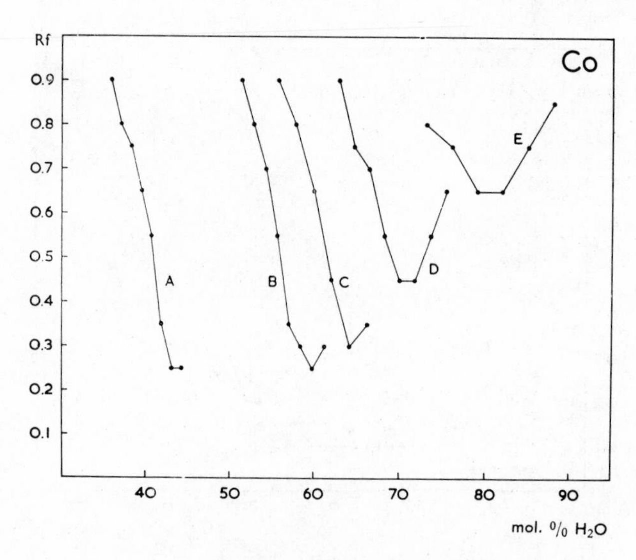

Fig. 12. (See page 248).

proposed by Lederer (24). He deduced for a number of cations a linear relationship between the R_M value and the log $[Cl^-]$ of the solvent mixture. Guedes de Carvalho (25) has studied the solvent mixture *n*-butyl alcohol-hydrochloric acid-water. During his experiments with this solvent mixture $[H_2O]$ (or [HCl]) was fixed at various levels and at each level [HCl] (or $[H_2O]$) varied. However, *n*-butyl alcohol strongly inhibits the movement of cations. In each experiment also the *n*-butyl alcohol content of the solvent mixture varied. Therefore, it is impossible to determine from the R_M curves of Guedes de Carvalho the exact influence of *n*-butyl alcohol, hydrochloric acid, and water on the chromatographic behavior of cations.

From the literature it appears that most investigators, using paper chromatography or thin-layer chromatography (on cellulose) for the separation of cations, have chosen a mixture of an organic solvent, hydrochloric acid, and water as solvent mixture. The solvent mixtures involved in the experiments of Merkus (21) were acetone-HCl-H_2O, methanol-HCl-H_2O, and *n*-butyl alcohol-HCl-H_2O. In order to investigate the influence of the individual components of the mixture in all series the organic solvent concentration was fixed, while the concentrations of hydrochloric acid and water varied inversely.

The connection between the R_f value of a cation and the partition coefficient (α) can be expressed by eq. (1). Bate-Smith and Westall (26) introduced the term R_M, which they find

$$R_M = \log [(1/R_f) - 1] \tag{25}$$

The formation of a chloro-complex of a cation, moving with the mobile phase (containing hydrochloric acid), can be described as follows

$$M(H_2O)x^{y+} + x\text{Cl}^- \rightleftharpoons M\text{Cl}x^{(y-x)} + xH_2O \tag{26}$$

The equilibrium constant K is found from the following equation:

$$K = \frac{[M\text{Cl}x^{(y-x)}]\,[H_2O]^x}{[M(H_2O)x^{y+}]\,[\text{Cl}^-]^x} \tag{27}$$

and its logarithm from:

$$\log K = \log \frac{[M\text{Cl}x^{(y-x)}]}{[M(H_2O)x^{y+}]} + x \log [H_2O] - x \log [\text{Cl}^-] \tag{28}$$

If in the mobile phase the concentration of water varies inversely with the concentration of hydrochloric acid, the effect of progressive dehydration and complex formation proceed to the same extent. Lederer writes:

$$x \log [H_2O] - x \log [\text{Cl}^-] = -2x \log [\text{Cl}^-] \tag{29}$$

This simplification has to be corrected because the difference between the two logarithms has to be written as the logarithm of the quotient:

$$x \log [H_2O] - x \log [Cl^-] = x \log \frac{[H_2O]}{[Cl^-]} \tag{30}$$

The partition coefficient of the cation, being present in the mobile phase as a chloro-complex and in the stationary phase as an aquo-ion, is expressed by Lederer in the form:

$$\alpha = \frac{[MCl_x^{(y-x)}]}{[M(H_2O)_x^{y+}]} \tag{31}$$

This equation has to be corrected because

$$\alpha = \frac{[M(H_2O)_x^{y+}]}{[MCl_x^{(y-x)}]} \tag{32}$$

When eq. (1) is substituted into eq. (25) the term R_M becomes

$$R_M = \log \alpha + \log (A_s/A_m) \tag{33}$$

Since

$$\log \alpha = \log \frac{[M(H_2O)_x^{y+}]}{[MCl_x^{(y-x)}]} \tag{34}$$

we may write

$$R_M = -\log K + \log \frac{A_s}{A_m} + \log \frac{[H_2O]}{[Cl^-]} \tag{35}$$

where K and A_s/A_m are constant. The concentration of water is very large in comparison with the concentration of hydrochloric acid. Therefore, log $[H_2O]$ may be also considered constant.

$$R_M = -x \log [Cl^-] \pm \text{a constant} \tag{36}$$

From eq. (36) it is possible to get information about the number of chloride ions ($= x$) in the chloro-complex. Plotting the R_M value of several cations against the log $[Cl^-]$ it is possible to calculate in which region the best separation of a given mixture of cations can be obtained. The graphs in Figures 13 and 14 demonstrate that the relation between the R_M value and the log $[Cl^-]$ is a straight line. The slope of the line is proportional to x. Variation of the transport mechanism of the cation can be revealed by the change of the slope of the straight line.

During the normal and inhibited normal transport the cation is moving with the mobile phase as an aquo-ion. This transport is inhibited by increasing the hydrochloric acid content and by decreasing the water content of the solvent mixture. The dehydration of the aquo-ion, which results from the hydration of the hydrogen ions and the chloride ions, may be written as follows:

$$M(H_2O)_n \rightleftharpoons M(H_2O)_{n-x} + xH_2O \tag{37}$$

The equilibrium constant K is expressed by

$$K = \frac{[M(H_2O)_{n-x}]\,[H_2O]^x}{[M(H_2O)_n]} \tag{38}$$

The dehydration of the cation is proportional to the concentration of hydrochloric acid. Therefore, we may write without introducing any serious error:

$$[H_2O]^x = [Cl^-]^x \tag{39}$$

As in the above described equations [HCl] is expressed by $[Cl^-]$. If we assume that the cations are present in the mobile phase as dehydrated aquo-ions and in the stationary phase as aquo-ions, we may write

$$\alpha = \frac{[M(H_2O)_n]}{[M(H_2O)_{n-x}]} \tag{40}$$

From eqs. (38,39,40) log K can be expressed in the form:

$$\log K = -\log \alpha + x \log [Cl^-] \tag{41}$$

or

$$\log K = -R_M + \log (A_s/A_m) + x \log [Cl^-] \tag{42}$$

or

$$R_M = x \log [Cl^-] + \text{a constant} \tag{43}$$

Figures 13 and 14 demonstrate the linear relationship between the R_M value and the log $[Cl^-]$. The slope of the straight line shows the ionic species and gives information about dehydration and complex formation of the cation, moving with the mobile phase (see also Tables I and II).

By comparing the properties of several cations it is possible to estimate the ionic hydration, the dehydration, or the type of the chloro-complex.

It could be possible that an exact measurement of the value of x enables us to calculate the hydration number of the cations or the stability constants of the chloro-complexes. Figure 15 gives a schematic representation of the R_M/log $[Cl^-]$ relationship for the four different previously described transport groups.

D. Results with Other Solvents

We have now seen how the influence of the components of the solvent mixture acetone-HCl-H_2O on the transport of the cation in TLC on cellu-

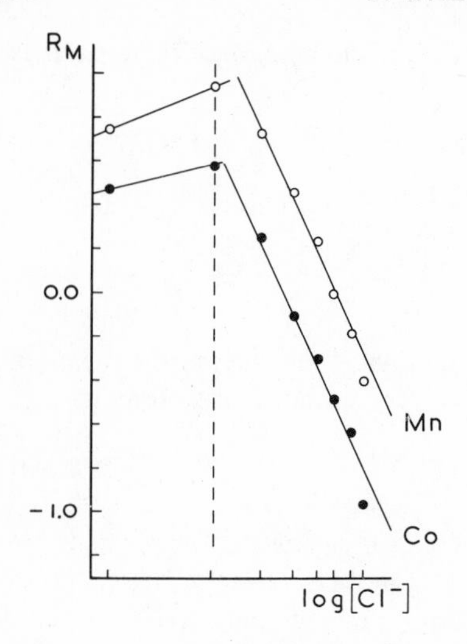

Fig. 13. Relationship between the R_M values of Mn and Co and the log [Cl^-] (*cf.* Table I) (21).

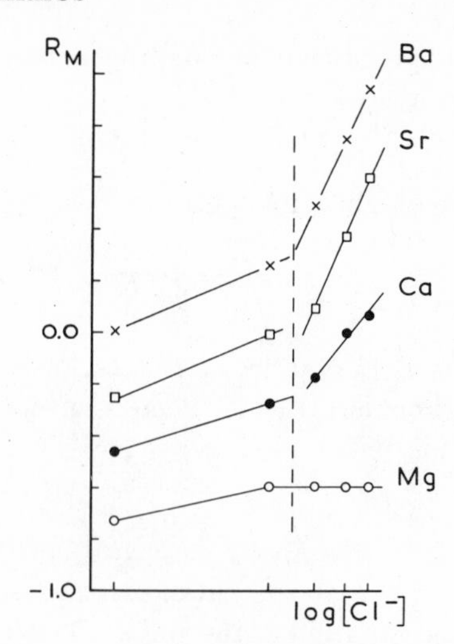

Fig. 14. Relationship between the R_M values of Ba, Sr, Ca, and Mg and the log [Cl^-] (*cf.* Table II) (9).

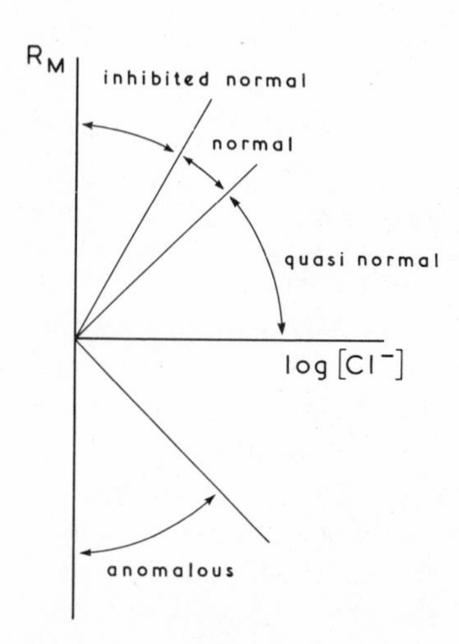

Fig. 15. Schematic representation of the relationship between the R_M value of the cations, classified in four transport groups and the log [Cl^-] of the solvent mixture [Merkus (19)].

lose layers could be elucidated. However, in some cases the use of other solvent mixtures, such as *n*-butyl alcohol-HCl-H_2O or methanol-HCl-H_2O, is justified because the separation of cations with these solvent mixtures can be better than with acetone-HCl-H_2O.

1. n-BUTYL ALCOHOL-HCl-H_2O

This solvent mixture requires a long development time. The solvent mixture is strongly separated as it rises up the cellulose layer. An increase of the HCl content in the solvent mixture decreases this separation. In other words, the higher the HCl content the closer the second front is to the first one. But a high HCl content gives very low R_f values by inhibition of the transport of the cations as aquo-ions. Furthermore, the use of *n*-butyl alcohol as a component of the solvent mixture auses cvery low R_f values because of

TABLE I

Composition of a Number of Solvent Mixtures and the Corresponding R_f Values of Mn^{2+} and Co^{2+} (*cf.* Figs. 13 and 16)

Acetone (ml)	10 N HCl (ml)	H_2O (ml)	Acetone (mol. %)	HCl (mol. %)	H_2O (mol. %)	Mn	Co
73	1	14	55·0	0·6	44·4	0·15	0·25
73	3	12	55·0	1·8	43·2	0·10	0·25
73	5	10	55·0	3·0	42·0	0·15	0·35
73	7	8	55·0	4·2	40·8	0·25	0·55
73	9	6	55·0	5·4	39·6	0·35	0·65
73	11	4	55·0	6·6	38·4	0·50	0·75
73	13	2	55·0	7·8	37·2	0·60	0·80
73	15	0	55·0	9·0	36·0	0·70	0·90

TABLE II

Composition of a Number of Solvent Mixtures and the Corresponding R_f Values of Ba^{2+}, Ca^{2+}, Sr^{2+}, and Mg^{2+} (*cf.* Fig. 14)

Methanol (ml)	10 N HCl (ml)	H_2O (ml)	Methanol (mol.%)	HCl (mol.%)	H_2O (mol.%)	Ba	Ca	Sr	Mg
50	5	45	31·1	1·2	67·7	0·50	0·75	0·65	0·85
50	20	30	31·1	5·0	63·9	0·35	0·65	0·50	0·80
50	30	20	31·1	7·5	61·4	0·25	0·60	0·45	0·80
50	40	10	31·1	10·0	58·9	0·15	0·50	0·30	0·80
50	50	0	31·1	12·5	56·4	0·10	0·45	0·20	0·80

the great difference in structure and properties between *n*-butyl alcohol and methyl alcohol (see below) and acetone. Using *n*-butyl alcohol-HCl-H_2O only a few cations have relatively high R_f values: As, Sb, Sn, Bi, Cd, Hg, Fe, and Zn. These cations are located in distinct zones near the second front or between the two fronts. Since these cations are difficult to separate with other solvents, it is understandable that many authors have chosen this solvent mixture for the separation of mixtures of these cations.

2. *METHANOL-HCl-H_2O*

The structural resemblance of water and methanol is responsible for the high R_f values of all cations using a methanol-containing solvent. The cations, classified under the normal, quasi-normal, and anomalous transport group have too high R_f values to obtain sharp separations. Only the

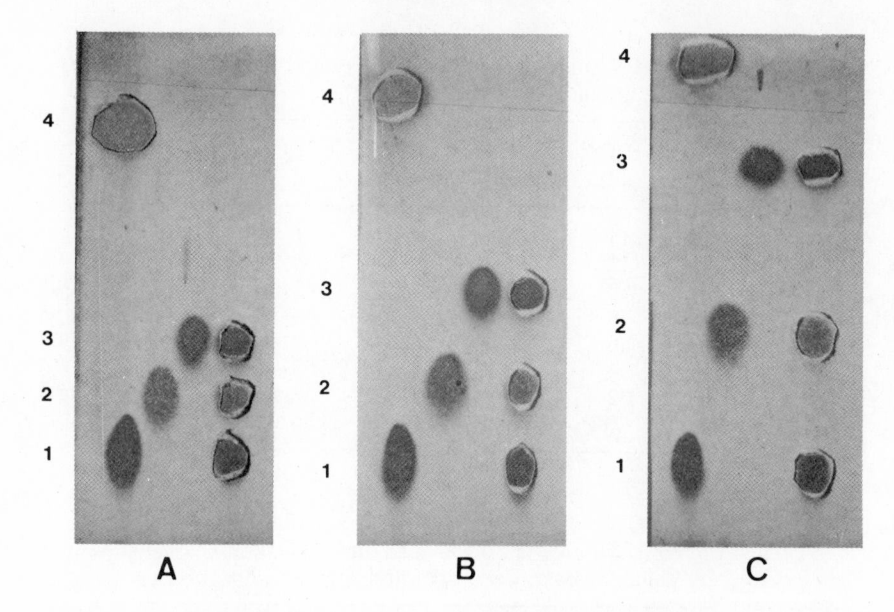

Fig. 16. The effect of an increasing content of HCl (increasing from A to C) in a solvent mixture on the R_f value of Ni (1), Mn (2), Co (3) and Cu (4). Solvent mixtures: acetone —HCl—H_2O (see Table I). Spray reagent: pyridylazonaphthol (0,2% in methanol), followed by treatment of the chromatoplates with NH_3 vapor [Merkus (19)]. Thin layer: procoated celluloseplates, Merck, Nr 5716.

alkaline and alkaline earth cations, classified under the inhibited normal transport group, can be easily separated using methanol, HCl, and H_2O as solvent mixture.

E. Experiments

In order to demonstrate these theoretical considerations in practice some experiments are presented.

Solutions of the cations were made up on the basis of 1 mg of cation in 1 ml of 4 *N* HCl. As sorbent cellulose powder MN 300 was used. A general spray reagent for the visualization of Ni, Mn, Co, and Cu was found in PAN (pyridylazonaphthol: 0.2% in methanol), for Ba, Ca, Sr, Mg, and Be in oxine-kojic acid (0.1 and 0.5% respectively in 60% ethanol). After spraying the spots were visualized by exposure to NH_3-vapor. Examination of the spots was executed in visual light or in UV light of 365 nm (oxine-kojic acid). The influence of the increasing content of HCl in the solvent mixture for the separation of Ni, Mn, Co, and Cu is demonstrated in Figure 16. It is possible to choose every desired distance between the R_f

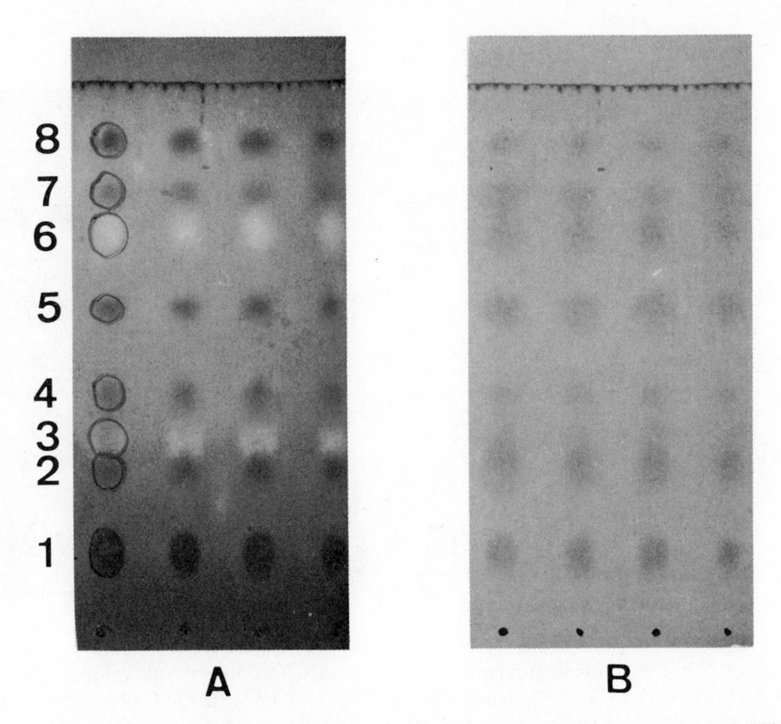

Fig. 17. Separation of Th (1), Ni (2), Al (3), Mn (4), Co (5), Be (6), Cu (7), UO_2(8). Thin layer: cellulose, precoated plates, Merck, Nr 5716. Solvent mixture: acetone-HCl (d = 1,19)-H_2O (40:20:15). Spray reagent: 0.5% oxine (8-hydroxyquinoline) in 60% ethanol, mixed with 0.2% pyridylazonaphthol in methanol (10 + 1v/v), followed by exposure to the vapor of ammonia. Visualization in (A) UV light (365 nm), or in (B) visual light [Merkus (27)].

values of the cations in order to separate more cations in one chromatogram. This is demonstrated in Figures 17A and B. A separation of Ba, Ca, Sr, and Mg can be achieved with methanol-HCl-H_2O in various compositions. The separation of these cations from Be is sharper when a solvent mixture containing acetone-HCl-H_2O is chosen. With methanol as organic solvent, Mg is transported well by the formation of solvo-ions (see quasi-normal transport).

In the case of acetone-HCl-H_2O the dehydration of Mg proceeds easily by an increase of the HCl content in the solvent mixture and the formation of solvo-ions can be neglected. In other words, in the case of acetone-HCl-H_2O there is a greater difference between the R_f values of Be and Mg, because Mg is transported as aquo-ion, according to the normal transport group. An increase in the HCl content of the solvent mixture causes a

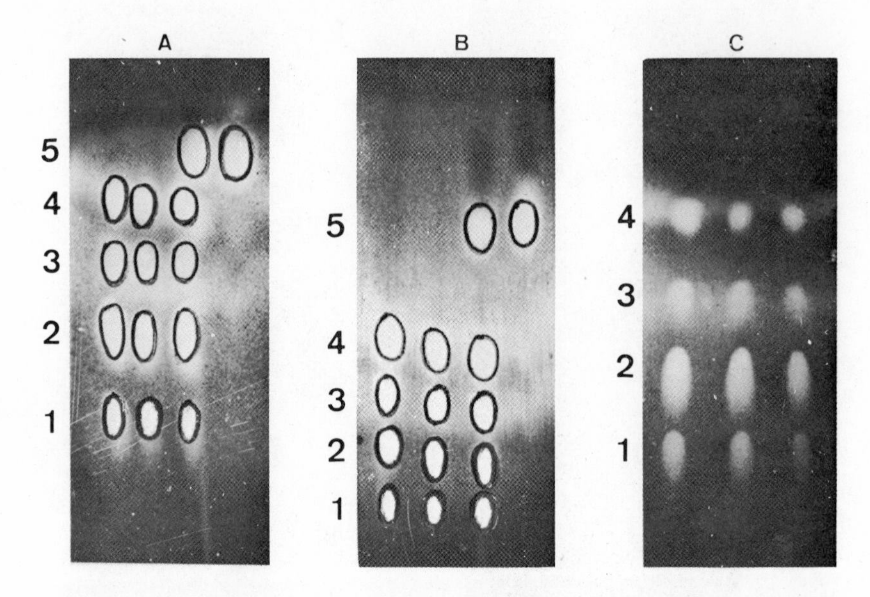

Fig. 18. Separation of Ba (1), Sr (2), Ca (3), Mg (4), and Be (5). Thin layer: cellulose powder MN 300. Solvent mixture: (*A*) acetone-HCl(10 *N*)-H_2O (6:1:4); (*B*) acetone-HCl(10 *N*)-H_2O (6:3:2); (*C*) methanol-HCl(d = 1.19)-H_2O (7:1:2). Spray reagent: oxine-kojic acid (see test). (*A*) and (*B*) show the relative strong decrease in the R_f value of Mg, Ca, Sr, and Ba in comparison with Be by an increase of the HCl content of the solvent mixture. Explanation of this phenomenon in text [Merkus (27)].

dehydration of the cation. This results in a decrease of the R_f value The decrease of the R_f value of Mg, Ca, Sr, and Ba is greater than the decrease of the R_f value of Be because the ionic radius of these ions is greater (see Figs. 18*A*, *B*, *C*).

The most important implications of these results are that:

1. Information about hydration, dehydration, solvation, and formation of a chloro-complex of the cation can be obtained;

2. For a given mixture of cations the composition of the solvent which will yield the best separation can be predicted.

III. DETECTION OF CATIONS

The identification of cations in TLC is not very difficult. Most reagents published in the past are sensitive and simply applicable. However, some results are not reproducible. In extensive research, Merkus (9) has criticized some of the authors who claimed to have located cations in paper or thin-layer chromatography. It seems useful to discuss briefly a number of the universal and the more specific reagents.

For each group of cations of the classical H_2S-system a number of universal reagents is summarized in Table III. A characteristic coloration of a cation with a universal reagent, coupled with a measurement of its R_f value, is often sufficient for its identification. Therefore, it is not always necessary to confirm the identity of a cation with a specific reagent. The more characteristic color-producing and sensitive reagents for each cation are listed in Table IV. It must be noted here that it can be of great value to use mixtures of reagents such as oxine, kojic acid, quercetin, or pyridylazonaphthol. These mixtures possess great versatility for the detection of cations. A description is given in Table V of the preparation of the reagents and the conditions, regarding pH, temperature, and exposure to NH_3-vapor, required for these color reactions. It is difficult to ascribe an exact color to the spots because the colors depend on pH and cation concentration. The most probable color reaction for a number of cations is given in Table VI.

TABLE III
Universal Reagents (9)

Ag, Pb, Tl	H_2S; KI; dithizone, diphenylcarbazide; K_2CrO_4 glyoxal-*bis*-(2-hydroxyanil)
Cu, Cd, Bi, Hg, Pb	H_2S; dithizone; diphenylcarbazone; KI
As, Sb, Sn	Dithizone; H_2S; KI
Fe, Al, Cr, Be, UO_2, Th	Oxine-quercetin-kojic acid (mixture)
Co, Ni, Mn, Zn	PAN(pyridylazonaphthol)
Ba, Ca, Mg, Sr, Be	Tetrahydroxyquinone, oxine-kojic acid; glyoxal-*bis*-(2-hydroxyanil)
Na, K, Li	Quercetin; zinc uranyl acetate

IV. SEPARATION OF CATIONS BY TLC ON CELLULOSE

1. *INTRODUCTION*

In this section a complete scheme of analysis is described as proposed by Merkus (9). Since the data of inorganic paper chromatography are applicable to the cellulose layers, the literature on inorganic paper chromatography (PC) will be referred to in the following sections. The first papers on inorganic PC are published by Lederer [47–50), Arden *et al.* (51,52), Burstall *et al.* (53,54), and Pollard *et al.* (32,33,56–58). Although the chromatographic separation took a long time and the sharpness of the separation on paper was not outstanding, the results were reliable. Subsequently hundreds of papers were published on the subject.

TABLE IV

Very Sensitive or Specific Reagents (9)

	Reagent	Color
Ag:	*p*-Dimethylbenzylidene-rhodanine	(red violet)
	K_2CrO_4	(red)
	Dithizone	(red)
Pb:	Dithizone	(red)
	Rhodizonic acid	(violet)
	K_2CrO_4	(yellow)
Tl:	KI	(yellow)
	Dithizone	(red)
	K_2CrO_4	(yellow)
	Rhodamine B	(blue)
Bi:	Cinchonine-KI	(orange)
	KI	(yellow)
Cd:	Dithizone	(red violet)
	Oxine	(yellow UV)
Cu:	Rubeanic acid	(green)
	α-Benzoinoxime	(green)
	Diphenylcarbazone	(red brown)
	Salicylaldoxime	(green)
Hg:	Dithizone	(yellow)
	KI	(red)
	Diphenylcarbazone	(red violet)
As:	KI	(brown)
	H_2S	(yellow)
	Dithizone	(yellow)
	Sodium hypophosphite	(brown)
Sb:	Dithizone	(red)
	H_2S	(orange)
	Phosphomolybdic acid	(blue)
Sn:	Oxine	(yellow UV)
	Dithizone	(violet)
	Phosphomolybdic acid	(blue)
	Quercetin	(yellow orange UV)
Al:	Oxine	(yellow UV)
	Aluminon	(red)
	Morin	(green UV)
Be:	Oxine	(yellow UV)
	Aluminon	(red)
	Morin	(yellow UV)

(*Continued*)

TABLE IV (*cont.*)

Cr:	Oxine-quercetin	(black UV)
	Diphenylcarbazide	(red violet after oxidation)
Fe:	Potassium ferrocyanide	(blue)
	KCNS	(red)
	Oxine-quercetin	(black UV)
UO_2:	PAN	(brown)
	Potassium ferrocyanide	(brown)
	Quercetin	(brown)
Co:	PAN	(green)
	Rubeanic acid	(brown)
	KCNS	(blue)
	Dimethylglyoxim	(brown)
	α-Nitroso-β-naphthol	(brown)
Mn:	PAN	(red)
	Oxine-quercetin	(brown UV)
Ni:	Dimethylglyoxim	(red)
	Rubeanic acid	(blue)
	PAN	(red violet)
Zn:	Dithizone	(red)
	PAN	(red)
	Oxine-quercetin	(yellow-orange UV)
Ba:	Rhodizonic acid	(red violet)
	Tetrahydroxyquinone	(red)
	Kojic acid	(yellow UV)
Ca:	Oxine-kojic acid	(yellow UV)
	GBHA	(red)
Mg:	Oxine-kojic acid	(yellow UV)
	GBHA	(red)
Sr:	Rhodizonic acid	(red)
	Tetrahydroxyquinone	(red)
	Oxine-kojic acid	(yellow UV)
Na:	Zinc uranyl acetate	(blue UV)
K:	Quercetin	(yellow UV)
	Dipicrylamine	(red)
	Lead-cobalt nitrite	(grey black)
Li:	GBHA	(orange)
	Zinc uranyl acetate	(blue UV)

TABLE V

A Description of the Major Reagents and Their Method of Application

Alizarin

After spraying with saturated solution in ethanol, the chromatoplate is treated with NH_3 vapor and warmed. The chromatoplate is decolorized by warming and the red violet spots become visible. Many cations can be detected by this reagent. The ammonium salt of alizarin is red violet.

Alizarin S

The chromatoplate is sprayed with a 0.1% solution in 60% ethanol. After spraying the chromatoplate is treated with NH_3 vapor. Like alizarin this reagent is suitable for many cations.

Aluminon

The chromatoplate is sprayed with a 0.1% solution of aluminon in a 1% solution of ammonium acetate in water. Al and Be can be detected by this reagent.

Benzoin-α-oxime (= cupron)

The chromatoplate is sprayed with a 5% solution in ethanol and exposed to NH_3 vapor and warmed. This reagent is specific for Cu (green).

Chloranilic acid

The chromatoplate is sprayed with a 0.1% solution in diethyl ether and examined under UV light (365 nm). This reagent is proposed by Barreto *et al.* (29), especially for the monovalent ions. Their results are not reproducible [Merkus (9), de Vries *et al.* (30)].

Cinchonine-potassium iodide

This reagent can be used for the detection of Bi. It must be freshly prepared. Cinchonine (100 mg) is dissolved in 10 ml warm water, to which 1 drop of concd HNO_3 was added. After cooling 200 mg potassium iodide is dissolved in this solution.

p-Dimethylaminobenzylidene-rhodanine

The chromatoplate is sprayed with a 0.03% solution in ethanol, held over ammonia and viewed under UV light (365 nm). This reagent is very sensitive for the detection of Ag but it reacts also with a great number of cations.

Dimethyl-glyoxim

The chromatoplate is sprayed with a 1% solution in ethanol and exposed to NH_3 vapor. This reagent can be used as a specific reagent for Ni.

Diphenylcarbazide

The chromatoplate is sprayed with a 1% solution in ethanol and treated with NH_3 vapor. Diphenylcarbazide can be used as a general reagent.

Diphenylcarbazone

The chromatoplate is sprayed with a saturated solution in ethanol and exposed to NH_3 vapor. This reagent gives good results as a general reagent.

Dipicrylamine

The chromatoplate is sprayed with a 1% solution of dipicrylamine in 0.1 *N* sodium carbonate and after 10 min is sprayed with 0.1 *N* HCl. *K* becomes visible as a red spot.

(Continued)

Dithizone

The chromatoplate is sprayed with a freshly prepared 0.05% solution in $CHCl_3$ or CCl_4. It must be noted here that the chromatoplate must not be dried before spraying with dithizone. After spraying the chromatoplate can be exposed to NH_3 vapor. Several cations show a characteristic change of color.

Glyoxal-bis-(2-hydroxyanil) (= GBHA)

The chromatoplate is sprayed with a solution of 1% GBHA and further with 3% KOH in methanol and subsequently warmed. This general reagent is studied by Möller and Zeller (31). It must be freshly prepared. The chromatoplate must not react acid (see Fig. 19).

8-Hydroxyquinoline (= oxine)

The chromatoplate is sprayed with a 0.5% solution in 60% ethanol, held over ammonia and viewed under UV light (365 nm). This reagent was first described for the detection of cations by Pollard *et al.* (32,33). It is the best and most used general reagent. Merkus (9) found that pyridylazonaphthol possesses an equal capacity for the detection of cations, except for the alkaline earth metals. Pollard *et al.* (32,33) have also recommended a mixture of 8-hydroxyquinoline and kojic acid, especially for the alkaline earth cations.

H_2S

After treatment with NH_3 vapor the chromatoplate is exposed to H_2S gas. Many other procedures are possible to locate the cations as sulfides on the chromatoplate, e.g., spraying with $(NH_4)_2S$ or a thioacetamide-solution.

Kojic acid

The chromatoplate is sprayed with a 0.1% solution in ethanol, held over ammonia, and viewed under UV light (365 nm). Many investigators use a mixture of the kojic acid and the 8-hydroxyquinoline reagent, as proposed by Pollard *et al.* (32,33).

Lead cobalt nitrite

The chromatoplate is sprayed with an aqueous solution of 5% cobalt nitrate, 5% lead nitrate, and two drops concd HCl. After drying the chromatoplate is sprayed with a saturated $NaNO_2$ solution and dried. *K* gives a grey spot [Heisig and Pollard (34) and Pollard *et al.* (33)].

Morin

Two solutions can be used. A 1% solution in glacial acetic acid gives a specific green fluorescence for Al under UV light. The 0.1% solution in methanol locates many cations on the chromatoplate. The chromatoplate must be sprayed with the methanolic solution, held over ammonia, and viewed under UV light. Except perhaps as specific reagent for Al, morin can be missed as general reagent, because 8-hydroxyquinoline, pyridylazonaphthol, etc., are more sensitive. The use of morin has been extensively investigated by Schneer-Erdey (35).

α-Nitroso-β-naphthol

The chromatoplate is sprayed with a 1% solution in ethanol and held over ammonia. This reagent locates a great number of cations but the sensitivity is inferior to 8-hydroxyquinoline and pyridylazonaphthol.

(*Continued*)

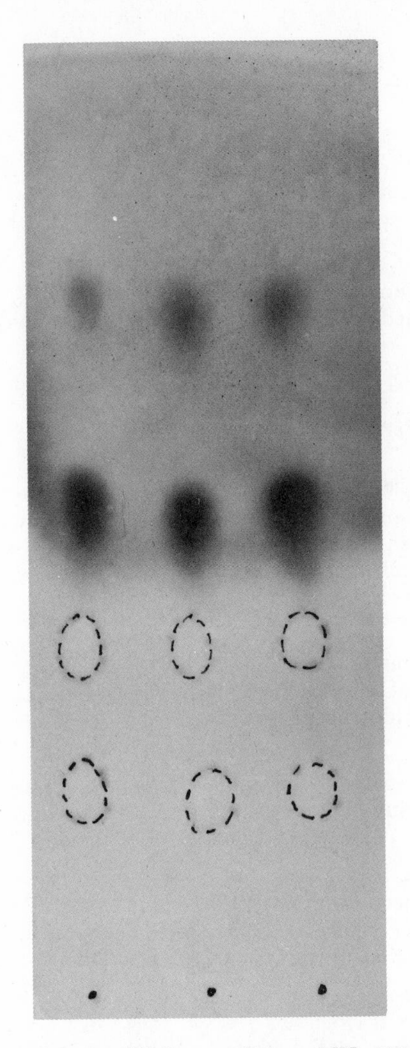

Fig. 19. TLC on cellulose MN 300. R_f values Ba < Sr < Ca < Mg. Solvent mixture: acetone-10 N HCl-H_2O (25:45:10). Spray reagent: GBHA. This photograph was made intentionally at the moment only Mg and Ca were colored. The bottom of the plate reacts too acid for the color reaction of GBHA and Ba and Sr. This photograph demonstrates the demixion of the solvent mixture, because the content of HCl, retained in the cellulose layer, is apparently decreasing from start to front.

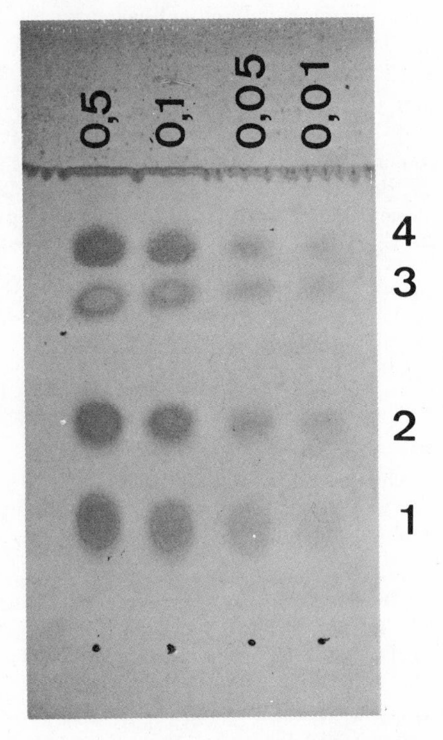

Fig. 20. TLC on a precoated cellulose plate, Merck (Nr 5716) of Ni (1), Mn (2), Co (3), and Cu (4). Solvent mixture: acetone-HCl(d = 1,19)-H_2O (40:20:10); Spray reagent: pyridylazonaphthol (0.2% in methanol), followed by exposure to NH_3 vapor. From left to right 0.5–0.01 μg of each cation was applied. Figure shows that PAN is extremely sensitive.

PAN (pyridylazonaphthol)

The chromatoplate is sprayed with a 0.2% solution in methanol and held over ammonia. This reagent is extremely sensitive and together with 8-hydroxyquinoline is the best universal reagent for the detection of inorganic cations (9) (see Fig. 20).

PAR (pyridylazoresorcinol)

The chromatoplate is sprayed with a 1% solution in water and exposed to NH_3 vapor. The results are similar to PAN. The use of PAR is recommended by Pollard *et al.* (36); Merkus found PAN more sensitive (9).

Phosphomolybdic acid

The chromatoplate is sprayed with a freshly prepared 5% solution in 60% ethanol. Subsequently the chromatoplate is warmed at 80–90°. Sn(II) and Sb(III) give blue spots.

Potassium chromate

The chromatoplate is sprayed with a 1% solution in water. This reagent can be used for the detection of Ag, Pb, and Tl.

Potassium ferrocyanide

The chromatoplate is sprayed with a 5% solution in water. This reagent is specific for Cu, UO_2, and Fe.

Potassium iodide

The chromatoplate is sprayed with a 1% solution in water and subsequently warmed. Some cations give characteristic colored spots.

Potassium thiocyanate

The chromatoplate is sprayed with a 5% solution in a mixture of equal volumes of water and acetone. This reagent detects Co (blue), Fe (red), and Bi (yellow).

Pyrocatecholviolet

The chromatoplate is sprayed with a 0.05% solution in ethanol and held over ammonia. This reagent was proposed by Macek and Moravek (37), but appeared to be not very sensitive (9).

Quercetin

The chromatoplate is sprayed with 0.2% solution in ethanol, held over ammonia, and viewed under UV light (365 nm). Before treatment with ammonia we can see also some cations under UV light. This reagent is very useful. It gives yellow, brown, or black spots with almost all divalent and trivalent cations. More details can be found in the papers of Weiss and Fallab (38) and Michal (39).

Quinalizarin (1,2,5,8-tetrahydroxyanthraquinone)

The chromatoplate is sprayed with a 0.05% solution in 0.1 *N* NaOH, exposed to NH_3 vapor and warmed. This reagent or a solution in ethanol is recommended in the literature [e.g., Johnson and Krause (40), Carleson (41)] for the detection of Be and Mg. However, the results are rather poor. Brinkman *et al.* (42) describes another procedure for the detection of Mg, Al, Sc, Y, rare earths, Ge, Ga, In, Ti, Zr, Hf, Th, and U. The chromatoplate is treated with NH_3 vapor and is left for 1 min in a 1% solution of quinalizarin in diethyl ether. Exposure to the vapor of glacial acetic acid enhances the constrast.

(*Continued*)

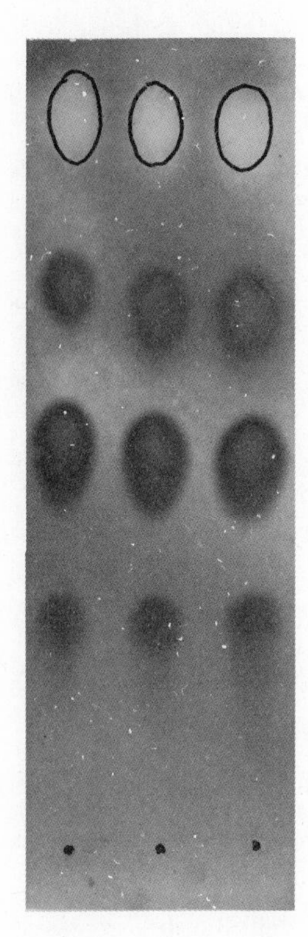

Fig. 21. TLC on cellulose MN 300. R_f values Ba < Sr < Ca < Mg. Solvent mixture: acetone-10N HCl-H_2O (25:45:10). Spray reagent: tetrahydroxyquinone (0.2% in ethanol), followed by exposure to NH_3 vapor. Before spraying and before treatment with ammonia vapor, the plate must be dried thoroughly.

Rhodamine B

The chromatoplate is sprayed with a 5% solution of sodium nitrite in water and subsequently sprayed with a 0.05% solution of rhodamine B in water. This reagent is specific for Sb and Tl.

Rhodizonic acid (sodium salt)

The chromatoplate is sprayed with 1% solution in water. The chromatoplate is treated before spraying with ammonia vapor. The reagent must be freshly prepared. For Ba and Sr this reagent can be used as a specific reagent.

Rubeanic acid (dithio-oxamide)

The chromatoplate is sprayed with a 0.5% solution in ethanol and subsequently treated with NH_3 vapor. This reagent is specific for Ni (blue). There are some investigators who use a mixture of rubeanic acid, salicylaldoxime, and alizarin as general reagent. There is no doubt that this reagent gives good results, but pyridylazonaphthol and oxine, for example, are more sensitive and universal.

Salicylaldoxime

The solution is prepared by dissolving 1 g salicylaldoxime in 5 ml ethanol. Subsequently this solution is diluted with water to 100 ml. This reagent is specific for Cu (yellow green).

Sodium hypophosphite

The chromatoplate is sprayed with a 5% solution in 4 N HCl and subsequently warmed; As becomes visible as a brown spot.

Tetracyanoquinodimethanide (TCNQ)

The use of this reagent has been investigated by Druding (43). TNCQ was found particularly sensitive for the detection of monovalent ions. This reagent is, however, not commercially available and must be synthesized.

Tetrahydroxyquinone

The chromatoplate is sprayed with a 0.2% solution in ethanol. Before and after spraying with this reagent the chromatoplate must be dried completely. After spraying and drying the chromatoplate is held over ammonia. This reagent was first proposed by

(*Continued*)

TABLE V (*cont.*)

Bock-Werthmann (44). It must be freshly prepared. If the chromatoplate is not completely dry the detection fails, because the entire chromatoplate darkens (see Fig. 21).

Violuric acid

The chromatoplate is sprayed with a freshly prepared 1.5% solution in water. Subsequently, the chromatoplate is warmed for 30 min at 100°. This reagent was first proposed by Erlenmeyer *et al.* (45) and later described by Seiler (46) for Ca, Ba, Sr, Na, K, Li in the form of their acetates. The published results with this reagent for the location of alkaline earth and alkali metals are not reproducible (9). This reagent is inferior to 8-hydroxyquinoline, GBHA, tetrahydroxyquinone, and so on.

Zinc uranyl acetate

The chromatoplate is sprayed with a saturated solution in 2 *N* acetic acid and viewed under UV light (365 nm). This reagent gives bright fluorescent spots for Na and Li. K is only visible in higher concentrations.

Two lines of approach to the use of PC for a complete analysis of cations can be distinguished here:

a. One is devoted to the separation of cations, grouped according to the orthodox qualitative analysis (H_2S-system): group I: Ag, Pb, Tl, Hg(I); group II: Pb, Hg, Cu, Cd, Bi and As, Sb, Sn; group III: Fe, Al, Cr, Be, U(VI); group IV: Ni, Co, Mn, Zn: group V: (Mg), K, Na, Li, or other groups. [Refs. Burstall *et al.* (53,54), Frierson and Ammons (59), Harasawa (60,61), Tamura (62), Surak (63), Pfeil *et al.* (64–67), Blasius and Göttling (68), Schulte and Henke (69), and Schneider and Patel (70)].

b. The second approach to the separation problem is based on the work of Pollard and co-workers. The separation of cations takes place without any previous separation of the cations in groups. Two different schemes of procedure have been evolved by Pollard *et al.* (32,55,56). The first consists of preparing chromatograms with each of three solvent mixtures and identifying the individual cations by their approximate R_f value and characteristic color reaction [see also the scheme of Merck (71)].

The second is based on a sequence of characteristic reactions applied to a number of chromatograms prepared with one solvent mixture. This method has been referred to as "differential spraying." For further details the reader is referred to the handbooks of Lederer and Lederer (72), Blasius (73), Hecht and Zacherl (74), and Hais and Macek (75).

TABLE VI
Color Reactions of Cations with the Reagents

	$K_4Fe(CN)_6$	Kojic Acid (UV)	KI	Rhodizonic Acid	α-Nitroso-β-naphthol	KCNS	Quercetin (UV)
Ag		green F[a]	yellow	brown	brown		dark
Hg (I)							
Pb			yellow	violet	brown		orange
Tl (I)		green F	yellow				
As (III)			yellow				
Sb (III)			yellow				brown
Sn (III)		yellow F					orange
Cd					yellow		
Cu	brown	grey	brown		brown		black
Bi			brown			yellow	
Hg (II)		green F	red				dark
Fe (III)	blue	black	brown		green	red	black
Cr (III)	yellow						dark
Al		pink F					yellow F
U (VI)	brown	brown					brown
Be		pink F					yellow F
Mn					brown		brown
Ni		grey			red brown		brown
Co					red brown	blue	brown
Zn					brown		yellow F
Ba		yellow green F		red			
Ca		yellow green F					
Sr		yellow green F		red			yellow
Mg							yellow
Na							
K							yellow F
Li							

[a] F = Fluorescence.

TABLE VI (*cont.*)

	Alizarin	Alizarin S	H_2S	PAR, Pyridylazo resorcinol	GBHA	Rhodanine	Diphenyl-carbazide
Ag			brown			dark brown	brown
Hg (I)			black				
Pb	violet	violet	brown	red	blue		red violet
Tl (I)			brown		red		red violet
As (III)			yellow				
Sb (III)			orange				
Sn (III)					violet		red violet
Cd	violet	violet	yellow	red	violet		violet
Cu	violet	violet	brown	red violet	violet	black	brown
Bi	red violet	violet	brown		blue		violet
Hg (II)	red violet	violet	black	red	brown	brown	blue
Fe (III)	red brown	brown	black	red brown	violet	brown	brown
Cr (III)						brown	
Al	red	red					
U (VI)	violet	violet		brown	blue		
Be	violet	violet					
Mn	violet	pink		red	black		red violet
Ni	violet	violet	black	red	violet	brown	red violet
Co	violet	violet	black	violet	black		red violet
Zn	violet	violet		red	violet		red violet
Ba		pink			red		
Ca	red violet				red		red violet
Sr		pink			red		
Mg	red violet				red		
Na							
K							
Li					orange		

TABLE VI (*cont.*)

	Diphenyl-carbazone	Dimethyl-glyoxim	Dithizone	Rubeanic Acid	PAN, Pyridylazo-naphthol	Tetrahydroxy-quinone	8-Hydroxy-quinoline (= oxine)
Ag			red	brown	red		brown
Hg (I)	violet		red				black
Pb	red violet		red	brown	red violet	grey blue	brown
Tl (I)	red violet		red	brown		grey	brown
As (III)			yellow				
Sb (III)			red				grey
Sn (III)			violet				yellow F
Cd	violet		violet (orange NH_3)	yellow	red	grey	yellow F
Cu	red brown	green	grey brown	green	red violet	grey	brown
Bi	violet		red violet	yellow brown		grey	brown
Hg (II)	violet		yellow orange (NH_3)	brown	red	brown	grey
Fe (III)	violet brown	brown		brown	red	brown	black
Cr (III)							black
Al	red						yellow F
U (VI)	red				violet		brown
Be	red						yellow F
Mn	red violet		pink	brown	red (NH_3)	grey brown	brown
Ni	red violet	red	brown violet	blue	red violet	grey brown	brown
Co	red violet	brown	violet	brown	green	grey brown	brown
Zn	red violet		red		red	grey brown	yellow F
Ba						red	blue F
Ca						red	yellow F
Sr						red	blue F
Mg	pink					red	yellow F
Na							
K							
Li							

2. FACTORS AFFECTING THE SEPARATION OF CATIONS IN INORGANIC TLC ON CELLULOSE

It is advisable to preseparate the cations to be analyzed according to the classical H_2S-system. If one has some experience in separating and locating cations or if one expects to find some special group of cations, the preseparation can be omitted.

The solution to be applied may be neutral or dilute acid. The best method of application is to spot several different concentrations of the solution on the plate. The best *concentration* is a solution containing between 0.1 and 10 mg of the cation per milliliter. It is advisable to run a standard mixture of cations on the same chromatogram. It must be noted here that TLC on cellulose is very sensitive. The best results are obtained with dilute solutions. Between 0.1 and 1 μl of the solution is chromatographed.

Equilibration of the chromatoplates is not necessary. The chromatoplates are air-dried after the sample solution is applied and immediately thereafter the development can start. Equilibration does not improve the separation because the cellulose-water-complex (stationary phase) would be imbibed by the absorbate of the mobile phase during such an equilibration. Thus the difference between the mobile phase and the stationary phase would decrease as the partition coefficient of the cation approaches unity.

In preparing the chromatogram some irregularities can occur. In organic chromatography most solvents are composed of an organic component, acid, and water. The solvent is separated as it rises on the sorbent layer. A double front can be distinguished after development (see Fig. 22). The first front is the organic component and the second is the aqueous-acid component. The higher the acid content in the solvent, the closer the second front will be to the first.

After spraying a cation may become visible in two (or more) spots. This phenomenon is called "double spot." The presence of double spots can be explained as follows:

a. The cation M, applied to the plate, is bound to anion X. When the mobile phase contains anion Y, it is possible that MX and MY are formed at the moment the mobile phase reaches the point of application. If MX and MY have different partition coefficients, "double spots" will appear.

b. If the cation forms more than one complex with components of the solvent double spots can also appear. It is also possible for the cation or cation-complex to be eluted stepwise when the concentration of the cation is too high, while finally it could be separated when the solvent demixes. In the experiments of Merkus double spots of Co and Co-complexes were sometimes obtained using solvents composed of acetone, HCl, and H_2O (see Fig. 22).

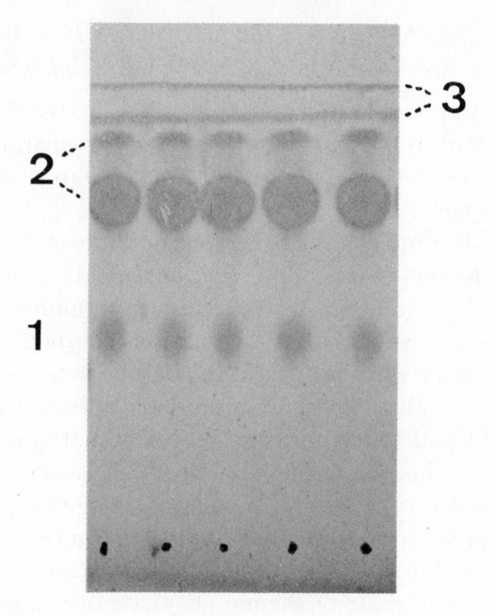

Fig. 22. Separation of Ni (1) and Cu (2) on precoated cellulose plates, Nr. 5716, Merck. Solvent mixture: acetone-HCl(d = 1,19)-H_2O (37:15:15). Spray reagent: pyridylazonaphthol. Clearly visible is the double spot of Cu and the double front (3) by demixion of the solvent mixture [Merkus (79)].

3. Ag, Hg(I), Pb, Tl

Most investigators have chosen for the paper chromatographic separation of these cations a solvent composed of n-butyl alcohol, saturated with HCl of various molarities. With this solvent, however, insoluble chlorides are formed causing excessive tailing of the spots. Therefore, solvent mixtures containing HCl have to be rejected. Some authors recommend n-butyl alcohol saturated with other acids, or solvents containing pyridine. Also isopropyl alcohol is used instead of n-butyl alcohol. All solvent mixtures described in the reviews on inorganic paper chromatography (72–75) gave an insufficient separation of Ag, Hg(I), Tl, and Pb in the experiments of Merkus (9) on cellulose layers. The R_f value of Pb and Tl were mostly similar. It is worthy to note the idea of Diller and Rex (76) to add H_2SO_4 to the solvent in order to separate Pb and Tl. Since $PbSO_4$ is insoluble, Pb is retained at the origin. Diller and Rex use a mixture of methanol-25% H_2SO_4-H_2O (7:1:4) for the separation of Pb < Tl < As < Hg < Cu < Zn.

In the literature on the separation of Ag, Pb, Hg(I), Tl, Merkus (9)noted a spot test described by Lederer in 1949 (49). He separated the insoluble chlorides of Hg(I), Ag, and Pb by elution on a piece of filter paper with two

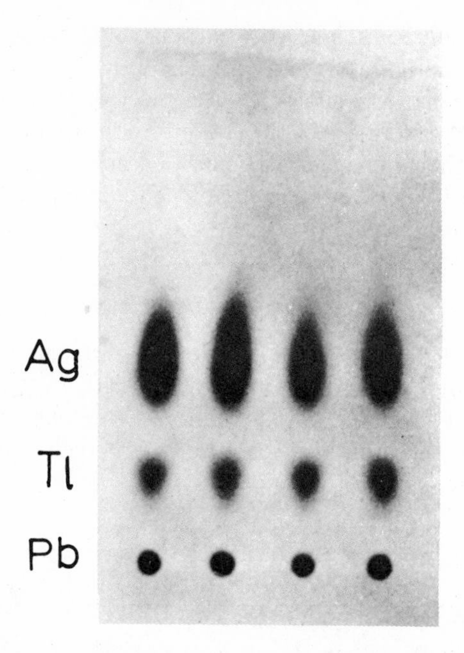

Fig. 23. TLC on cellulose Mn 300 of Ag, Tl, Pb. For detailed information see text. Spray reagent: the chromatogram is not dried after developing but is held above a bottle, containing saturated ammonium sulphide. Solvent mixture: 10% ammonia [Merkus (78)].

drops of water and subsequently with two drops 6N ammonia. After detection of the cations with H_2S, Lederer found three concentric rings: Hg(I) < Ag < Pb. This sequence can be explained as follows: $PbCl_2$ is first solubilized as $Ag \cdot 2NH_3^+$ by the elution with ammonia, while Hg(I) is insoluble in both liquids.

Merkus (9) tried this technique to separate Ag, Pb, Tl, and Hg(I) on cellulose layers using 10% ammonia as solvent. The result was excellent, especially because the separation was not only possible after application of a solution of these ions but also after application of a suspension of the insoluble chlorides. However, Pb and Hg(I) are not separated by 10% ammonia, but the identification of Hg offers no problem. Hg_2Cl_2 is colored black by ammonia. Furthermore, only Hg(II) is present after solubilization of a sample by HNO_3 or other oxidizing agents.

The separation of Pb, Tl, and Ag is presented in Figure 23. It appears that the hydroxide of Pb is insoluble and that Tl is transported over a short distance since the hydroxide is slightly soluble in the mobile phase. The movement of Ag is better since the amine complex of Ag is easily transported by ammonia. Sometimes the sharpness of the separation will be not as good as shown in Figure 23. The variation of the R_f value with the concentration of the cations, as well as the occurrence of tailing can be explained if one accepts that adsorption of the cations by the cellulose molecules is the major factor in their movement. It must be noted here that in all other separations discussed below there is no such adsorption effect, because if the mobile phase contains free acid, the hydroxonium ions are preferentially adsorbed by the cellulose.

4. Cu, Pb, Hg(II), Cd, Bi

In 1949 Lederer (50) described the paper chromatographic separation of these cations by *n*-butyl alcohol, saturated with 1 *N* HCl. In later years Burstall *et al.* (54) examined the influence of the molarity of the HCl for the saturation of the *n*-butyl alcohol, on the R_f value of the cations (see Table VII), using descending paper chromatography.

TABLE VII
R_f Values

	Cu	Pb	Bi	Cd	Second Front	Hg
H_2O	0.04	0.05	0.50	0.61	0.66	0.74
1 *N*·HCl	0.08	0.12	0.54	0.65	0.69	0.77
2 *N*·HCl	0.12	0.15	0.56	0.68	0.71	0.76
2,5 *N*·HCl	0.19	0.25	0.62	0.73	—	0.77
3 *N*·HCl	0.20	0.27	0.59	0.77	0.79	0.81

From Table VII it can be concluded that *n*-butyl alcohol saturated with 3 *N* HCl gives a relatively good result. But the use of *n*-butyl alcohol is not generally recommended because higher alcohols inhibit the transport of the cations and cause long development times. Methanol-HCl-H_2O and acetone-HCl-H_2O as solvents for the separation of these cations on cellulose layers (and paper) also give rather poor results. Therefore, Merkus (9) also preferred *n*-butyl alcohol saturated with 3 *N* HCl. An advantage in the use of this solvent mixture is the fact that only a few cations have high R_f values (Hg, Bi, Cd, Zn, Fe, As, Sb, Sn) and that consequently their location and identification will not be disturbed by the presence of traces of most of the other cations.

Figure 23 shows the separation of these cations on cellulose. The visualization of the cations was achieved with H_2S. The yellow spot of Cd (localized in the second front line) is not visible in Figure 24. Other detection methods give also good results.

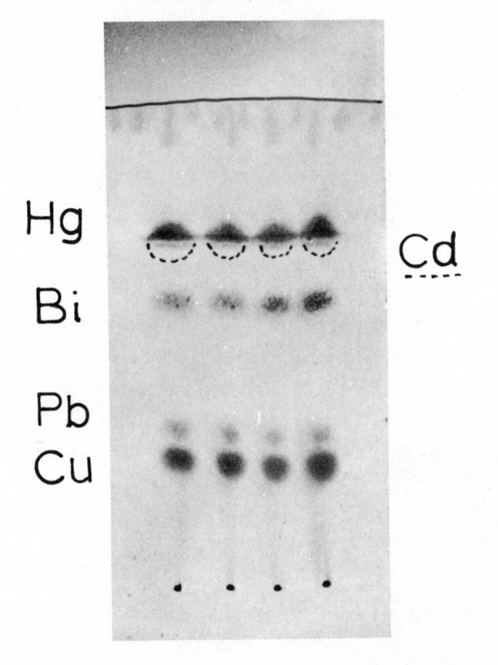

Fig. 24. TLC on cellulose MN 300 of Cu, Pb, Bi, Cd, Hg. Spray reagent: ammonium sulphide (see Fig. 23). Solvent mixture: *n*-butanol, saturated with 3 *N* HCl [Merkus (78)].

Not only *n*-butyl alcohol-HCl mixtures but also solvents of different composition are recommended in the literature on PC. But better results than those shown in Figure 24 are not likely to be obtained with these other solvent mixtures (9).

5. *As, Sb, Sn*

The great number of different methods that have been proposed in the literature for the separation of these three ions indicates that this is a difficult separation. Lederer has described (50) a separation using *n*-butyl alcohol, saturated with 1 *N* HCl as solvent mixture. The sequence of the R_f values was As < Sb < second front < Sn. Many authors have adopted this solvent mixture.

The separation of As(III), Sb(III), and Sn(II) is also possible on cellulose layers using *n*-butyl alcohol, saturated with 1, 2, or 3 *N* HCl. The identification of the cations gives no problem since they give a characteristic color with dithizone:

As: yellow
Sb: red

and

Sn: violet

A good separation can also be obtained with the solvent mixture of Blasius and Göttling (68): *t*-butyl alcohol-acetone-water-HNO_3 (d = 1,40)-acetylacetone (49:33:10:3:5): As < Sb < Sn. These and other separations of these cations (see Refs. 72–75) have been previously reviewed by Merkus (9).

6. Al, Cr(III), Be, U(VI), Fe, Th

Many authors have described the paper chromatographic separation of a number of these cations. The elucidation of the transport mechanism of these cations, using acetone-HCl-H_2O as a solvent, enables us to compose a suitable solvent for the separation of this group of cations on cellulose layers.

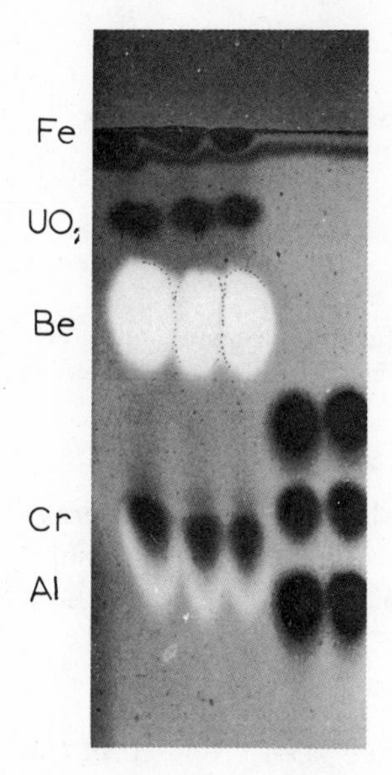

Fig. 25. TLC on cellulose MN 300 of Al, Cr, Be, UO_2, Fe. Spray reagent: mixture of oxine (0.5%), quercetin (0.2%), and pyridylazonaphtol (0.2%) in ethanol, followed by exposure to NH_3 vapor. (On the right Ni, Mn, and Co are also visible). Visualization in UV light (365 nm). Solvent mixture: acetone-10 N HCl-H_2O (37:18:12) [Merkus (78)].

Al and Cr(III) can be classified in the normal transport group, while U(VI) and Fe belong in the anomalous transport group. The transport of Be is different from the four described groups. The R_f values of Be are mostly between 0.50–0.70. The R_f curves of Be resemble the quasi-normal transport curves, but another process than solvation probably is the major factor in their movement (see Ref. 77).

The separation of these cations on cellulose layers gives excellent results (see Fig. 25). The R_f values of Cr(III) and Al are rather similar, but the detection causes no problem since Al gives a fluorescent and Cr a dark spot in UV light with 8-hydroxyquinoline. The separation of Th from U and many other cations is presented in the Figures 26 to 28.

7. Ni, Mn, Co, Zn

The R_f values of these cations on cellulose layers, using mixtures of acetone-HCl-H_2O, are different enough to get a sharp separation. Every desired difference can be obtained, using the data of the transport mechanism of these cations. Ni is classified

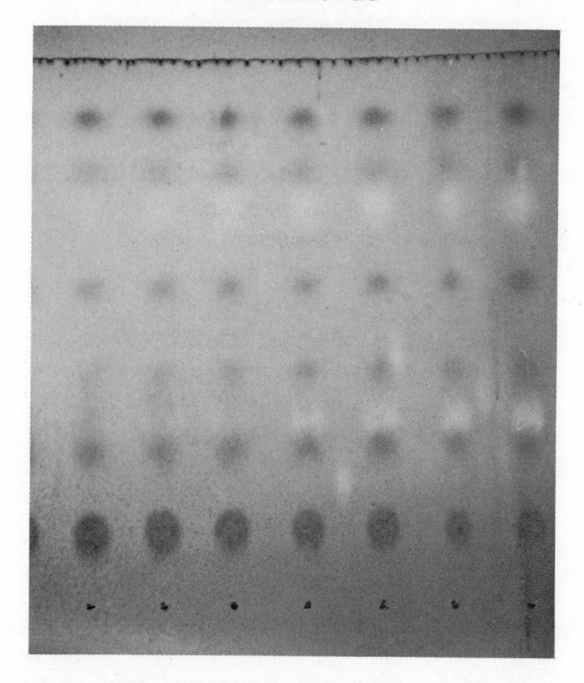

Fig. 26. Separation of Th < Ni < Al < Mn < Co < Be < Cu < UO_2 < (Zn and Fe in the front). Thin layer: precoated cellulose plates, Merck, Nr. 5716. Solvent mixture: acetone-HCl(d = 1,19)-H_2O (40:20:15). Spray reagent: 0.5% 8-hydroxyquinoline in 60% ethanol, mixed with 0.2% pyridylazonaphthol in methanol (10:1, v/v). After spraying, the plate is exposed to NH_3 and examined under UV (365 nm) [Merkus (78)].

in the normal and Co and Mn in the anomalous transport group with Zn. The R_f value of Zn always is near 1.0.

Figure 29 is an example of the separation of Ni, Mn, Co, and Zn while in Figures 26 to 28 the separation of these cations from a mixture with some other cations is presented. These figures show that the elucidation of the mechanism of the transport enables us to separate many cations in one chromatogram by changing the composition of the solvent mixture in such a way that the R_f values of the cations, classified in the four transport groups, become different. Bearing the R_f curves of the four transport groups in mind, the composition of a suitable solvent will be easy. From this point of view it seems somewhat strange when an author claims to have found a "new" solvent mixture.

8. Ba, Ca, Sr, Mg, Be

Good separations of the alkaline earth cations can be obtained with solvents containing methanol or acetone mixed with water and hydrochloric

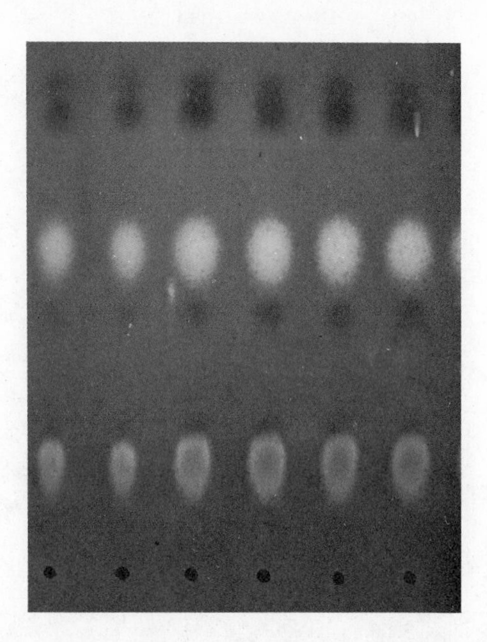

Fig. 27. This picture was intentionally taken as a demonstration of the visualization of Al and Be under UV light. Thin layer: see legend to Fig. 26. Solvent mixture: see legend to Fig. 26 (40:20:10). Spray reagent: 8-hydroxyquinoline [Merkus (37)].

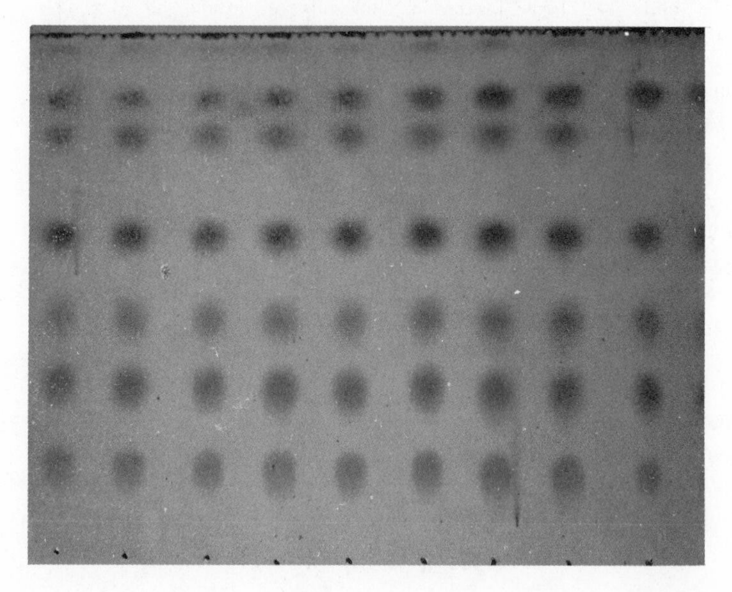

Fig. 28. Separation of Th $<$ Ni $<$ Mn $<$ Co $<$ Cu $<$ UO_2 $<$ (Zn and Fe in the front). Thin layer and solvent mixture: see Fig. 26. Spray reagent: pyridylazonaphthol [Merkus (78)].

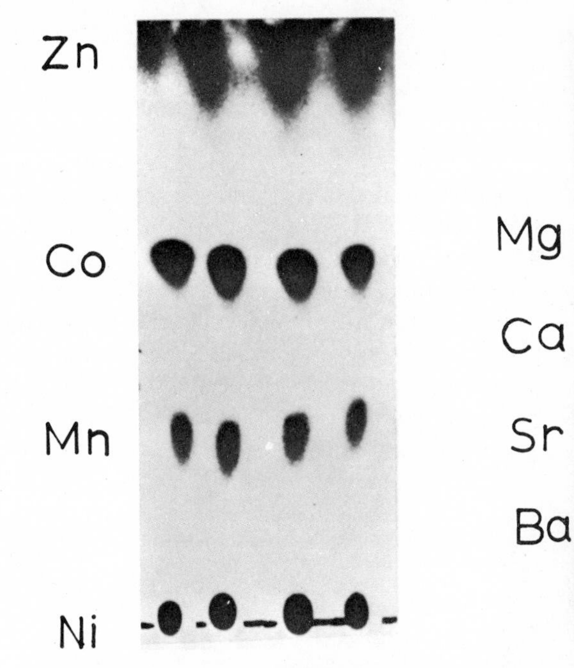

Fig. 29. TLC on cellulose MN 300 of Ni, Mn, Co, Zn. Spray reagent: pyridylazonaphtol. Solvent mixture: acetone-10 N HCl-H_2O (90:5:5) [Merkus (78)].

Fig. 30. TLC on cellulose MN 300 of Ba, Sr, Ca, Mg. Spray reagent: 8-hydroxyquinoline. Solvent mixture: methanol-HCl(d = 1,19)-H_2O (7:1:2) [Merkus (78)].

acid. For example, in the case of acetone-HCl-H_2O good results can be achieved when the solvent contains 30–60 vol % acetone and 70–40 vol % of a mixture of HCl and H_2O (these data are taken from Merkus' (9) experiments on cellulose MN 300 layers). It has already been pointed out that for the separation of Be from the other alkaline earth cations, acetone-HCl-H_2O is preferable to methanol-HCl-H_2O because of the similarity in the R_f values of Be and Mg with the methanol containing solvent. The solvation of Be and Mg is responsible for the good transport of both cations (see quasi-normal transport).

The separation of Ba, Ca, Sr, and Mg with methanol-HCl-H_2O is excellent (see Fig. 30). The number of solvents yielding a sharp separation of the alkaline earth cations is enormous. The R_f value depends on the dehydration of the cation under the influence of the content of HCl and H_2O in the solvent. Since the ionic radii of these ions differ, it is possible to choose a suitable solvent for their separation (see inhibited normal transport). The best general spray reagent for these cations is 8-hydroxyquinoline (= oxine) or a mixture of this reagent with kojic acid. Tetrahydroxyquinone and glyoxal-*bis*-(2-hydroxyanil) also give satisfactory results.

9. Na, K, Li

Na and K can be classified in the inhibited normal transport group while Li is in the quasi-normal group, using a solvent composed of methanol-HCl-H_2O. This division enables us to choose a composition of this solvent to get a separation of these three cations. In 1950, Burstall *et al.* (54) found that pure methanol could separate the alkali metals in the sequence Li (0.8), Na(0.5), and K (0.1). Later many investigators used methanol containing solvent mixtures (see the reviews in Refs. 21, 72–75). The successful separation with so many different solvents is explicable by the difference between the transport mechanisms of Li and Na-K and by the difference in ionic radius between Na and K. Thus, it is possible to obtain any desired separation simply by changing the HCl-H_2O content in the solvent.

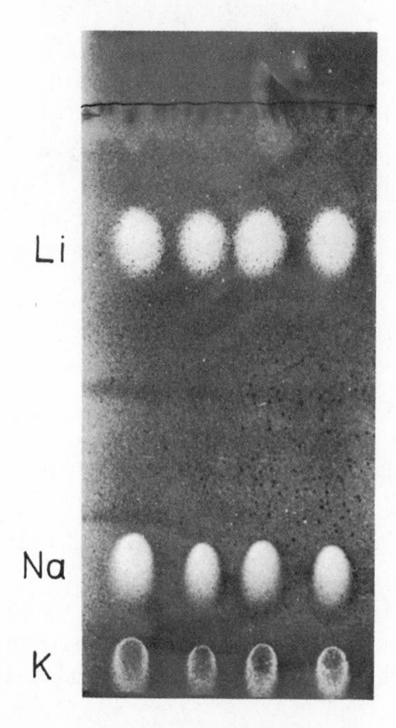

Fig. 31. TLC on cellulose MN 300 of Li, Na, K. Spray reagent: zinc uranyl acetate (saturated solution in 2 *N* acetic acid). Visualization in UV light (365 nm). Solvent mixture: methanol-10 *N* HCl-H_2O (80:12:8) [Merkus (78)].

The detection of the alkali metal ions is still a question of some dispute. The reagents violuric acid and tetracyanoquinodimethanide are discussed above. In Figure 31 the alkali metal ions are visualized by zinc uranyl acetate in 2 *N* acetic acid. In UV light bright fluorescent spots become visible. This reagent, however, is not very sensitive for K. In the paper chromatographic literature some reagents are described for the detection of the anions bound to the alkali metals. For instance, $AgNO_3$-fluoresceine for chlorides, or $BaCl_2$-sodium rhodizonate for sulphates are mentioned. The use of pH indicators is sometimes recommended to detect the buffer formed by the acid of the solvent mixture and the alkali salt. However, these reagents may give misleading results when traces of other cations or anions are present. For K a number of specific reagents are described: dipicrylamine, $Na_3Co(NO_2)_6$, and Na-Pb-Co-hexanitrite. The best detection method is probably described by Pollard *et al.* (33). First, the plate is sprayed with zinc uranyl acetate and examined under UV light (for Na and Li). The plate is then sprayed with a lead-cobalt nitrate solution. The formation of a greyish-black triple nitrite of Pb-Co-K indicates K. In case SO_4 ions are present, a preliminary spray of the K area with a saturated solution of barium nitrate is advisable in order to precipitate all sulphate ions.

10. INORGANIC MICRO—TLC

While preparing the manuscript for this chapter, we became aware of the excellent results of Brinkman, de Vries and van Dalen (42) using a micro-technique for reversed-phase TLC. We therefore investigated the possibility of separating cations on cellulose layers, within a few minutes and with a length of run of 2–3 cm. The results were very good. The separations could be achieved in this short time, on very small plates. For more details and photographs of actual chromatograms, one is referred to reference 80.

Experimental: Precoated cellulose plates containing no fluorescence indicator (Merck Nr 5716) are cut into pieces about the size of *microscope slides.* It is not advisable to use any other cellulose layer, because other layers are often too thin or too loose to get a sharp separation within a few minutes. Samples of the solution to be analyzed are spotted on the plates using a pointed paper strip, impregnated with the solution. By this means it is possible to apply very small spots on the plate having a diameter of about 1 mm. After application of the sample solution, containing for instance 0.1 to 1 mg of the cation in 1 ml, the plate is air-dried and developed in a small bottle without equilibration. In 5–10 minutes the solvent mixture travels a distance of 2–3 cm. The plate is air-dried or dried with warm air and subsequently treated with the ammonia vapor. After spraying with the reagent, the plate is treated again with ammonia vapor.

References

1. H. Seiler, in E. Stahl, *Dünnschichtchromatographie (ein Laboratoriumshandbuch)* 1st and 2nd eds., Springer Verlag, Berlin.
2. K. Randerath, *Dünnschichtchromatographie,* 1st and 2nd eds., Verlag Chemie, Weinheim.
3. F. H. Pollard, G. Nickless, K. Burton, and J. Hubbard, *Microchem. J.*, **10,** 131 (1966).
4. F. H. Pollard, K. W. C. Burton, and D. Lyons, *Lab. Pract.*, 505 (1964).
5. J. P. Garel, *Bull. Soc. Chim. France,* 1899 (1965).

6. M. Lesigang-Buchtela, *Oesterr. Chem. Ztg.*, **67**, 115 (1966).
7. T. Takeuchi, *Kagaku No Ryoiki, Zokan*, **64**, 197 (1964).
8. S. Takitana and K. Kawanabe, *Kagaku No Ryoiki, Zokan*, **64**, 221 (1964).
9. F. W. H. M. Merkus, Thesis, Amsterdam, 1966.
10. M. Lederer, *Chromatog. Rev.*, **9**, 115 (1967).
11. P. Wollenweber, in Hais en Macek, *Stationary phase in paper and thin-layer chromatography*, Elsevier, Amsterdam, 1965, p. 98.
12. R. Consden, in Hais en Macek, *Stationary phase in paper and thin-layer chromatography*, Elsevier, Amsterdam, 1965, p. 155.
13. J. B. Schute, Thesis, Leiden, 1953.
14. J. B. Schute, *Meded. Vlaamse Chem. Ver.*, **15**, 1 (1953).
15. J. B.Schute, *Nature*, **171**, 839 (1953).
16. G. Ackermann and J. Michal, *Talanta*, **11**, 441 and 451 (1964).
17. G. Ackermann and J. Michal, *Talanta*, **12**, 171 (1965).
18. G. Ackermann and J. Michal, in Hais en Macek, *Stationary phase in paper and thin-layer chromatography*, Elsevier, Amsterdam, 1965, p. 167.
19. F. W. H. M. Merkus, *Pharm. Weekblad*, **103**, 1037 (1968).
20. H. Hartkamp and H. Specker, *Z. Anal. Chem.*, **152**, 107 (1956).
21. H. Hartkamp and H. Specker, *Z. Anal. Chem.*, **158**, 92, 161 (1957).
22. P. W. J. M. Boumans, *Chem. Weekblad*, **54**, 729 (1958).
23. L. Horner, W. Emrich, and A. Kirschner, *Z. Electrochem.*, **56**, 987 (1952).
24. M. Lederer, *J. Chromatog.*, **1**, 172 (1958).
25. R. A. Guedes de Carvalho, *J. Chromatog.*, **4**, 353 (1960).
26. E. C. Bate-Smith and R. G. Westall, *Biochim. Biophys. Acta*, **4**, 427 (1950).
27. F. W. H. M. Merkus, unpublished results.
28. F. W. H. M. Merkus, *Intern. Symp. V, Chromatograpnie et Electroforèse* (Presses Académiques Européenes), Brussels, 1969, in press.
29. H. S. R. Barreto *et al.*, *J. Chromatog.*, **4**, 153 (1960) and **5**, 5 (1961).
30. G. de Vries, G. P. Schütz, and E. van Dalen, *J. Chromatog.*, **13**, 119 (1964).
31. H. G. Möller and N. Zeller, *J. Chromatog.*, **14**, 560 (1964).
32. F. H. Pollard, J. F. W. McOmie, and I. I. M. Elbeih, *Faraday Soc. Discussions*, No. 7, 183 (1949); *J. Chem. Soc.*, 466 (1951).
33. F. H. Pollard, J. F. W. McOmie, and H. M. Stevens, *J. Chem. Soc.*, **771** (1951).
34. G. B. Heisig and F. H. Pollard, *Anal. Chim. Acta*, **16**, 234 (1957).
35. A. Schneer-Erdey, *Talanta*, **10**, 591 (1963).
36. F. H. Pollard, P. Hanson, and W. J. Geary, *Anal. chim. Acta*, **20**, 26 (1959).
37. K. Macek and L. Moravek, *Nature* **178**, 102 (1956).
38. A. Weiss and S. Fallab, *Helv. chim. Acta*, **37**, 1253 (1954).
39. J. Michal, *Chem. Losty*, **50**, 77 (1956).
40. O. H. Johnson and H. H. Krause, *Anal. Chim. Acta*, **11**, 128 (1954).
41. G. Carleson, *Acta Chem. Scand.*, **8**, 1673 (1954).
42. U. A. Th. Brinkman, *et al.*, *J. Chromatog.*, **25**, 447 (1966).
43. L. F. Druding, *Anal. Chem.*, **35**, 1582 (1963).
44. W. Bock-Werthmann, *Anal. Chim. Acta*, **28**, 519 (1963).
45. H. Erlenmeyer, H. Von Hahn, and E. Sorkin, *Helv. chim. Acta*, **34**, 1419 (1951).
46. H. Seiler, in E. Stahl, *Dünschichtchromatographie*, 2nd ed., Springer Verlag, Berlin, 1967.
47. M. Lederer, *Anal. Chim. Acta*, **2**, 261 (1948).
48. M. Lederer, *Nature*, **162**, 776 (1948).
49. M. Lederer, *Nature*, **163**, 598 (1949).

50. M. Lederer, *Anal. Chim. Acta*, **3,** 476 (1949).
51. T. V. Arden, F. H. Burstall, G. R. Davies, J. A. Lewis, and R. P. Linstead, *Nature*, **162,** 691 (1948).
52. T. V. Arden, F. H. Burstall, and R. P. Linstead, *J. Chem. Soc.*, 311 (1949).
53. F. H. Burstall, G. R. Davies, R. P. Linstead, and R. A. Wells, *Nature*, **163,** 64 (1949).
54. F. H. Burstall, G. R. Davies, R. P. Linstead, and R. A. Wells, *J. Chem. Soc.*, 516 (1950).
55. F. H. Pollard, J. R. W. McOmie, and I. I. M. Elbeih, *Nature*, **163,** 292 (1949).
56. F. H. Pollard and J. F. W. McOmie, *Endeavour*, **10,** 213 (1951).
57. F. H. Pollard, J. F. W. McOmie, and H. M. Stevens, *J. Chem. Soc.*, 1863 (1951).
58. F. H. Pollard and J. F. W. McOmie, *Chromatographic Methods of Inorganic Analysis*, Butterworths, London, 1953.
59. W. J. Frierson and M. J. Ammons, *J. Chem. Educ.*, **27,** 37 (1950).
60. S. Harasawa, *J. Chem. Soc. Japan*, **72,** 107, 236, 423 (1951).
61. S. Harasawa, *Kagaku no Ryoiki*, **5,** 461 (1951).
62. Z. Tamura, *Japan Analyst*, **1,** 117 (1952).
63. J. G. Surak, N. Leffler, and R. J. Martinovich, *J. Chem. Educ.*, **30,** 20 (1953).
64. E. Pfeil, *Z. anal. Chem.*, **146,** 241 (1955).
65. E. Pfeil, *Chemie für Labor u. Betrieb*, **5,** 177 (1957).
66. E. Pfeil, *Chemie für Labor u. Betrieb*, **6,** 123 (1955).
67. E. Pfeil, A. Friedrich, and Th. Wachsmann, *Z. anal. Chem.*, **158,** 429 (1957).
68. E. Blasius and W. Göttling, *Z. anal. Chem.*, **162,** 423 (1958).
69. K. E. Schulte and G. Henke, *Pharmazie*, **18,** 601 (1963).
70. W. Schneider and B. Patel, *Arch. Pharmaz.*, **297,** 97 (1964).
71. E. Merck, *Chromatographie*, E. Merck A.G., Darmstadt.
72. E. Lederer and M. Lederer, *Chromatography*, Elsevier, Amsterdam, 1957.
73. E. Blasius, *Chromatographische Methoden in der analytischen und präparativen anorganischen Chemie unter besonderer Berücksichtigung der Ionenaustauscher*, Enke-Verlag, Stuttgart, 1958.
74. M. Lederer in Hecht en Zacherl, *Handbuch der mikrochemischen Methoden*, Bd. III, Springer-Verlag, Wenen, 1961.
75. I. M. Hais and K. Macek, *Handbuch der Papierchromatographie*, Bd. I (1958); II (1960); III (1963), G. Fischer Verlag, Jena.
76. H. Diller and O. Rex, *Z. anal. Chem.*, **137,** 241 (1952/53).
77. H. Hartkamp and H. Specker, *Z. anal. Chem.*, **152,** 107 (1956); **158,** 161 (1957).
78. F. W. H. M. Merkus, Intern. Symp. V., Chromatographie et Electroforese, P.A.E., Brussels, 1969, in press.
79. F. W. H. M. Merkus, unpublished results.
80. F. W. H. M. Merkus, *J. Chromatog.*, **41,** 497 (1969).

Separation of Nickel and Cobalt by an Electrolytic Process*

T. A. SULLIVAN

U.S. Department of the Interior, Bureau of Mines, Boulder City Metallurgy Research Laboratory, Boulder City, Nevada

I. INTRODUCTION

A. Nickel and Cobalt

Nickel and cobalt occur together in laterite and serpentine ores and are generally extracted together from such ores because of their physical and chemical similarities. Laterite-type deposits constitute the largest known potential resource of nickel in the world. Typical composition of these nickeliferous ores is 40% iron, 1% nickel, and 0.05% cobalt. Recovery of nickel and cobalt together as metals would give a product containing about 95% nickel and 5% cobalt.

* Major features of this manuscript were presented before the 124th meeting of The Electrochemical Society at New York during September 1963.

Nickel containing 1% or less cobalt is usually suitable for most metallurgical applications. However, for use in atomic energy applications, the maximum tolerable amount of cobalt in nickel is 0.2% (12).

B. Separation Methods

Several methods have been reported for treating nickel products extracted from laterite ores to obtain nickel that would meet the cobalt specifications for metallurgical or atomic energy usage. Mahan (8) converted nickel oxide from laterite ore into a nickel metal by first reducing the oxide to sponge nickel and then melting the sponge to remove gangue and part of the cobalt. He was able to produce nickel containing about 1% cobalt. Brooks and Rosenbaum (2) reported a method for the separation of nickel and cobalt by solvent extraction and their recovery by electrolysis, using a nickel oxide starting material similar to that used by Mahan. Ferrante and Butler (5) developed a combination electrolytic-leaching method for producing nickel from basic carbonates extracted from laterite ore to meet metallurgical specifications. Their method depended on the selective dissolutions of the carbonates and required an electrolyte containing a maximum of 0.1 g per liter of cobalt to produce nickel containing 1% or less of cobalt.

The direct separation of nickel and cobalt by electrolysis in aqueous solution has not been possible because of their chemical similarity. This is illustrated by their nearly identical standard electrode potentials in aqueous solutions. Schlain (10) lists the standard single electrode potentials of nickel and cobalt, relative to the standard hydrogen electrode at 25°C, as a −0.25 volt for Ni (II)-Ni(0) and a −0.27 volt for Co (II)-Co (0). The more noble or electronegative element will normally be deposited first by electrolysis. When two or more species are quite similar, as is the case with cobalt and nickel, their ions may be discharged simultaneously at the cathode. Thus, to separate nickel and cobalt by aqueous electrolysis to produce nickel low in cobalt, it has been necessary to resort to chemical means for separating the bulk of the cobalt from the nickel prior to electrolysis.

The possibility of using molten salt electrolysis for the separation of cobalt and nickel was suggested by the Bureau of Mines' success in purification of metals such as beryllium (13), chromium (3), molybdenum (4), titanium (9), and vanadium (11) by molten salt electrorefining. The use of molten salt electrolyses to separate nickel and cobalt appeared feasible from reported standard electrode potentials on the electromotive force series of metals in molten salts. The standard electrode potentials of nickel and cobalt in some molten salt electrolytes differ sufficiently to offer the possibility of separating nickel and cobalt by electrolysis in such electrolytes. The standard electrode potentials of the metals were determined by Laitinen and Liu (7) using as the solvent the eutectic mixture of LiCl-KCl at 450°C. They reported a value of −0.795 volt for Ni (II)-Ni(0) and a value of −0.991 volt for Co (II)-Co (0) using the Pt (II)-Pt (0) system as the reference electrode. A similar spread in the standard electrodepotentials of the two metals in molten 1 to 1 *M* KCl-NaCl at 700°C was reported by Flengas and Ingraham (6), who also calculated the activity coefficients of the metal chlorides in that solvent as 0.046 for $CoCl_2$ and 0.34 for $NiCl_2$. Using the values reported by Flengas, the activity coefficients for the metals in a eutectic mixture of KCl-LiCl at 450°C are 0.049 for $CoCl_2$ and 0.31 for $NiCl_2$. The activity coefficients indicate that the cobalt is complexed to a much greater degree than the nickel in these two molten solvents. The difference in the electrode potentials and the greater complexing of the cobalt in the molten solvents indicate the possibility of selectively electrodepositing nickel from nickel-cobalt materials. Atkinson (1), in a paper on the electrolytic transfer of platinum metals using fused chloride electrolytes, found that nickel nearly free of cobalt could be prepared from nickel containing 0.41% cobalt. Molten salt electrorefining was investigated as a means of refining nickel products containing up to 5% cobalt to produce nickel to meet metallurgical or atomic energy specifications.

II. MOLTEN-SALT ELECTROLYTIC CELLS

Two types of cells were utilized for the electrolytic separation of cobalt and nickel. One was a controlled-atmosphere cell which was designed to permit electrolysis in molten salt under a protective atmosphere of an inert gas. Helium was used for the inert gas. The second cell, which will be called the normal atmosphere cell, did not use any special means of protecting the electrolyte or deposit from air oxidation.

The inert-atmosphere cell was constructed so that the consumable anode and the cathode could be inserted or removed from the cell without exposing to air either the molten electrolyte or the hot, refined metal product. A helium atmosphere was kept over the electrolyte at all times. A schematic drawing of the cell is shown in Figure 1. The cell consisted of an electrolyte compartment, a slide valve, and a water-cooled lock. The electrolyte compartment was a nickel crucible constructed from a 24-in. length of 6-in.-diameter, schedule 40 nickel pipe with a welded nickel bottom. A water-cooled, mild steel flange was welded to the top of the crucible for connection to the slide valve. A high-density graphite liner was inserted in the nickel crucible to contain the molten electrolyte. The slide valve was inserted

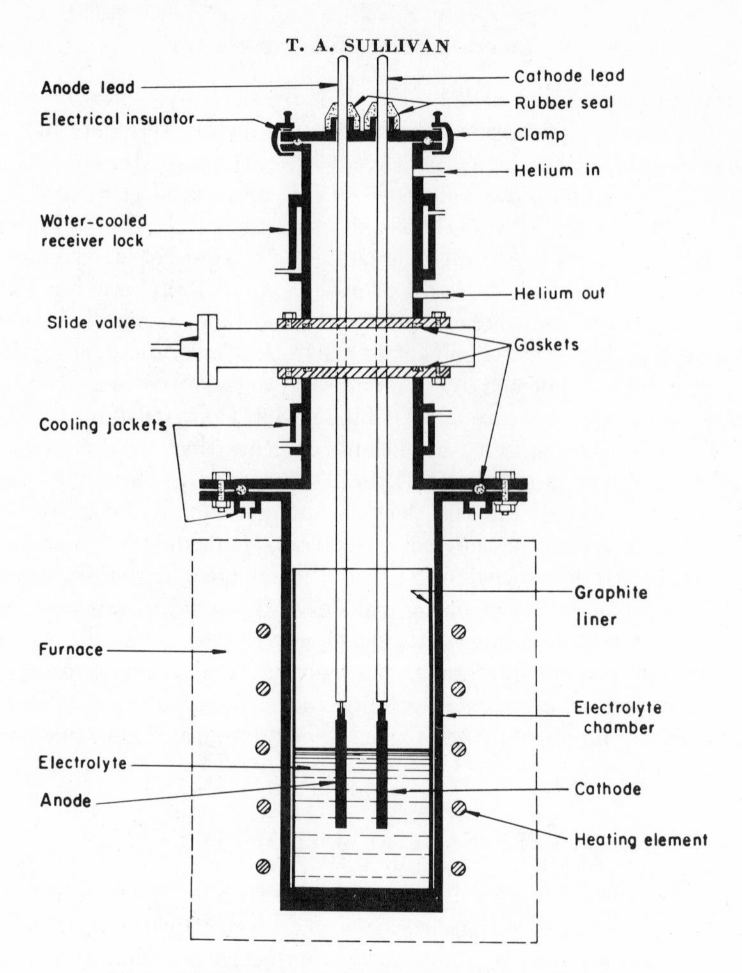

Fig. 1. Helium atmosphere electrolytic cell drawing.

between the compartment and the lock. When this valve was opened, it provided a full 6-in. access to the electrolyte, and when it was closed, it provided a gas and vacuum seal for isolating the lock from the electrolyte chamber. The lock was water-jacketed for rapid cooling of metal deposits and had valved inlets for helium and vacuum connections. A clamp-on cover was used to close the lock. The cover was electrically insulated from the lock by a rubber gasket and insulators at the clamps and had insulated fittings for the anode and cathode lead rods. These rods were sealed in the fittings with rubber sleeves that allowed the raising or lowering of the electrodes without contamination of the helium atmosphere. Sight glasses were provided in the cover for observation and manipulation of the electrodes and deposits.

The normal atmosphere cell consisted of a flanged electrolyte compartment similar to that used in the inert-atmosphere cell. A fused-silica liner was used to contain the electrolyte instead of the graphite liner. A cover plate with fittings for the anode and cathode leads was placed directly on top of the electrolyte compartment. No slide valve or lock was used.

III. ELECTROLYTE

The molten electrolyte selected for use in the electrolytic separation of nickel and cobalt was the eutectic mixture of LiCl-KCl (44 wt. % Li-Cl–56 wt. % KCl) as the solvent and $NiCl_2$ as the solute. The nickel content of the starting electrolytes was between 3 and 6%. Electrolyses were performed at 450°C. This electrolyte had several advantages over other molten electrolytes which have higher operating temperatures, such as KCl-NaCl-$NiCl_2$ (750°C) and NaCl-$NiCl_2$ (850°C). One advantage was the possibility of electrolysis without the need for providing a protective atmosphere over the molten electrolyte to prevent the loss of $NiCl_2$ by air oxidation. Another reason for selecting the lower temperature electrolyte was to diminish the corrosive effects of molten electrolytes containing $NiCl_2$. Requirements for materials of construction for an electrolytic cell to contain the molten salts and their vapors are less severe with an electrolyte operating at 450°C than for electrolytes which require operating temperatures of 750–850°C. Vapor losses of $NiCl_2$ are lower, and less power is required to maintain the molten electrolyte at the lower temperature.

IV. OPERATIONAL PROCEDURES

The electrolytes used in the helium-atmosphere cell were prepared by melting in the electrolyte compartment of the cell, at 400°C under vacuum, a mixture of the dried component salts. Melting of the salts was accompanied by a large evolution of moisture not removed by drying and, possibly, from reaction of the salts with moisture. When evolution of gases from the molten electrolyte ceased, the cell was slightly pressurized with helium. In the work being described, 3000 to 5000 g of electrolyte were used.

The anode materials used in the electrolytic separation of nickel and cobalt were prepared by arc-melting the desired amounts of nickel and cobalt into 300-g bars. Figure 2 shows two new and two used anodes. Nickel anodes containing 2.7 to 5.5% cobalt were investigated.

To prepare the cell for electrolysis, the slide valve was closed to maintain the helium atmosphere over the electrolyte, the cover was removed from the lock, and a nickel-cobalt alloy bar was attached to the anode electrode lead. A nickel cathode was attached to the cathode electrode lead.

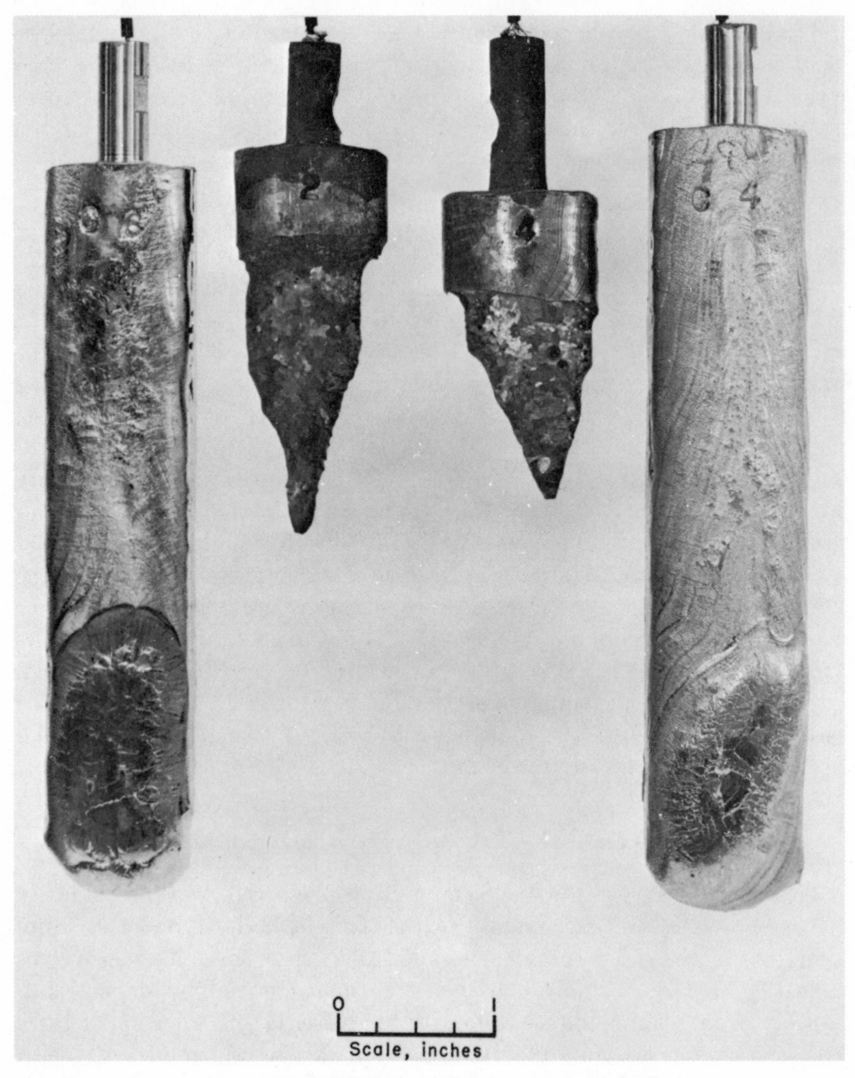

Fig. 2. Nickel-cobalt anode bars, new and used.

The cover was clamped on the lock which was alternately evacuated and back-flushed three times with helium to remove the air. A slight positive pressure of helium was maintained by bubbling the exit gas through an oil bubbler. The slide valve was opened, and the anode and cathode were lowered into position in the electrolyte. Electrolysis was started using direct current from a selenium rectifier. When the desired amount of product had been deposited on the cathode, electrolysis was terminated and both the anode and cathode were raised just out of the molten electrolyte. After draining for 10 min to remove excess electrolyte, they were withdrawn into the lock and the slide valve was closed. The anode and the nickel deposits were removed from the lock after they had been cooled to 25°C. The anode bar was leached free of salt, dried, and weighed. The anode and a new cathode were attached to their respective leads and a new operating cycle started. Cathode configurations used were various diameter nickel rods 6 in. long, and one was a nickel plate 3 in. wide by 6 in. long. One complete deposition cycle will be referred to as a test, and a group of tests made using the same nickel-cobalt alloy anode materials and the same electrolyte will be called a series.

The electrolyte was prepared for the normal atmosphere cell by melting the dried salt mixture and raising the temperature to 450°C. Electrolysis in the cell was similar to that in the helium atmosphere cell. The major difference was that the electrolyte and the hot deposit of refined nickel were exposed to air when the deposit was removed at the conclusion of each test.

The deposits from both types of cells consisted of dendritic nickel crystals attached to the cathode. When the metal deposits were withdrawn from the molten electrolyte, they were coated with a thin film of salt. This coating protected the hot, dendritic metal crystals from oxidation when they were exposed to air on removal from the normal atmosphere cell.

The refined metal was stripped from the cathode and leached in dilute hydrochloric acid (HCl) 1 to 50 to remove the electrolyte. It was then rinsed free of chloride and dried.

V. ELECTROREFINING IN A HELIUM ATMOSPHERE CELL

The selective electrodeposition of nickel using a molten KCl-$LiCl$-$NiCl_2$ electrolyte and a nickel-cobalt alloy anode is illustrated by the following series of tests: The electrolyte contained about 3000 g of a mixture of 1550 gm KCl, 1220 g LiCl, and 220 g $NiCl_2$. The nickel content of the molten electrolyte was 3.66%. The nickel-cobalt alloy for refining contained 2.7% cobalt. Electrolysis was with constant current at an initial cathode current density of 210 amp/ft^2. Approximately 800 g of metal was refined in 21 tests in the series with about 40 g nickel deposited per test. Figure 3 shows a typical nickel deposit from the series. It weighed 41 g, contained 0.02% cobalt, and was the thirteenth deposit in the series. The metal from individual tests showed a cobalt content of less than 0.01% at the start of the series and 0.04% at the end of the series.

Fig. 3. Unleached nickel deposit.

The average cobalt concentration was 0.02% for all products of the series. The nickel content of the electrolyte dropped from 3.66 to 3.00%, while the cobalt content increased from 0 to 0.67% during the series.

This series of tests demonstrated that a good separation of nickel from cobalt was possible by molten salt electrolysis. The series also indicated that the cobalt content of the electrolyte increased as more nickel was refined and that the cobalt content of the refined nickel was dependent, in part, on the amount of cobalt in the electrolyte.

A series of investigations was made to determine what factors influenced the selective deposition of nickel from nickel-cobalt alloys.

A. Factors Involved in the Selective Deposition of Nickel

1. ANODE DISSOLUTION

A series of tests was made to determine what happened at the nickel-cobalt anode during electrolysis. A nickel-cobalt anode was used that contained 5% cobalt. Anode current efficiencies for each test were always close to 100%. Inspection and weighing of the anode indicated that it was dissolving in a nonselective manner. No sludge or sponge residue was left on the anode at the conclusion of each refining test to indicate selective solution of the nickel. In addition, analyses of the electrolyte showed an increase in cobalt content after each test. Nickel was selectively deposited at the cathode. With an increase in the cobalt content of the electrolyte, there was a proportional decrease in its nickel content. Cathode current efficiencies for the tests averaged over 96%, which indicated that the excess nickel deposited at the cathode over that being dissolved at the anode came from the nickel chloride content of the electrolyte. The loss in nickel from the electrolyte was balanced by its gain in cobalt content. From these tests, it was evident that both the cobalt and nickel were being dissolved at the anode in a nonselective manner.

2. COBALT CONCENTRATION OF THE ELECTROLYTE

The effect of the cobalt concentration in the electrolyte on the selective deposition of nickel was determined in a series of 61 consecutive tests. The starting electrolyte contained 5.75% nickel as $NiCl_2$; this was equivalent to approximately a 0.1 mole concentration of the $NiCl_2$. The nickel-cobalt anode contained 5% cobalt. Cathode deposits of nickel averaging 40 g per deposit were made using initial cathode current densities of 100 to 400 amp/ft^2. Table I shows the results of representative tests in the series. It was evident that as the series progressed there was a buildup of cobalt in the electrolyte. The first eight nickel deposits contained $< 0.01\%$ cobalt. The product of the ninth test contained 0.012% cobalt. The electrolyte, at this point, contained 0.34% cobalt. As the $CoCl_2$ content of the electrolyte increased in successive tests, so did the cobalt content of the refined nickel products. However, even with a 2:1 mole ratio of nickel to cobalt in the electrolyte, only 0.067% cobalt was found in the product, compared to the 5% present in the starting anode.

TABLE I
Effect of Electrolyte Composition on Cobalt Concentration in Nickel Product

Test	Electrolyte (%)		Mole ratio, Ni:Co	Product, % cobalt
	Nickel	Cobalt		
[a]	5.75	<0.01		
9	5.51	.34	16.26:1	0.012
16	4.98	.90	5.55:1	.024
21	4.37	1.35	3.25:1	.045
28	4.04	2.00	2.03:1	.067
33	3.56	2.33	1.55:1	.120
38	3.01	2.80	1.07:1	.170
44	6.12	3.04	2.02:1	.072
51	5.86	3.61	1.63:1	.090
61	4.92	4.20	1.18:1	.110

[a] Original electrolyte, 95% nickel, 5% cobalt anode.

An addition of $NiCl_2$ was made to the electrolyte after test 40 to determine the effect of increasing the nickel content of the electrolyte to its original value, because a substantial amount of the original $NiCl_2$ had been replaced with $CoCl_2$. Enough $NiCl_2$ was added to bring its concentration to 0.1 mole which resulted in a nickel-to-cobalt mole ratio of 2:1. The effect of the added $NiCl_2$ was to reduce the rate of cobalt deposition on the cathode product. The nickel product from test 38, made with a nickel-to-cobalt mole ratio of 1.07:1, contained 0.17% cobalt, while that from test 44, made after the $NiCl_2$ addition and with a nickel-to-cobalt mole ratio of 2.02:1, contained only 0.072% cobalt. The amount of cobalt in the products after test 44 again increased with an increasing cobalt content in the electrolyte. The rate of increase was not as great as it was for similar nickel-to-cobalt ratios in the first part of the series.

From this series of tests it was found that the selective deposition of nickel in a molten salt electrolyte is a function of at least two variables: (1) the cobalt content of the electrolyte and (2) the nickel content of the electrolyte.

Two tests were made to determine the effect of replacing the $NiCl_2$ content of the electrolyte with $CoCl_2$. With an electrolyte that contained 4.76% cobalt and < 0.01% nickel, electrolysis products from a nickel-cobalt anode containing 4.85% cobalt contained 4.42 and 4.48% cobalt. The transfer of cobalt under these conditions was approximately equal to its concentration in the anode material.

While the primary objective in the development of the process for the separation of cobalt and nickel was to obtain nickel that would meet specifications for metallurgical or nuclear-grade nickel from nickel containing up to 5% cobalt, the possibility of preparing high-purity nickel was also investigated. The results are illustrated by the analyses of the nickel feed material and refined product in Table II. The refined nickel product was from test 9, Table I and was of high purity except for 120 ppm of cobalt. A good reduction of the impurity elements present in the feed material was achieved.

TABLE II
Analysis of Anode Material and Refined Nickel (%)

Element	Anode Material	Refined Nickel
Aluminum	0.05 –0.5	0.0020
Calcium	.01 –0.10	.0007
Chromium	.001–0.01	.0007
Cobalt	5.05	.0120
Copper	.005–0.05	.0008
Iron	.50	.0006
Lead	[a]	.0003
Manganese	.01 –0.10	.0001
Molybdenum	.001–0.01	<.0001
Tin	[a]	.0003
Titanium	[a]	.0001
Vanadium	[a]	.0001
Zinc	[a]	.0002

[a] Not determined.

Cobalt-free nickel shot instead of the nickel-cobalt alloy was used as anode material in another series of tests. The electrorefined nickel products were similar in analyses to the product shown in Table II, with the exception that no cobalt was detected in the nickel products. Molten salt electrorefining to prepare high-purity nickel was shown to be feasible.

3. EFFECT OF CATHODE CURRENT DENSITY

The proportion of cobalt transferring to the nickel cathode deposit was influenced by the cathode current density used in electrolysis. The effect of varying the cathode current densities in consecutive tests is shown in Table III. The group of tests shown was made during a series, using a nickel-cobalt alloy anode containing 5.5% cobalt. Electrorefining had

Fig. 4. Nickel deposited with a cathode current density of 30 amp/ft^2.

resulted in a cobalt concentration of 1.84% in the electrolyte at the time this group of tests was started. The amount of nickel deposited in each test was held at approximately 40 g. The results show that, in general, the higher the initial cathode current density, the greater was the transfer of cobalt to the product. The influence of current density on the physcial appearance of nickel deposits is shown in Figures 4 and 5. A cathode current density range of 200 to 500 amp/ft^2 was chosen as the most practical to use on the basis of both transfer of cobalt and deposition rate.

4. *COMPOSITION OF THE ANODE MATERIAL*

Two successive series of tests were made to determine the role of the anode composition on the transfer of cobalt to the nickel product. The series were made using the same electrolyte, first with an anode containing 5% cobalt, followed by a series with a pure nickel anode.

When an electrolyte composition of 4.38% nickel and 4.40% cobalt was reached in the refining series with a nickel-cobalt anode containing 5% cobalt, a nickel product containing 0.19% cobalt was obtained. At this

Fig. 5. Nickel deposited with a cathode current density of 835 amp/ft^2.

point the nickel-cobalt alloy anode was removed and a pure nickel anode was substituted for it. The second series of refining tests was then made which consisted of 13 tests with cathode products averaging 30 g each. The cobalt content of the products did not decrease, varying between 0.17 and 0.21%. This indicated that the cobalt content of the anode material was not a significant factor in the transfer of cobalt to the nickel product, except for its effect in increasing the cobalt concentration of the electrolyte.

TABLE III
Variation of Cobalt in Nickel Products with Cathode Current Densities

Test	Volts	Initial Cathode Current Density (amp/ft²)	Deposit, % cobalt	Electrolyte (%)	
				Nickel	Cobalt
1	2.0	1,160	0.15	3.34	1.84
2	2.0	1,735	.16		
3	2.2	2,090	.19	3.21	1.94
4	2.3	2,295	.19		
5	2.3	325	.14	3.00	1.99
6	.05	30	.10		
7	1.0	835	.16	2.89	2.14
8	.2	80	.12		
9	.5	205	.15	2.76	2.26
10	.5	390	.16		
11	2.2	1,380	.17	2.68	2.36

With an electrolyte composition that was fairly constant in both cobalt and nickel contents, nickel products were obtained that contained approximately the same proportions of cobalt. This reconfirmed the fact that one of the major factors controlling the deposition of cobalt in the refined nickel was the cobalt concentration of the electrolyte.

B. Recovery of Cobalt from the Electrolyte

Electrorefining of nickel-cobalt alloys leaves the bulk of the cobalt dissolved in the electrolyte as $CoCl_2$ or a complexed salt. Recovery of the cobalt was effected by first removing any nickel in the electrolyte by electrolysis using a cobalt anode instead of a nickel-cobalt alloy anode. As nickel is the more noble of the two, it will deposit preferentially at the cathode. When the nickel content of the electrolyte was depleted, a carbon anode was substituted for the cobalt anode and the cobalt was electrowon from the electrolyte. The results of one series of tests made to recover cobalt from the electrolyte are given in Table IV.

The 3200 g of electrolyte contained 0.20% nickel and 1.68% cobalt at the start of the cobalt removal tests. Four depositions using a cobalt anode and an initial cathode current density of 200 amp/ft² stripped the nickel from the electrolyte which then contained 1.72% cobalt, or 55 g. Two electrowinning depositions were made with a carbon anode and an initial cathode current density of 350 amp/ft² to remove the cobalt from the electrolyte. The final cobalt concentration in the electrolyte was 0.03%. A 98% recovery of cobalt was achieved.

TABLE IV

Recovery of the Nickel and Cobalt Content of the Electrolyte Using Cobalt and Carbon Anodes

Test	Anode	Deposit (g)	Deposit (%) Nickel	Deposit (%) Cobalt	Electrolyte (%)[a] Nickel	Electrolyte (%)[a] Cobalt	Electrolyte Depletion (g) Nickel	Electrolyte Depletion (g) Cobalt
1	Cobalt	65.8	8.21	91.4	0.20	1.68	5.40	—
2	Cobalt	48.2	1.40	98.2	.03	1.68	.67	—
3	Cobalt	51.0	.39	99.2	.01	1.69	.20	—
4	Cobalt	50.4	.09	>99.5	<.01	1.70	.05	—
5	Carbon	49.2	<.01	>99.5	<.01	1.72	—	49.0
6	Carbon	5.0	<.01	>99.5	<.01	.16	—	5.0

[a] Values for electrolytes are before start of each test.

This method not only served to recover the cobalt from the electrolyte but also made possible reuse of the molten electrolyte. All that was necessary to reuse the electrolyte was to add $NiCl_2$ in the desired amount.

VI. ELECTROREFINING IN A NORMAL ATMOSPHERE CELL

The use of a cell without a protective atmosphere for the separation of cobalt and nickel was investigated to compare with results obtained in the helium atmosphere cell. Several advantages are possible by using a normal atmosphere cell. The need for a pressure- and vacuum-tight system with a slide valve would be eliminated. The equipment cost would be lower with an overall simplicity in construction and operation. Two main points that had to be determined were (1) whether exposure of the hot electrolyte to air would result in a loss of $NiCl_2$ from the electrolyte by oxidation and (2) the effect of air on the electrorefined product on its removal from the electrolyte.

A series of tests was made in a cell using a nickel-cobalt alloy containing 4.2% cobalt. Separation of nickel and cobalt was accomplished with results similar to those obtained by refining in a helium atmosphere cell. The major difference in the performance of the two types of cells was in a slightly greater loss in the nickel content of the normal atmosphere cell by oxidation or sublimation. Using similar amounts of nickel and cobalt in the electrolyte, and with similar current densities, the concentrations of cobalt in the refined nickel products were similar for both cells. As the nickel products were removed from the molten electrolyte, a film of the electrolyte froze

Fig. 6. Unleached nickel deposit.

on the metal crystals and protected them from air oxidation. A typical unleached deposit from the normal atmosphere cell is shown in Figure 6. The major difference in the appearance of this deposit, when compared to deposits produced in the helium atmosphere cell, was that the electrolyte film on the surface of the refined nickel products was darker when cooled in contact with air than when cooled in a helium atmosphere. A small amount of nickel oxide was found in the electrolyte leached from the metal crystals of the normal atmosphere cell. No oxidation of the metal crystals was observed.

VII. COMPARISON OF REFINING IN NORMAL AND HELIUM ATMOSPHERE CELLS

Two parallel series of refining tests were made to compare refining in both types of cells. Testing was performed under similar conditions in each cell. Electrolyses were made at 450°C with initial cathode current densities ranging from 25 to 2,600 amp/ft^2 with potentials of 0.05 to 2.3 volts. The nickel-cobalt anode alloy contained 4.1% cobalt. The anode bars were similar to those shown in Figure 2. Two anode bars were used with a single $\frac{7}{16}$-in.-diameter cathode and were hung from the same anode lead. The anodes were immersed 4 in. into the electrolyte, and the cathode depth was varied between 1 and 4 in., depending on the cathode current density desired. In each cell 5000 g of KCl-$LiCl$-$NiCl_2$ electrolyte was used. The starting nickel content of the electrolyte was 5.08% in the helium atmosphere cell and 4.99% in the normal atmosphere cell. Electrolysis was continued for a long enough period in each test to produce approximately 40 g of refined product. The performance in each cell is summarized in Table V. The table groups consecutive deposits into three groups which contained up to 0.05%, between 0.05 and 0.10%, and between 0.10 and

TABLE V

Summary of Data from Comparison Tests with Normal and Helium Atmosphere Cells [a]

	Cobalt in Refined Nickel (%)		
	0–0.05	0.05–0.10	0.10–0.20
Normal Atmosphere Cell			
Refined nickel (g)	520	530	570
Electrolyte, range (%)			
Nickel	4.99–4.19	4.19–3.30	3.30–2.18
Cobalt	0 –0.56	.56–1.16	1.16–1.92
Helium Atmosphere Cell			
Refined nickel (g)	530	670	420
Electrolyte, range (%)			
Nickel	5.08–4.36	4.36–3.58	3.58–2.68
Cobalt	0 –0.74	.74–1.46	1.46–2.36

[a] Ranges show percent at start and finish, respectively, of electrolysis.

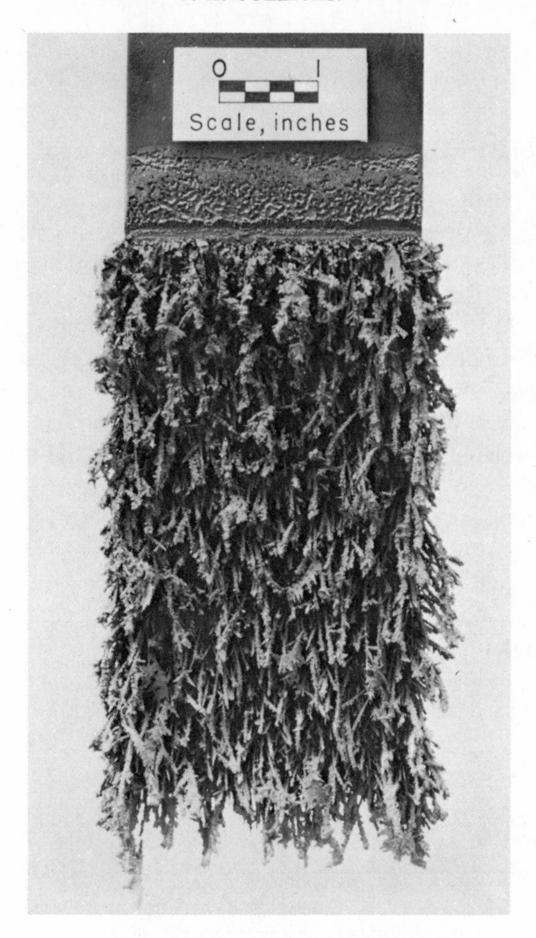

Fig. 7. Nickel deposited on a plate cathode.

0.20% cobalt. Little difference was found in the refining in either type of cell. Electrolyte dragout was higher in the normal atmosphere cell with a metal-to-salt ratio of 2.5:1, compared to 3.3:1 in the helium atmosphere cell. The average cobalt content of the nickel produced in the normal atmosphere cell was 0.09%, and it was 0.08% in the helium atmosphere cell. The electrolyte of the helium atmosphere cell contained 5.08% nickel at the start and ended with a combined nickel and cobalt content of 5.04%. The nickel content of the normal atmosphere cell started at 4.99% and ended with a nickel-cobalt content of 4.10%. The loss of nickel-cobalt chlorides from the normal atmosphere cell was only slightly larger than from the helium atmosphere cell. Refining of nickel containing 4.1%

cobalt to a product containing less than 0.20% cobalt was achieved in both cells.

After the comparison tests were concluded, a test was made in the normal atmosphere cell using a plate instead of a rod cathode. Figure 7 shows the type of deposit obtained. The deposit was confined to the side of the plate facing the anode. A cell utilizing plate cathodes would require alternate anode and cathode plates for uniform depositions.

VIII. SUMMARY

The electrolytic separation of nickel and cobalt by electrorefining in a molten salt electrolyte was shown to be technically feasible. Nickel-cobalt alloys containing up to 5.5% cobalt were electrorefined in a KCl-LiCl-$NiCl_2$ electrolyte to a high-purity nickel metal meeting atomic energy specifications for cobalt content. The transfer of cobalt to the refined nickel product was found to be dependent on the nickel content of the electrolyte, the cobalt content of the electrolyte, the ratio of nickel-to-cobalt in the electrolyte, and the cathode current densities.

Nickel metal containing $<0.10\%$ cobalt was routinely prepared from nickel-cobalt alloys containing up to 5.5% cobalt by controlling the nickel-cobalt content of the electrolyte. The nickel content of the electrolyte was gradually replaced by the solution of cobalt from the alloy in the process.

A method of electrowinning the cobalt from the electrolyte was developed which not only recovered the cobalt but also prepared the electrolyte for reuse on the addition of more $NiCl_2$.

Either a normal atmosphere or a protected atmosphere cell was found to be usable for the electrorefining process. The major difference in the performance of the two types of cells was that in the normal atmosphere cell slightly greater quantities of the nickel and cobalt were lost from the electrolyte by oxidation or sublimation.

References

1. R. H. Atkinson, "Research on the Electrolytic Transfer of Platinum Metals Using Fused Chloride Electrolytes," *Trans. Faraday Soc.*, **26**(8), 1930, pp. 490–496.
2. P. T. Brooks and J. B. Rosenbaum, "Separation and Recovery of Cobalt and Nickel by Solvent Extraction and Electrorefining," *Bu. Mines Rept. of Inv.*, 6159, 1963, 30 pp.
3. F. R. Cattoir and D. H. Baker, Jr., "Electrorefining Chromium," *Bu. Mines Rept. of Inv.*, 5682, 1960, 15 pp.
4. R. E. Cumings, F. R. Cattoir, and T. A. Sullivan, "Preparation of High Purity Molybdenum by Molten Salt Electrorefining," *Bu. Mines Rept. of Inv.*, 6850, 1966, 24 pp.

5. M. J. Ferrante and M. O. Butler, "An Electrolytic Method for Separating Nickel and Cobalt," *Bu. Mines Rept. of Inv.*, 5543, 1959, 23 pp.
6. S. N. Flengas and T. R. Ingraham, "Electromotive Force Series of Metals in Fused Salts and Activities of Metal Chlorides in 1:1 Molar KCL-NaCl Solutions," *J. Electrochem. Soc.*, V, **106,** (8), 1959, pp. 714–721.
7. H. A. Laitinen and C. H. Liu, "An Electromotive Force Series in Molten Lithium Chloride-Potassium Chloride Eutectic," *J. Am. Chem. Soc.*, **80**(5), 1958, pp. 1015–1020.
8. W. M. Mahan, N. B. Melcher, J. P. Riott, and E. J. Ostrawski, "Conversion of Nicaro Nickel Oxide to Nickel Metal," *Bu. Mines Rept. of Inv.*, 5465, 1959, 36 pp.
9. J. R. Nettle, D. H. Baker, Jr., and F. S. Wartman, "Electrorefining Titanium Metal," *Bu. Mines Rept. of Inv.*, 5315, 1957, 43 pp.
10. David Schlain, "Preparation of Primary Purified Metals by Electrowinning and Electrorefining from Aqueous and Nonaqueous Electrolytic," in *Techniques of Metal Research*, R. F. Bunshah, Interscience, New York, Vol. 1, Part 2, 1968, pp. 493–548.
11. T. A. Sullivan, "Electrorefining Vanadium," *J. Metals*, **17**(1), 1965, pp. 45–48.
12. Glen C. Ware, "Nickel," in *Mineral Facts and Problems, Bu. Mines Bull.*, 630, 1965, p. 610.
13. M. M. Wong, F. R. Cattoir, and D. H. Baker, Jr., "Electrorefining Beryllium," *Bu. Mines Rept. of Inv.*, 5581, 1960, 9 pp.

Author Index

Numbers in parentheses are reference numbers and show that an author's work is referred to although his name is not mentioned in the text. Numbers in *italics* indicate the pages on which the full references appear.

Subject Index